Numerical Analysis and Computational Mathematics

Numerical Analysis and Computational Mathematics

Editors

Jesús Martín Vaquero
Deolinda Dias Rasteiro
Araceli Queiruga-Dios
Fatih Yilmaz

MDPI • Basel • Beijing • Wuhan • Barcelona • Belgrade • Manchester • Tokyo • Cluj • Tianjin

Editors

Jesús Martín Vaquero
Department of Applied
Mathematics, Institute of
Fundamental Physics and
Mathematics, Universidad de
Salamanca
Spain

Deolinda Dias Rasteiro
Department of Physics and
Mathematics, ISEC—Coimbra
Institute of Engineering,
Polytechnic Institute of Coimbra
(IPC)
Portugal

Araceli Queiruga-Dios
Department of Applied
Mathematics, Institute of
Fundamental Physics and
Mathematics, Universidad de
Salamanca
Spain

Fatih Yilmaz
Department of Mathematics,
Ankara Hacı Bayram Veli
University
Turkey

Editorial Office
MDPI
St. Alban-Anlage 66
4052 Basel, Switzerland

This is a reprint of articles from the Special Issue published online in the open access journal *Axioms* (ISSN 2075-1680) (available at: http://www.mdpi.com).

For citation purposes, cite each article independently as indicated on the article page online and as indicated below:

LastName, A.A.; LastName, B.B.; LastName, C.C. Article Title. *Journal Name* **Year**, *Volume Number*, Page Range.

ISBN 978-3-0365-3258-5 (Hbk)
ISBN 978-3-0365-3259-2 (PDF)

© 2022 by the authors. Articles in this book are Open Access and distributed under the Creative Commons Attribution (CC BY) license, which allows users to download, copy and build upon published articles, as long as the author and publisher are properly credited, which ensures maximum dissemination and a wider impact of our publications.

The book as a whole is distributed by MDPI under the terms and conditions of the Creative Commons license CC BY-NC-ND.

Contents

About the Editors vii

Preface to "Numerical Analysis and Computational Mathematics" ix

Somayeh Nemati and Delfim F. M. Torres
Application of Bernoulli Polynomials for Solving Variable-Order Fractional Optimal Control-Affine Problems
Reprinted from: *Axioms* **2020**, *9*, 114, doi:10.3390/axioms9040114 1

Wiyada Kumam and Kanikar Muangchoo
Inertial Iterative Self-Adaptive Step Size Extragradient-Like Method for Solving Equilibrium Problems in Real Hilbert Space with Applications
Reprinted from: *Axioms* **2020**, *9*, 127, doi:10.3390/axioms9040127 19

Wiyada Kumam and Kanikar Muangchoo
Approximation Results for Equilibrium Problems Involving Strongly Pseudomonotone Bifunction in Real Hilbert Spaces
Reprinted from: *Axioms* **2020**, *9*, 137, doi:10.3390/axioms9040137 41

Bruno Carbonaro and Marco Menale
Towards the Dependence on Parameters for the Solution of the Thermostatted Kinetic Framework
Reprinted from: *Axioms* **2021**, *10*, 59, doi:10.3390/axioms10020059 61

Chainarong Khanpanuk, Nuttapol Pakkaranang, Nopparat Wairojjana and Nattawut Pholasa
Approximations of an Equilibrium Problem without Prior Knowledge of Lipschitz Constants in Hilbert Spaces with Applications
Reprinted from: *Axioms* **2021**, *10*, 76, doi:10.3390/axioms10020076 83

Nataša Kontrec, Stefan Panić, Biljana Panić, Aleksandar Marković and Dejan Stošović
Mathematical Approach for System Repair Rate Analysis Used in Maintenance Decision Making
Reprinted from: *Axioms* **2021**, *10*, 96, doi:10.3390/axioms10020096 103

Emmanuel A. Bakare, Snehashish Chakraverty and Radovan Potucek
Numerical Solution of an Interval-Based Uncertain SIR (Susceptible–Infected–Recovered) Epidemic Model by Homotopy Analysis Method
Reprinted from: *Axioms* **2021**, *10*, 114, doi:10.3390/axioms10020114 113

Eliana Costa e Silva, Aldina Correia and Ana Borges
Unveiling the Dynamics of the European Entrepreneurial Framework Conditions over the Last Two Decades: A Cluster Analysis
Reprinted from: *Axioms* **2021**, *10*, 149, doi:10.3390/axioms10030149 133

Ioannis K. Argyros, Stepan Shakhno, Roman Iakymchuk, Halyna Yarmola and Michael I. Argyros
Gauss–Newton–Secant Method for Solving Nonlinear Least Squares Problems under Generalized Lipschitz Conditions
Reprinted from: *Axioms* **2021**, *10*, 158, doi:10.3390/axioms10030158 147

Asuka Ohashi and Tomohiro Sogabe
Numerical Algorithms for Computing an Arbitrary Singular Value of a Tensor Sum
Reprinted from: *Axioms* **2021**, *10*, 211, doi:10.3390/axioms10030211 . . . . **161**

Víctor Gayoso Martínez, Luis Hernández Encinas, Agustín Martín Muñoz and Araceli Queiruga Dios
Using Free Mathematical Software in Engineering Classes
Reprinted from: *Axioms* **2021**, *10*, 253, doi:10.3390/axioms10040253 . . . . **175**

About the Editors

Jesús Martín-Vaquero is a Full Professor at Salamanca University, where he graduated in Mathematics (2002). He completed his PhD (2006) at University Rey Juan Carlos. He is the co-author of over 40 papers, and has made numerous contributions to conferences. He is a member of the Editorial Board of several journals and has served as an Invited Editor for others. He has participated in numerous funding projects, and obtained mobility grants to universities worldwide (Milwaukee, Minnesota, Cambridge and Simon Fraser). His current research interests include the development of numerical methods for solving nonlinear parabolic PDEs, approximation theory and mathematical education based on projects and competencies.

Deolinda Dias Rasteiro is a Full Professor at Department of Physics and Mathematics, ISEC—Coimbra Institute of Engineering and received a Doctorate in Mathematics, in the field of Network Optimization. She is the co-coordinator of SPEE (Portuguese Society for Engineering Education) and the chair of SEFI MSIG (European Society for Engineering Education Mathematics Special Interest Group). She is the co-author of over 50 papers, has made numerous contributions to conferences and has participated as an Invited Editor in several journals. In addition to this, she has participated in funding projects and obtained mobility grants to international universities (Spain, Poland, Romania, and The Fields Institute). Her current research interests include network-optimization algorithms and mathematical education based on projects and competencies.

Araceli Queiruga-Dios teaches at the Universidad de Salamanca, is a graduate in Physics at the University of Salamanca (Spain) and obtained a Ph.D. in Mathematics at the same university, in 2006. She now teaches mathematics at the Department of Applied Mathematics in the Higher Technical School of Industrial Engineering at the University of Salamanca. Additionally, she has participated in several research projects, is the author of several research papers and has made contributions to workshops and conferences about teaching and learning methodologies as well as teaching innovation. Her major fields of study include cryptography, number theory, applications of mathematics to education and innovation in education.

Fatih Yılmaz obtained his PhD from Selçuk University (Turkey) in Mathematics. He is currently studying at Ankara Hacı Bayram Veli University as an Associate Professor and has served his university in various administrative positions. His major field of study is matrix theory, number theory, graph theory and combinatorics. He is co-author of over 30 papers, and has made numerous contributions to conferences. He is a member of the Editorial Board of several journals and has participated as an Invited Editor in others. Moreover, he has participated in several funding projects.

Preface to "Numerical Analysis and Computational Mathematics"

This Special Issue includes (but is not limited to) papers that were accepted at the International Conference on Mathematics and its Applications in Science and Engineering (ICMASE 2020), held in Ankara Hacı Bayram Veli University, Turkey, between 9th and 10th July, 2020 and conducted online due to the COVID-19 pandemic. The aim of this conference was to exchange ideas, discuss developments in mathematics, develop collaborations and interact with professionals and researchers from all over the world. The following topics of interest were explored: Functional Analysis, Approximation Theory, Real Analysis, Complex Analysis, Harmonic and non-Harmonic Analysis, Applied Analysis, Numerical Analysis, Geometry, Topology and Algebra, Modern Methods in Summability and Approximation, Operator Theory, Fixed Point Theory and Applications, Sequence Spaces and Matrix Transformation, Spectral Theory and Differential Operators, Boundary Value Problems, Ordinary and Partial Differential Equations, Discontinuous Differential Equations, Convex Analysis and its applications, Optimization and its application, Mathematics Education, the application of Variable Exponent Lebesgue Spaces, applications of Differential Equations and Partial Differential Equations, Fourier Analysis, Wavelet and Harmonic Analysis Methods in Function Spaces, applications of Computer Engineering, and Flow Dynamics.

However, the talks were not restricted to these subjects alone, and we expect to include more topics in the future, since we will now be hosting the conference annually.

In 2020, this conference was also organized as a final multiplier event of the Rules Math Project, supported by the EU, and in the future, we aim to dedicate a considerable level of attention to educational studies (in mathematics).

Jesús Martín Vaquero, Deolinda Dias Rasteiro, Araceli Queiruga-Dios, Fatih Yilmaz
Editors

Article

Application of Bernoulli Polynomials for Solving Variable-Order Fractional Optimal Control-Affine Problems

Somayeh Nemati [1] and Delfim F. M. Torres [2,*]

[1] Department of Mathematics, Faculty of Mathematical Sciences, University of Mazandaran, Babolsar 47416-13534, Iran; s.nemati@umz.ac.ir
[2] Center for Research and Development in Mathematics and Applications (CIDMA), Department of Mathematics, University of Aveiro, 3810-193 Aveiro, Portugal
* Correspondence: delfim@ua.pt; Tel.: +351-234-370-668

Received: 4 September 2020; Accepted: 6 October 2020; Published: 13 October 2020

Abstract: We propose two efficient numerical approaches for solving variable-order fractional optimal control-affine problems. The variable-order fractional derivative is considered in the Caputo sense, which together with the Riemann–Liouville integral operator is used in our new techniques. An accurate operational matrix of variable-order fractional integration for Bernoulli polynomials is introduced. Our methods proceed as follows. First, a specific approximation of the differentiation order of the state function is considered, in terms of Bernoulli polynomials. Such approximation, together with the initial conditions, help us to obtain some approximations for the other existing functions in the dynamical control-affine system. Using these approximations, and the Gauss—Legendre integration formula, the problem is reduced to a system of nonlinear algebraic equations. Some error bounds are then given for the approximate optimal state and control functions, which allow us to obtain an error bound for the approximate value of the performance index. We end by solving some test problems, which demonstrate the high accuracy of our results.

Keywords: variable-order fractional calculus; Bernoulli polynomials; optimal control-affine problems; operational matrix of fractional integration

MSC: 34A08; 65M70 (Primary); 11B68 (Secondary)

1. Introduction

The Bernoulli polynomials, named after Jacob Bernoulli (1654–1705), occur in the study of many special functions and, in particular, in relation with fractional calculus, which is a classical area of mathematical analysis whose foundations were laid by Liouville in a paper from 1832 and that is nowadays a very active research area [1]. One can say that Bernoulli polynomials are a powerful mathematical tool in dealing with various problems of dynamical nature [2–6]. Recently, an approximate method, based on orthonormal Bernoulli's polynomials, has been developed for solving fractional order differential equations of Lane–Emden type [7], while in [8] Bernoulli polynomials are used to numerical solve Fredholm fractional integro-differential equations with right-sided Caputo derivatives. Here we are interested in the use of Bernoulli polynomials with respect to fractional optimal control problems.

An optimal control problem refers to the minimization of a functional on a set of control and state variables (the performance index) subject to dynamic constraints on the states and controls. When such dynamic constraints are described by fractional differential equations, then one speaks of fractional optimal control problems (FOCPs) [9]. The mathematical theory of fractional optimal control has born in 1996/97 from practical problems of mechanics and began to be developed in the context of the fractional calculus of variations [10–12]. Soon after, fractional optimal control theory became a mature research area, supported with many applications in engineering and physics. For the state-of-the-art, see [13–15] and references therein. Regarding the use of Bernoulli polynomials to numerically solve FOCPs, we refer to [2], where the operational matrices of fractional Riemann–Liouville integration for Bernoulli polynomials are derived and the properties of Bernoulli polynomials are utilized, together with Newton's iterative method, to find approximate solutions to FOCPs. The usefulness of Bernoulli polynomials for mixed integer-fractional optimal control problems is shown in [16], while the practical relevance of the methods in engineering is illustrated in [17]. Recently, such results have been generalized for two-dimensional fractional optimal control problems, where the control system is not a fractional ordinary differential equation but a fractional partial differential equation [18]. Here we are the first to develop a numerical method, based on Bernoulli polynomials, for FOCPs of variable-order.

The variable-order fractional calculus was introduced in 1993 by Samko and Ross and deals with operators of order α, where α is not necessarily a constant but a function $\alpha(t)$ of time [19]. With this extension, numerous applications have been found in physics, mechanics, control, and signal processing [20–24]. For the state-of-the-art on variable-order fractional optimal control we refer the interested reader to the book [25] and the articles [26,27]. To the best of our knowledge, numerical methods based on Bernoulli polynomials for such kind of FOCPs are not available in the literature. For this reason, in this work we focus on the following variable-order fractional optimal control-affine problem (FOC-AP):

$$\min J = \int_0^1 \phi(t, x(t), u(t)) dt \quad (1)$$

subject to the control-affine dynamical system

$${}_0^C D_t^{\alpha(t)} x(t) = \varphi\left(t, x(t), {}_0^C D_t^{\alpha_1(t)} x(t), \ldots, {}_0^C D_t^{\alpha_s(t)} x(t)\right) + b(t) u(t) \quad (2)$$

and the initial conditions

$$x^{(i)}(0) = x_0^i, \quad i = 0, 1, \ldots, n, \quad (3)$$

where ϕ and φ are smooth functions of their arguments, $b \neq 0$, n is a positive integer number such that for all $t \in [0,1]$, $0 < \alpha_1(t) < \alpha_2(t) < \ldots < \alpha_s(t) < \alpha(t) \leq n$, and ${}_0^C D_t^{\alpha(t)}$ is the (left) fractional derivative of variable-order defined in the Caputo sense. We employ two spectral methods based on Bernoulli polynomials in order to obtain numerical solutions to problem (1)–(3). Our main idea consists of reducing the problem to a system of nonlinear algebraic equations. To do this, we introduce an accurate operational matrix of variable-order fractional integration, having Bernoulli polynomials as basis vectors.

The paper is organized as follows. In Section 2, the variable-order fractional calculus is briefly reviewed and some properties of the Bernoulli polynomials are recalled. A new operational matrix of variable-order is introduced for the Bernoulli basis functions in Section 3. Section 4 is devoted to two new numerical approaches based on Bernoulli polynomials for solving problem (1)–(3). In Section 5, some error bounds are proved. Then, in Section 6, some FOC-APs are solved using the proposed methods. Finally, concluding remarks are given in Section 7.

2. Preliminaries

In this section, a brief review on necessary definitions and properties of the variable-order fractional calculus is presented. Furthermore, Bernoulli polynomials and some of their properties are recalled.

2.1. The Variable-Order Fractional Calculus

The two most commonly used definitions in fractional calculus are the Riemann–Liouville integral and the Caputo derivative. Here, we deal with generalizations of these two definitions, which allow the order of the fractional operators to be of variable-order.

Definition 1 (See, e.g., [25]). *The left Riemann—Liouville fractional integral of order $\alpha(t)$ is defined by*

$$_0I_t^{\alpha(t)} y(t) = \frac{1}{\Gamma(\alpha(t))} \int_0^t (t-s)^{\alpha(t)-1} y(s) ds, \quad t > 0,$$

where Γ is the Euler gamma function, that is,

$$\Gamma(t) = \int_0^\infty \tau^{t-1} \exp(-\tau) d\tau, \quad t > 0.$$

Definition 2 (See, e.g., [25]). *The left Caputo fractional derivative of order $\alpha(t)$ is defined by*

$$_0^C D_t^{\alpha(t)} y(t) = \frac{1}{\Gamma(n-\alpha(t))} \int_0^t (t-s)^{n-\alpha(t)-1} y^{(n)}(s) ds, \quad n-1 < \alpha(t) < n,$$

$$_0^C D_t^{\alpha(t)} y(t) = y^{(n)}(t), \quad \alpha(t) = n.$$

For $0 \leq \beta(t) < \alpha(t) \leq n$, $n \in \mathbb{N}$, $\gamma > 0$, and $\nu > -1$, some useful properties of the Caputo derivative and Riemann–Liouville fractional integral are as follows [25]:

$$_0I_t^{\alpha(t)} t^\nu = \frac{\Gamma(\nu+1)}{\Gamma(\nu+1+\alpha(t))} t^{\nu+\alpha(t)}, \tag{4}$$

$$_0I_t^\gamma (_0^C D_t^\gamma y(t)) = y(t) - \sum_{i=0}^{\lceil \gamma \rceil - 1} y^{(i)}(0) \frac{t^i}{i!}, \quad t > 0, \tag{5}$$

$$_0I_t^{n-\alpha(t)} (y^{(n)}(t)) = {}_0^C D_t^{\alpha(t)} y(t) - \sum_{i=\lceil \alpha(t) \rceil}^{n-1} y^{(i)}(0) \frac{t^{i-\alpha(t)}}{\Gamma(i+1-\alpha(t))}, \quad t > 0, \tag{6}$$

$$_0I_t^{\alpha(t)-\beta(t)} (_0^C D_t^{\alpha(t)} y(t)) = {}_0^C D_t^{\beta(t)} y(t) - \sum_{i=\lceil \beta(t) \rceil}^{\lceil \alpha(t) \rceil - 1} y^{(i)}(0) \frac{t^{i-\beta(t)}}{\Gamma(i+1-\beta(t))}, \quad t > 0, \tag{7}$$

where $\lceil \cdot \rceil$ is the ceiling function.

2.2. Bernoulli Polynomials

The set of Bernoulli polynomials, $\{\beta_m(t)\}_{m=0}^\infty$, consists of a family of independent functions that builds a complete basis for the space $L^2[0,1]$ of all square integrable functions on the interval $[0,1]$. These polynomials are defined as

$$\beta_m(t) = \sum_{i=0}^m \binom{m}{i} b_{m-i} t^i, \tag{8}$$

where b_k, $k = 0, 1, \ldots, m$, are the Bernoulli numbers [28]. These numbers are seen in the series expansion of trigonometric functions and can be given by the following identity [29]:

$$\frac{t}{e^t - 1} = \sum_{i=0}^\infty b_i \frac{t^i}{i!}.$$

Thus, the first few Bernoulli numbers are given by

$$b_0 = 1, \quad b_1 = -\frac{1}{2}, \quad b_2 = \frac{1}{6}, \quad b_3 = 0, \quad b_4 = -\frac{1}{30}, \quad b_5 = 0, \quad b_6 = \frac{1}{42}.$$

Furthermore, the first five Bernoulli polynomials are

$$\beta_0(t) = 1,$$
$$\beta_1(t) = t - \frac{1}{2},$$
$$\beta_2(t) = t^2 - t + \frac{1}{6},$$
$$\beta_3(t) = t^3 - \frac{3}{2}t^2 + \frac{1}{2}t,$$
$$\beta_4(t) = t^4 - 2t^3 + t^2 - \frac{1}{30}.$$

For an arbitrary function $x \in L^2[0,1]$, we can write

$$x(t) = \sum_{m=0}^{\infty} a_m \beta_m(t).$$

Therefore, an approximation of the function x can be given by

$$x(t) \simeq x_M(t) = \sum_{m=0}^{M} a_m \beta_m(t) = A^T B(t), \qquad (9)$$

where

$$B(t) = [\beta_0(t), \beta_1(t), \ldots, \beta_M(t)]^T \qquad (10)$$

and

$$A = [a_0, a_1, \ldots, a_M]^T.$$

The vector A in (9) is called the coefficient vector and can be calculated by the formula (see [2])

$$A = D^{-1} \langle x(t), B(t) \rangle,$$

where $\langle \cdot, \cdot \rangle$ is the inner product, defined for two arbitrary functions $f, g \in L^2[0,1]$ as

$$\langle f(t), g(t) \rangle = \int_0^1 f(t) g(t) dt,$$

and $D = \langle B(t), B(t) \rangle$ is calculated using the following property of Bernoulli polynomials [29]:

$$\int_0^1 \beta_i(t) \beta_j(t) dt = (-1)^{i-1} \frac{i! j!}{(i+j)!} b_{i+j}, \quad i, j \geq 1.$$

It should be noted that

$$X = \text{span} \{\beta_0(t), \beta_1(t), \ldots, \beta_M(t)\}$$

is a finite dimensional subspace of $L^2[0,1]$ and x_M, given by (9), is the best approximation of function x in X.

3. Operational Matrix of Variable-Order Fractional Integration

In this section, we introduce an accurate operational matrix of variable-order fractional integration for Bernoulli functions. To this aim, we rewrite the Bernoulli basis vector B given by (10) in terms of the Taylor basis functions as follows:

$$B(t) = Q\mathbb{T}(t), \qquad (11)$$

where $\mathbb{T}$ is the Taylor basis vector given by

$$\mathbb{T}(t) = \begin{bmatrix} 1, t, t^2, \ldots, t^M \end{bmatrix}^T$$

and Q is the change-of-basis matrix, which is obtained using (8) as

$$Q = \begin{bmatrix} 1 & 0 & 0 & 0 & 0 & \cdots & 0 \\ -\frac{1}{2} & 1 & 0 & 0 & 0 & \cdots & 0 \\ \frac{1}{6} & -1 & 1 & 0 & 0 & \cdots & 0 \\ 0 & \frac{1}{2} & -\frac{3}{2} & 1 & 0 & \cdots & 0 \\ \vdots & \vdots & \vdots & \vdots & \vdots & & \vdots \\ b_M & \binom{M}{1} b_{M-1} & \binom{M}{2} b_{M-2} & \binom{M}{3} b_{M-3} & \binom{M}{4} b_{M-4} & \cdots & 1 \end{bmatrix}.$$

Since Q is nonsingular, we can write

$$\mathbb{T}(t) = Q^{-1} B(t). \tag{12}$$

By considering (11) and applying the left Riemann–Liouville fractional integral operator of order $\alpha(t)$ to the vector $B(t)$, we get that

$$_0 I_t^{\alpha(t)} B(t) = {}_0 I_t^{\alpha(t)} (Q \mathbb{T}(t)) = Q({}_0 I_t^{\alpha(t)} \mathbb{T}(t)) = Q S_t^{\alpha(t)} \mathbb{T}(t), \tag{13}$$

where $S_t^{\alpha(t)}$ is a diagonal matrix, which is obtained using (4) as follows:

$$S_t^{\alpha(t)} = \begin{bmatrix} \frac{1}{\Gamma(1+\alpha(t))} t^{\alpha(t)} & 0 & 0 & 0 & \cdots & 0 \\ 0 & \frac{1}{\Gamma(2+\alpha(t))} t^{\alpha(t)} & 0 & 0 & \cdots & 0 \\ 0 & 0 & \frac{2}{\Gamma(3+\alpha(t))} t^{\alpha(t)} & 0 & \cdots & 0 \\ \vdots & \vdots & \vdots & \vdots & & \vdots \\ 0 & 0 & 0 & 0 & \cdots & \frac{\Gamma(M+1)}{\Gamma(M+1+\alpha(t))} t^{\alpha(t)} \end{bmatrix}.$$

Finally, by using (12) in (13), we have

$$_0 I_t^{\alpha(t)} B(t) = Q S_t^{\alpha(t)} Q^{-1} B(t) = P_t^{\alpha(t)} B(t), \tag{14}$$

where $P_t^{\alpha(t)} = Q S_t^{\alpha(t)} Q^{-1}$ is a matrix of dimension $(M+1) \times (M+1)$, which we call the operational matrix of variable-order fractional integration $\alpha(t)$ for Bernoulli functions. Since Q and Q^{-1} are lower triangular matrices and $S_t^{\alpha(t)}$ is a diagonal matrix, $P_t^{\alpha(t)}$ is also a lower triangular matrix. In the particular case of $M = 2$, one has

$$P_t^{\alpha(t)} = \begin{bmatrix} \frac{1}{\Gamma(\alpha(t)+1)} t^{\alpha(t)} & 0 & 0 \\ \left(\frac{1}{2\Gamma(\alpha(t)+2)} - \frac{1}{2\Gamma(\alpha(t)+1)} \right) t^{\alpha(t)} & \frac{1}{\Gamma(\alpha(t)+2)} t^{\alpha(t)} & 0 \\ \left(\frac{1}{6\Gamma(\alpha(t)+1)} - \frac{1}{2\Gamma(\alpha(t)+2)} + \frac{2}{3\Gamma(\alpha(t)+3)} \right) t^{\alpha(t)} & \left(\frac{2}{\Gamma(\alpha(t)+3)} - \frac{1}{\Gamma(\alpha(t)+2)} \right) t^{\alpha(t)} & \frac{2}{\Gamma(\alpha(t)+3)} t^{\alpha(t)} \end{bmatrix}.$$

4. Methods of Solution

In this section, we propose two approaches for solving problem (1)–(3). To do this, first we introduce

$$n = \max_{0 < t \leq 1} \{ \lceil \alpha(t) \rceil \}.$$

Then, we may use the following two approaches to find approximations for the state and control functions, which optimize the performance index.

4.1. Approach I

In our first approach, we consider an approximation of the nth order derivative of the unknown state function x using Bernoulli polynomials. Set

$$x^{(n)}(t) = A^T B(t), \tag{15}$$

where A is a $1 \times (M+1)$ vector with unknown elements and B is the Bernoulli basis vector given by (10). Then, using the initial conditions given in (3), and Equations (5), (14), and (15), we get

$$\begin{aligned} x(t) &= {}_0I_t^n(x^{(n)}(t)) + \sum_{i=0}^{n-1} x^{(i)}(0)\frac{t^i}{i!} \\ &= A^T({}_0I_t^n B(t)) + \sum_{i=0}^{n-1} x_0^i \frac{t^i}{i!} \\ &= A^T P_t^n B(t) + \sum_{i=0}^{n-1} x_0^i \frac{t^i}{i!}. \end{aligned} \tag{16}$$

Moreover, using (6), (14), and (15), we get

$$_0^C D_t^{\alpha(t)} x(t) = A^T P_t^{n-\alpha(t)} B(t) + \sum_{i=\lceil \alpha(t) \rceil}^{n-1} x_0^i \frac{t^{i-\alpha(t)}}{\Gamma(i+1-\alpha(t))} := F[A,t] \tag{17}$$

and

$$_0^C D_t^{\alpha_j(t)} x(t) = A^T P_t^{n-\alpha_j(t)} B(t) + \sum_{i=\lceil \alpha_j(t) \rceil}^{n-1} x_0^i \frac{t^{i-\alpha_j(t)}}{\Gamma(i+1-\alpha_j(t))} := F_j[A,t], \quad j=1,\ldots,s. \tag{18}$$

By substituting (16)–(18) into the control-affine dynamical system given by (2), we obtain an approximation of the control function as follows:

$$u(t) = \frac{1}{b(t)}\left[F[A,t] - \varphi\left(t, A^T P_t^n B(t) + \sum_{i=0}^{n-1} x_0^i \frac{t^i}{i!}, F_1[A,t], \ldots, F_s[A,t]\right)\right]. \tag{19}$$

Taking into consideration (16) and (19) in the performance index J, we have

$$J[A] = \int_0^1 \phi\left(t, A^T P_t^n B(t) + \sum_{i=0}^{n-1} x_0^i \frac{t^i}{i!}, \frac{1}{b(t)}\left[F[A,t] - \varphi\left(t, A^T P_t^n B(t) + \sum_{i=0}^{n-1} x_0^i \frac{t^i}{i!}, F_1[A,t], \ldots, F_s[A,t]\right)\right]\right) dt.$$

For the sake of simplicity, we introduce

$$G[A,t] = \phi\left(t, A^T P_t^n B(t) + \sum_{i=0}^{n-1} x_0^i \frac{t^i}{i!}, \frac{1}{b(t)}\left[F[A,t] - \varphi\left(t, A^T P_t^n B(t) + \sum_{i=0}^{n-1} x_0^i \frac{t^i}{i!}, F_1[A,t], \ldots, F_s[A,t]\right)\right]\right).$$

In many applications, it is difficult to compute the integral of function $G[A,t]$. Therefore, it is recommended to use a suitable numerical integration formula. Here, we use the Gauss–Legendre quadrature formula to obtain

$$J[A] \simeq \frac{1}{2}\sum_{i=1}^N w_i G\left[A, \frac{t_i+1}{2}\right], \tag{20}$$

where t_i, $i=1,2,\ldots,N$, are the zeros of the Legendre polynomial of degree N, $P_N(t)$, and w_i are the corresponding weights [30], which are given by

$$\omega_i = \frac{2}{\left(\frac{d}{dt}P_N(t_i)\right)^2 (1-t_i^2)}, \quad i=1,\ldots,N. \tag{21}$$

Finally, the first order necessary condition for the optimality of the performance index implies

$$\frac{\partial J[A]}{\partial A} = 0,$$

which gives a system of $M+1$ nonlinear algebraic equations in terms of the $M+1$ unknown elements of the vector A. By solving this system, approximations of the optimal state and control functions are, respectively, given by (16) and (19).

4.2. Approach II

In our second approach, we set

$$^C_0 D^{\alpha(t)}_t x(t) = A^T B(t). \tag{22}$$

Then, using (7) with $\beta(t) \equiv 0$, we obtain that

$$\begin{aligned} x(t) &= {_0I^{\alpha(t)}_t}({^C_0 D^{\alpha(t)}_t} x(t)) + \sum_{i=0}^{\lceil \alpha(t) \rceil - 1} x^{(i)}(0) \frac{t^i}{\Gamma(i+1)} \\ &= A^T({_0I^{\alpha(t)}_t} B(t)) + \sum_{i=0}^{\lceil \alpha(t) \rceil - 1} x_0^i \frac{t^i}{i!} \\ &= A^T P^{\alpha(t)}_t B(t) + \sum_{i=0}^{\lceil \alpha(t) \rceil - 1} x_0^i \frac{t^i}{i!}. \end{aligned} \tag{23}$$

Furthermore, we get

$$^C_0 D^{\alpha_j(t)}_t x(t) = A^T P^{\alpha(t)-\alpha_j(t)}_t B(t) + \sum_{i=\lceil \alpha_j(t) \rceil}^{\lceil \alpha(t) \rceil -1} x_0^i \frac{t^{i-\alpha_j(t)}}{\Gamma(i+1-\alpha_j(t))} := F_j[A,t], \quad j=1,\ldots,s. \tag{24}$$

Taking (22)–(24) into consideration, Equation (2) gives

$$u(t) = \frac{1}{b(t)} \left[A^T B(t) - \varphi\left(t, A^T P^{\alpha(t)}_t B(t) + \sum_{i=0}^{\lceil \alpha(t) \rceil -1} x_0^i \frac{t^i}{i!}, F_1[A,t],\ldots,F_s[A,t]\right) \right]. \tag{25}$$

By substituting the approximations given by (23) and (25) into the performance index, we get

$$J[A] = \int_0^1 \phi\left(t, A^T P^{\alpha(t)}_t B(t) + \sum_{i=0}^{\lceil \alpha(t) \rceil -1} x_0^i \frac{t^i}{i!}, \right.$$
$$\left. \frac{1}{b(t)}\left[A^T B(t) - \varphi\left(t, A^T P^{\alpha(t)}_t B(t) + \sum_{i=0}^{\lceil \alpha(t) \rceil -1} x_0^i \frac{t^i}{i!}, F_1[A,t],\ldots,F_s[A,t]\right)\right]\right) dt.$$

By introducing

$$G[A, t] = \phi\left(t, A^T P_t^{\alpha(t)} B(t) + \sum_{i=0}^{\lceil \alpha(t) \rceil - 1} x_0^i \frac{t^i}{i!}, \right.$$

$$\left. \frac{1}{b(t)} \left[A^T B(t) - \varphi\left(t, A^T P_t^{\alpha(t)} B(t) + \sum_{i=0}^{\lceil \alpha(t) \rceil - 1} x_0^i \frac{t^i}{i!}, F_1[A, t], \ldots, F_s[A, t] \right) \right] \right),$$

then this approach continues in the same way of finding the unknown parameters of the vector A as in Approach I.

5. Error Bounds

The aim of this section is to give some error bounds for the numerical solution obtained by the proposed methods of Section 4. We present the error discussion for Approach II, which can then be easily extended to Approach I.

Suppose that x^* is the optimal state function of problem (1)–(3). Let $f(t) := {}_0^C D_t^{\alpha(t)} x^*(t)$ with $f(t) \in H^\mu(0,1)$ ($H^\mu(0,1)$ is a Sobolev space [31]). According to our numerical method, $f_M(t) = A^T B(t)$ is the best approximation of function f in terms of the Bernoulli polynomials, that is,

$$\forall g \in X, \ \|f - f_M\|_2 \leq \|f - g\|_2.$$

We recall the following lemma from [31].

Lemma 1 (See [31]). *Assume that $f \in H^\mu(0,1)$ with $\mu \geq 0$. Let $L_M(f) \in X$ be the truncated shifted Legendre series of f. Then,*

$$\|f - L_M(f)\|_2 \leq CM^{-\mu} |f|_{H^{\mu;M}(0,1)},$$

where

$$|f|_{H^{\mu;M}(0,1)} = \left(\sum_{j=\min\{\mu, M+1\}}^\mu \|f^{(j)}\|_2^2 \right)^{\frac{1}{2}}$$

and C is a positive constant independent of function f and integer M.

Since the best approximation of function f in the subspace X is unique and f_M and $L_M(f)$ are both the best approximations of f in X, we have $f_M = L_M(f)$. Therefore, we get that

$$\|f - f_M\|_2 \leq CM^{-\mu} |f|_{H^{\mu;M}(0,1)}. \tag{26}$$

Hereafter, C denotes a positive constant independent of M and n.

Theorem 1. *Suppose x^* to be the exact optimal state function of problem (1)–(3) such that $f(t) := {}_0^C D_t^{\alpha(t)} x^*(t) \in H^\mu(0,1)$, with $\mu \geq 0$, and $\tilde{x}$ be its approximation given by (23). Then,*

$$\|x^*(t) - \tilde{x}(t)\|_2 \leq CM^{-\mu} |f|_{H^{\mu;M}(0,1)}. \tag{27}$$

Proof. Let $Y = L^2[0,1]$ and ${}_0 I_t^{\alpha(t)} : Y \to Y$ be the variable-order Riemann–Liouville integral operator. By definition of the norm for operators, we have

$$\|{}_0 I_t^{\alpha(t)}\|_2 = \sup_{\|g\|_2 = 1} \|{}_0 I_t^{\alpha(t)} g\|_2.$$

In order to prove the theorem, first we show that the operator $_0I_t^{\alpha(t)}$ is bounded. Since $\|g\|_2 = 1$, using Schwarz's inequality, we get

$$\left\|_0I_t^{\alpha(t)} g\right\|_2 = \left\|\frac{1}{\Gamma(\alpha(t))} \int_0^t (t-s)^{\alpha(t)-1} g(s) ds \right\|_2$$

$$\leq \|g\|_2 \left\|\frac{1}{\Gamma(\alpha(t))} \int_0^t (t-s)^{\alpha(t)-1} ds \right\|_2$$

$$= \left\|\frac{t^{\alpha(t)}}{\Gamma(\alpha(t)+1)}\right\|_2$$

$$\leq C,$$

where we have used the assumption $\alpha(t) > 0$, which gives $t^{\alpha(t)} < 1$ for $0 < t \leq 1$, and a particular property of the Gamma function, which is $\Gamma(t) > 0.8$. Therefore, $_0I_t^{\alpha(t)}$ is bounded. Now, using (26), and taking into account (7) and (23), we obtain that

$$\|x^*(t) - \tilde{x}(t)\|_2 = \left\|_0I_t^{\alpha(t)} f(t) + \sum_{i=0}^{\lceil\alpha(t)\rceil-1} x^{(i)}(0)\frac{t^i}{\Gamma(i+1)} - \left(_0I_t^{\alpha(t)}(A^T B(t)) + \sum_{i=0}^{\lceil\alpha(t)\rceil-1} x_0^i \frac{t^i}{i!}\right)\right\|_2$$

$$= \left\|_0I_t^{\alpha(t)}(f(t) - A^T B(t))\right\|_2$$

$$\leq \left\|_0I_t^{\alpha(t)}\right\|_2 \|f(t) - A^T B(t)\|_2$$

$$\leq CM^{-\mu} |f|_{H^{\mu;M}(0,1)}.$$

The proof is complete. □

Remark 1. *Since we have $\alpha(t) - \alpha_j(t) > 0$, $j = 1, 2, \ldots, s$, with a similar argument it can be shown that*

$$\left\|_0^C D_t^{\alpha_j(t)} x^*(t) - \left(A^T P_t^{\alpha(t)-\alpha_j(t)} B(t) + \sum_{i=\lceil\alpha_j(t)\rceil}^{\lceil\alpha(t)\rceil-1} x_0^i \frac{t^{i-\alpha_j(t)}}{\Gamma(i+1-\alpha_j(t))}\right)\right\|_2 \leq CM^{-\mu}|f|_{H^{\mu;M}(0,1)}.$$

With the help of Theorem 1, we obtain the following result for the error of the optimal control function. For simplicity, suppose that in the control-affine dynamical system given by (2) the function φ appears as $\varphi := \varphi(t,x)$ (cf. Remark 2).

Theorem 2. *Suppose that the assumptions of Theorem 1 are fulfilled. Let u^* and $\tilde{u}$ be the exact and approximate optimal control functions, respectively. If $\varphi : \mathbb{R}^2 \longrightarrow \mathbb{R}$ satisfies a Lipschitz condition with respect to the second argument, then*

$$\|u^*(t) - \tilde{u}(t)\|_2 \leq CM^{-\mu}|f|_{H^{\mu;M}(0,1)}. \tag{28}$$

Proof. Using Equation (2), the exact optimal control function is given by

$$u^*(t) = \frac{1}{b(t)}\left(f(t) - \varphi(t, x^*(t))\right) \tag{29}$$

and the approximate control function obtained by our Approach II is given by

$$\tilde{u}(t) = \frac{1}{b(t)}\left(A^T B(t) - \varphi(t, \tilde{x}(t))\right). \tag{30}$$

By subtracting (30) from (29), we get

$$u^*(t) - \tilde{u}(t) = \frac{1}{b(t)}\left(f(t) - \varphi(t, x^*(t)) - A^T B(t) + \varphi(t, \tilde{x}(t))\right). \quad (31)$$

Since the function φ satisfies a Lipschitz condition with respect to the second variable, there exists a positive constant K such that

$$|\varphi(t, \mathbf{x}_1) - \varphi(t, \mathbf{x}_2)| < K|\mathbf{x}_1 - \mathbf{x}_2|.$$

Therefore, using (26) and (27), and also taking into account $b(t) \neq 0$, we have

$$\|u^*(t) - \tilde{u}(t)\|_2 \leq \frac{1}{\|b(t)\|_2}\left(\left\|f(t) - A^T B(t)\right\|_2 + K\|x^*(t) - \tilde{x}(t)\|_2\right) \leq CM^{-\mu}|f|_{H^{\mu;M}(0,1)},$$

which yields (28). □

Remark 2. *For the general case $\varphi := \varphi(t, x, x_1, \ldots, x_s)$, the same result of Theorem 2 can be easily obtained by assuming that φ satisfies Lipschitz conditions with respect to the variables $x, x_1, \ldots, x_s$.*

As a result of Theorems 1 and 2, we obtain an error bound for the approximate value of the optimal performance index J given by (20). First, we recall the following lemma in order to obtain the error of the Gauss–Legendre quadrature rule.

Lemma 2 (See [30]). *Let g be a given sufficiently smooth function. Then, the Gauss–Legendre quadrature rule is given by*

$$\int_{-1}^{1} g(t)dt = \sum_{i=1}^{N} \omega_i g(t_i) + E_N(g), \quad (32)$$

where t_i, $i = 1, \ldots, N$, are the roots of the Legendre polynomial of degree N, and ω_i are the corresponding weights given by (21). In (32), $E_N(g)$ is the error term, which is given as follows:

$$E_N(g) = \frac{2^{2N+1}(N!)^4}{(2N+1)[(2N)!]^3} g^{2N}(\eta), \quad \eta \in (-1,1).$$

Now, by considering the assumptions of Theorems 1 and 2, we prove the following result.

Theorem 3. *Let J^* be the exact value of the optimal performance index J in problem (1)–(3) and $\tilde{J}$ be its approximation given by (20). Suppose that the function $\phi : \mathbb{R}^3 \longrightarrow \mathbb{R}$ is a sufficiently smooth function with respect to all its variables and satisfies Lipschitz conditions with respect to its second and third arguments, that is,*

$$|\phi(t, \mathbf{x}_1, u) - \phi(t, \mathbf{x}_2, u)| \leq K_1|\mathbf{x}_1 - \mathbf{x}_2| \quad (33)$$

and

$$|\phi(t, x, \mathbf{u}_1) - \phi(t, x, \mathbf{u}_2)| \leq K_1|\mathbf{u}_1 - \mathbf{u}_2|, \quad (34)$$

where K_1 and K_2 are real positive constants. Then, there exist positive constants C_1 and C_2 such that

$$|J^* - \tilde{J}| \leq C_1 M^{-\mu}|f|_{H^{\mu;M}(0,1)} + C_2 \frac{(N!)^4}{(2N+1)[(2N)!]^3}. \quad (35)$$

Proof. Using (20) and (32), we have

$$\tilde{J} = \frac{1}{2}\sum_{i=1}^{N} \omega_i \phi\left(\frac{t_i+1}{2}, \tilde{x}\left(\frac{t_i+1}{2}\right), \tilde{u}\left(\frac{t_i+1}{2}\right)\right) = \int_0^1 \phi(t, \tilde{x}(t), \tilde{u}(t))\,dt - \xi_N, \quad (36)$$

where

$$\xi_N = \left(\frac{1}{2}\right) \frac{2^{2N+1}(N!)^4}{(2N+1)[(2N!)]^3} \left(\frac{1}{2}\right)^{2N} \left.\frac{\partial^{2N}\phi(t,\tilde{x}(t),\tilde{u}(t))}{\partial t^{2N}}\right|_{t=\eta} = \frac{(N!)^4}{(2N+1)[(2N!)]^3} \left.\frac{\partial^{2N}\phi(t,\tilde{x}(t),\tilde{u}(t))}{\partial t^{2N}}\right|_{t=\eta}$$

for $\eta \in (0,1)$. Therefore, taking into consideration (33)–(36), we get

$$|J^* - \tilde{J}| = \left|\int_0^1 \phi(t,x^*(t),u^*(t))dt - \int_0^1 \phi(t,\tilde{x}(t),\tilde{u}(t))\,dt + \xi_N\right|$$

$$= \left|\int_0^1 \phi(t,x^*(t),u^*(t))dt - \int_0^1 \phi(t,\tilde{x}(t),u^*(t))dt + \int_0^1 \phi(t,\tilde{x}(t),u^*(t))dt - \int_0^1 \phi(t,\tilde{x}(t),\tilde{u}(t))\,dt + \xi_N\right|$$

$$\leq K_1 \int_0^1 |x^*(t) - \tilde{x}(t)|\,dt + K_2 \int_0^1 |u^*(t) - \tilde{u}(t)|\,dt + \max_{0<t<1}|\xi_N|$$

$$\leq C_1 M^{-\mu}|f|_{H^{\mu;M}(0,1)} + C_2 \frac{(N!)^4}{(2N+1)[(2N!)]^3},$$

where we have used the property of equivalence of L^1 and L^2-norms and

$$C_2 = \max_{0<t<1} \left|\frac{\partial^{2N}\phi(t,\tilde{x}(t),\tilde{u}(t))}{\partial t^{2N}}\right|.$$

The proof is complete. □

Remark 3. *A similar error discussion can be considered for Approach I by setting $f(t) := x^{*(n)}(t)$ with $f(t) \in H^\mu(0,1)$ and taking into account the fact that the operators I^n, $I^{\alpha(t)}$ and $I^{\alpha_j(t)}$, for $j = 1,2,\ldots,s$, are bounded.*

Remark 4. *In practice, since the exact control and state functions that minimize the performance index are unknown, in order to reach a given specific accuracy ϵ for these functions, we increase the number of basis functions (by increasing M) in our implementation, such that*

$$\max_{1\leq i \leq M} |F[A,t_i] - \varphi(t_i,\tilde{x}(t_i),F_1[A,t_i],\ldots,F_s[A,t_i]) - b(t_i)\tilde{u}(t_i)| < \epsilon \quad \text{(Approach I)},$$

and

$$\max_{1\leq i \leq M} \left|A^T B(t_i) - \varphi(t_i,\tilde{x}(t_i),F_1[A,t_i],\ldots,F_s[A,t_i]) - b(t_i)\tilde{u}(t_i)\right| < \epsilon \quad \text{(Approach II)},$$

where

$$t_i = \frac{i}{M+1}, \quad i = 1,2,\ldots,M.$$

6. Test Problems

In this section, some FOC-APs are included and solved by the proposed methods, in order to illustrate the accuracy and efficiency of the new techniques. In our implementation, the method was carried out using Mathematica 12. Furthermore, we have used $N = 14$ in employing the Gauss–Legendre quadrature formula.

Example 1. *As first example, we consider the following variable-order FOC-AP:*

$$\min J = \int_0^1 \left[\left(x(t) - t^2\right)^2 + \left(u(t) - \frac{1}{\Gamma(3-\alpha(t))}t^{2-\alpha(t)}e^{-t} + \frac{1}{2}e^{t^2-t}\right)^2\right]dt \quad (37)$$

subject to
$$^C_0D_t^{\alpha(t)}x(t) = e^{x(t)} + 2e^t u(t), \quad 0 < \alpha(t) \le 1,$$
$$x(0) = 0.$$

The exact optimal state and control functions are given by
$$x(t) = t^2, \quad u(t) = \frac{1}{\Gamma(3-\alpha(t))}t^{2-\alpha(t)}e^{-t} - \frac{1}{2}e^{t^2-t},$$

which minimize the performance index J with the minimum value $J = 0$. In [26], a numerical method based on the Legendre wavelet has been used to solve this problem with $\alpha(t) = 1$. For solving this problem with $\alpha(t) = 1$, according to our methods, we have $n = 1$. In this case, both approaches introduced in Section 4 give the same result. By setting $M = 1$, we suppose that
$$x'(t) = A^T B(t) = a_1\left(t - \frac{1}{2}\right) + a_0,$$

where
$$A = [a_0, a_1]^T \text{ and } B(t) = \left[1, t - \frac{1}{2}\right]^T.$$

The operational matrix of variable-order fractional integration is given by
$$P_t^1 = \begin{bmatrix} t & 0 \\ -\frac{t}{4} & \frac{t}{2} \end{bmatrix}.$$

Therefore, we have
$$x(t) = A^T P_t^1 B(t) = a_0 t + \frac{1}{2}a_1(t-1)t. \tag{38}$$

Moreover, using the control-affine dynamical system, we get
$$u(t) = \frac{1}{2}e^{-t}\left(A^T B(t) - e^{A^T P_t^1 B(t)}\right) = \frac{1}{2}e^{-t}\left(a_1\left(t - \frac{1}{2}\right) + a_0 - e^{a_0 t + \frac{1}{2}a_1(t-1)t}\right). \tag{39}$$

By substituting (38) and (39) into (37), using the Gauss–Legendre quadrature for computing J, and, finally, setting
$$\frac{\partial J}{\partial a_0} = 0, \quad \frac{\partial J}{\partial a_1} = 0,$$

we obtain a system of two nonlinear algebraic equations in terms of a_0 and a_1. By solving this system, we find
$$a_0 = 1, \quad a_1 = 2,$$

which gives the exact solution
$$x(t) = t^2 \text{ and } u(t) = te^{-t} - \frac{1}{2}e^{t^2-t}.$$

As it is seen, in the case of $\alpha(t) = 1$, our approaches give the exact solution with $M = 1$ (only two basis functions) compared to the method introduced in [26] based on the use of Legendre wavelets with $\hat{m} = 6$ (six basis functions).

Since the optimal state function is a polynomial of degree 2, Approach I gives the exact solution with $M = 1$ for every admissible $\alpha(t)$. On the other hand, if $\alpha(t) \ne 1$, then $^C_0D_t^{\alpha(t)}x(t) \in H^1(0,1)$. Therefore, according to the theoretical discussion and the error bound given by (35), the numerical solution given by Approach II converges to the exact solution, very slowly, that can be confirmed by the results reported in Table 1 obtained with $\alpha(t) = \sin(t)$ and different values of M. Furthermore, by considering a different $\alpha(t)$, and by applying the two proposed approaches with $M = 5$, the numerical results for the functions x and u are displayed in Figures 1 and 2. Figure 1 displays the numerical results obtained by Approach I, while Figure 2 shows the numerical results given by Approach II. For these results, we have used

$$\alpha_1(t) = 1, \quad \alpha_2(t) = \sin(t), \quad \alpha_3(t) = \frac{t}{2}, \quad \alpha_4(t) = \frac{t}{3}. \tag{40}$$

Moreover, the numerical results for the performance index, obtained by our two approaches, are shown in Table 2. It can be easily seen that, in this case, Approach I gives higher accuracy results than Approach II. This is caused by the smoothness of the exact optimal state function x.

Table 1. (Example 1.) Numerical results obtained by Approach II for the performance index with different M and $\alpha(t) = \sin(t)$.

M	1	2	3	2	5
J	6.80×10^{-3}	2.33×10^{-3}	1.76×10^{-3}	1.57×10^{-3}	1.56×10^{-3}

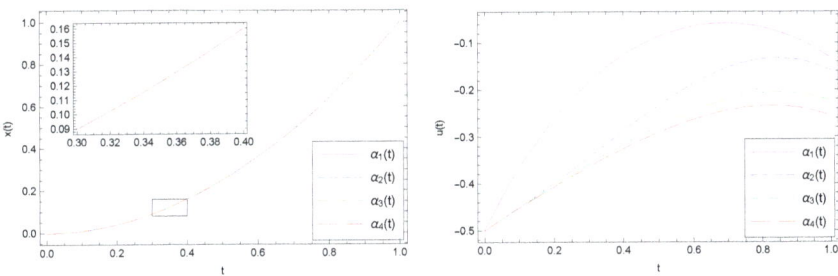

Figure 1. (Example 1.) Comparison between the approximate state (**left**) and control (**right**) functions obtained by Approach I with $M = 5$ and different $\alpha(t)$ (40).

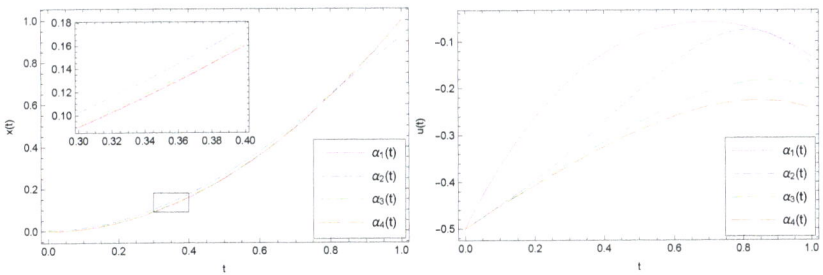

Figure 2. (Example 1.) Comparison between the approximate state (**left**) and control (**right**) functions obtained by Approach II with $M = 5$ and different $\alpha(t)$ (40).

Table 2. (Example 1.) Numerical results for the performance index with $M = 5$ and different $\alpha(t)$ (40).

Method	$\alpha_1(t)$	$\alpha_2(t)$	$\alpha_3(t)$	$\alpha_4(t)$
Approach I	3.05×10^{-33}	3.26×10^{-33}	6.89×10^{-33}	2.08×10^{-33}
Approach II	2.74×10^{-33}	1.56×10^{-3}	1.71×10^{-4}	2.50×10^{-5}

Example 2. Consider now the following FOC-AP borrowed from [32]:

$$\min J = \int_0^1 \left[\left(x(t) - t^{\frac{5}{2}}\right)^4 + (1+t^2)\left(u(t) + t^6 - \frac{15\sqrt{\pi}}{8}t\right)^2 \right] dt \qquad (41)$$

subject to

$${}_0^C D_t^{\frac{3}{2}} x(t) = tx^2(t) + u(t) \qquad (42)$$

and the initial conditions $x(0) = x'(0) = 0$. For this problem, the state and control functions

$$x(t) = t^{\frac{5}{2}}, \quad u(t) = -t^6 + \frac{15\sqrt{\pi}}{8}t$$

minimize the performance index with the optimal value $J = 0$. We have solved this problem by both approaches. The numerical results of applying Approach I to this problem, with different values of M, are presented in Figure 3 and Table 3. Figure 3 displays the approximate state (left) and control (right) functions obtained by $M = 1, 3, 5, 7$, together with the exact ones, while Table 3 reports the approximate values of the performance index. Here, we show that Approach II gives the exact solution by considering $M = 1$. To do this, we suppose that

$$_0^C D_t^{\frac{3}{2}} x(t) = A^T B(t) = a_1 \left(t - \frac{1}{2}\right) + a_0$$

with

$$A = [a_0, a_1]^T \text{ and } B(t) = \left[1, t - \frac{1}{2}\right]^T.$$

Therefore, we have

$$x(t) = A^T P_t^{\frac{3}{2}} B(t) = \frac{2}{3\sqrt{\pi}}(2a_0 - a_1)t^{\frac{3}{2}} + \frac{8}{15\sqrt{\pi}} a_1 t^{\frac{5}{2}}, \qquad (43)$$

where

$$P_t^{\frac{3}{2}} = \begin{bmatrix} \frac{4}{3\sqrt{\pi}} t^{\frac{3}{2}} & 0 \\ -\frac{2}{5\sqrt{\pi}} t^{\frac{3}{2}} & \frac{8}{15\sqrt{\pi}} t^{\frac{3}{2}} \end{bmatrix}.$$

Using the dynamical control-affine system given by (42), we get

$$u(t) = a_1\left(t - \frac{1}{2}\right) + a_0 - t\left(\frac{2}{3\sqrt{\pi}}(2a_0 - a_1)t^{\frac{3}{2}} + \frac{8}{15\sqrt{\pi}} a_1 t^{\frac{5}{2}}\right)^2$$

$$= -\frac{64a_1^2}{225\pi} t^6 + \left(\frac{32a_1^2}{45\pi} - \frac{64a_0 a_1}{45\pi}\right) t^5 + \left(\frac{16a_0 a_1}{9\pi} - \frac{16a_0^2}{9\pi} - \frac{4a_1^2}{9\pi}\right) t^4 + a_1 t + a_0 - \frac{a_1}{2}. \qquad (44)$$

By substituting (43) and (44) into (41), the value of the integral can be easily computed. Then, by taking into account the optimality condition, a system of nonlinear algebraic equations is obtained. Finally, by solving this system, we obtain

$$a_0 = \frac{15\sqrt{\pi}}{16}, \quad a_1 = \frac{15\sqrt{\pi}}{8}.$$

By taking into account these values in (43) and (44), the exact optimal state and control functions are obtained. Lotfi et al. have solved this problem using an operational matrix technique based on the Legendre orthonormal functions combined with the Gauss quadrature rule. In their method, the approximate value of the minimum performance index with five basis functions has been reported as 7.82×10^{-9} while our suggested Approach II gives the exact value only with two basis functions.

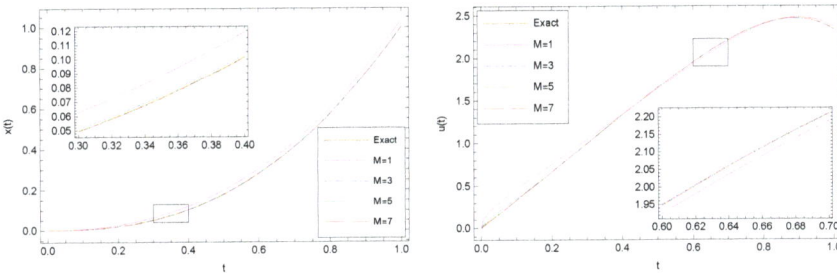

Figure 3. (Example 2.) Comparison between the exact and approximate state (**left**) and control (**right**) functions obtained by Approach I with different values of M.

Table 3. (Example 2.) Numerical results for the performance index obtained by Approach I with different M.

M	1	3	5	7
J	5.24×10^{-4}	7.59×10^{-6}	4.65×10^{-7}	5.86×10^{-8}

As we see, in this example, Approach II yields the exact solution with a small computational cost, while the precision of the results of Approach I increases by enlarging M. Note that here the optimal state function is not an infinitely smooth function.

Example 3. *As our last example, we consider the following FOC-AP [32]:*

$$\min J = \int_0^1 \left[e^t \left(x(t) - t^4 + t - 1 \right)^2 + (1+t^2) \left(u(t) + 1 - t + t^4 - \frac{8000}{77\Gamma\left(\frac{1}{10}\right)} t^{\frac{21}{10}} \right)^2 \right] dt$$

subject to

$${}_0^C D_t^{1.9} x(t) = x(t) + u(t),$$
$$x(0) = 1, \quad x'(0) = -1.$$

For this example, the following state and control functions minimize the performance index J with minimum value J = 0:

$$x(t) = t^4 - t + 1, \quad u(t) = -t^4 + \frac{8000}{77\Gamma\left(\frac{1}{10}\right)} t^{\frac{21}{10}} + t - 1.$$

This problem has been solved using the proposed methods with different values of M. By considering $M = 1$, the numerical results of Approach I are shown in Figure 4. In this case, an approximation of the performance index is obtained as $J = 7.21 \times 10^{-1}$. By choosing $M = 2$, according to our numerical method, we have $n = 2$. Therefore, we set

$$x''(t) = A^T B(t),$$

where

$$A = [a_0, a_1, a_2], \quad B(t) = \left[1, t - \frac{1}{2}, t^2 - t + \frac{1}{6} \right]^T.$$

Hence, using the initial conditions, the state function can be approximated by

$$x(t) = A^T P_t^2 B(t) - t + 1 = \frac{a_2}{12} t^4 + \frac{a_1 - a_2}{6} t^3 + \left(\frac{a_0}{2} - \frac{a_1}{4} + \frac{a_2}{12} \right) t^2 - t + 1,$$

where

$$P_t^2 = \begin{bmatrix} \frac{t^2}{2} & 0 & 0 \\ -\frac{t^2}{6} & \frac{t^2}{6} & 0 \\ \frac{t^2}{36} & -\frac{t^2}{12} & \frac{t^2}{12} \end{bmatrix}.$$

In the continuation of the method, we find an approximation of the control function u using the control-affine dynamical system. Then, the method proceeds until solving the resulting system, which yields

$$a_0 = 4, \quad a_1 = 12, \quad a_2 = 12.$$

These values give us the exact solution of the problem. This problem has been solved in [32] with five basis functions and the minimum value was obtained as $J = 5.42 \times 10^{-7}$ while our suggested Approach I gives the exact value with only three basis functions.

In the implementation of Approach II, we consider different values of M and report the results in Table 4 and Figure 5. These results confirm that the numerical solutions converge to the exact one by increasing the value of M. Nevertheless, we see that since the exact state function x is a smooth function, it takes much less computational effort to solve this problem by using Approach I.

Table 4. (Example 3.) Numerical results for the performance index obtained by Approach II with different M.

M	2	4	6	8
J	3.79×10^{-4}	5.42×10^{-7}	1.21×10^{-8}	7.36×10^{-10}

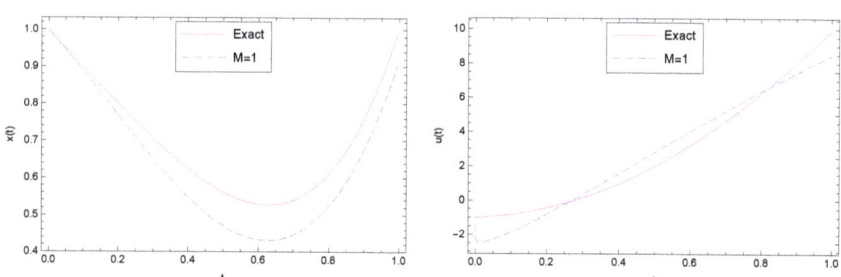

Figure 4. (Example 3.) Comparison between the exact and approximate state (**left**) and control (**right**) functions obtained by Approach I with $M = 1$.

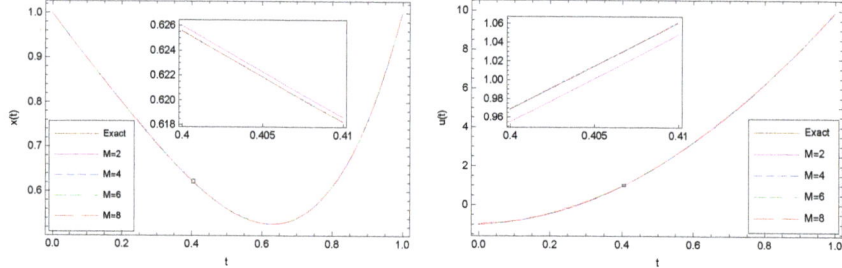

Figure 5. (Example 3.) Comparison between the exact and approximate state (**left**) and control (**right**) functions obtained by Approach II with different values of M.

7. Conclusions

Two numerical approaches have been proposed for solving variable-order fractional optimal control-affine problems. They use an accurate operational matrix of variable-order fractional integration for Bernoulli polynomials, to give approximations of the optimal state and control functions. These approximations, along with the Gauss–Legendre quadrature formula, are used to reduce the original problem to a system of algebraic equations. An approximation of the optimal performance index and an error bound were given. Some examples have been solved to illustrate the accuracy and applicability of the new techniques. From the numerical results of Examples 1 and 3, it can be seen that our Approach I leads to very high accuracy results with a small number of basis functions for optimal control problems in which the state function that minimizes the performance index is an infinitely smooth function. Moreover, from the results of Example 2, we conclude that Approach II may give much more accurate results than Approach I in the cases that the smoothness of ${}_0^C D_t^{\alpha(t)} x(t)$ is more than $x^{(n)}(t)$.

Author Contributions: Conceptualization, S.N. and D.F.M.T.; investigation, S.N. and D.F.M.T.; software, S.N.; validation, D.F.M.T.; writing—original draft, S.N. and D.F.M.T.; writing—review and editing, S.N. and D.F.M.T. All authors have read and agreed to the published version of the manuscript.

Funding: Torres was funded by *Fundação para a Ciência e a Tecnologia* (FCT, the Portuguese Foundation for Science and Technology) through grant UIDB/04106/2020 (CIDMA).

Acknowledgments: This research was initiated during a visit of Nemati to the Department of Mathematics of the University of Aveiro (DMat-UA) and to the R&D unit CIDMA, Portugal. The hospitality of the host institution is here gratefully acknowledged. The authors are also grateful to three anonymous reviewers for several questions and comments, which helped them to improve the manuscript.

Conflicts of Interest: The authors declare no conflict of interest.

References

1. Machado, J.T.; Kiryakova, V.; Mainardi, F. Recent history of fractional calculus. *Commun. Nonlinear Sci. Numer. Simul.* **2011**, *16*, 1140–1153. [CrossRef]
2. Keshavarz, E.; Ordokhani, Y.; Razzaghi, M. A numerical solution for fractional optimal control problems via Bernoulli polynomials. *J. Vib. Control* **2016**, *22*, 3889–3903. [CrossRef]
3. Bhrawy, A.H.; Tohidi, E.; Soleymani, F. A new Bernoulli matrix method for solving high-order linear and nonlinear Fredholm integro-differential equations with piecewise intervals. *Appl. Math. Comput.* **2012**, *219*, 482–497. [CrossRef]
4. Tohidi, E.; Bhrawy, A.H.; Erfani, K. A collocation method based on Bernoulli operational matrix for numerical solution of generalized pantograph equation. *Appl. Math. Model.* **2013**, *37*, 4283–4294. [CrossRef]
5. Toutounian, F.; Tohidi, E. A new Bernoulli matrix method for solving second order linear partial differential equations with the convergence analysis. *Appl. Math. Comput.* **2013**, *223*, 298–310. [CrossRef]
6. Bazm, S. Bernoulli polynomials for the numerical solution of some classes of linear and nonlinear integral equations. *J. Comput. Appl. Math.* **2015**, *275*, 44–60. [CrossRef]
7. Sahu, P.K.; Mallick, B. Approximate solution of fractional order Lane-Emden type differential equation by orthonormal Bernoulli's polynomials. *Int. J. Appl. Comput. Math.* **2019**, *5*, 89. [CrossRef]
8. Loh, J.R.; Phang, C. Numerical solution of Fredholm fractional integro-differential equation with right-sided Caputo's derivative using Bernoulli polynomials operational matrix of fractional derivative. *Mediterr. J. Math.* **2019**, *16*, 28. [CrossRef]
9. Rosa, S.; Torres, D.F.M. Optimal control of a fractional order epidemic model with application to human respiratory syncytial virus infection. *Chaos Solitons Fractals* **2018**, *117*, 142–149. [CrossRef]
10. Malinowska, A.B.; Torres, D.F.M. *Introduction to the Fractional Calculus of Variations*; Imperial College Press: London, UK, 2012. [CrossRef]
11. Malinowska, A.B.; Odzijewicz, T.; Torres, D.F.M. Advanced methods in the fractional calculus of variations. In *Springer Briefs in Applied Sciences and Technology*; Springer: Cham, Switzerland, 2015.
12. Almeida, R.; Pooseh, S.; Torres, D.F.M. *Computational Methods in the Fractional Calculus of Variations*; Imperial College Press: London, UK, 2015. [CrossRef]

13. Ali, M.S.; Shamsi, M.; Khosravian-Arab, H.; Torres, D.F.M.; Bozorgnia, F. A space-time pseudospectral discretization method for solving diffusion optimal control problems with two-sided fractional derivatives. *J. Vib. Control* **2019**, *25*, 1080–1095. [CrossRef]
14. Nemati, S.; Lima, P.M.; Torres, D.F.M. A numerical approach for solving fractional optimal control problems using modified hat functions. *Commun. Nonlinear Sci. Numer. Simul.* **2019**, *78*, 104849. [CrossRef]
15. Salati, A.B.; Shamsi, M.; Torres, D.F.M. Direct transcription methods based on fractional integral approximation formulas for solving nonlinear fractional optimal control problems. *Commun. Nonlinear Sci. Numer. Simul.* **2019**, *67*, 334–350. [CrossRef]
16. Rabiei, K.; Ordokhani, Y.; Babolian, E. Numerical solution of 1D and 2D fractional optimal control of system via Bernoulli polynomials. *Int. J. Appl. Comput. Math.* **2018**, *4*, 7. [CrossRef]
17. Behroozifar, M.; Habibi, N. A numerical approach for solving a class of fractional optimal control problems via operational matrix Bernoulli polynomials. *J. Vib. Control* **2018**, *24*, 2494–2511. [CrossRef]
18. Rahimkhani, P.; Ordokhani, Y. Generalized fractional-order Bernoulli-Legendre functions: An effective tool for solving two-dimensional fractional optimal control problems. *IMA J. Math. Control Inf.* **2019**, *36*, 185–212. [CrossRef]
19. Samko, S.G.; Ross, B. Integration and differentiation to a variable fractional order. *Integral Transform. Spec. Funct.* **1993**, *1*, 277–300. [CrossRef]
20. Lorenzo, C.F.; Hartley, T.T. Variable order and distributed order fractional operators. *Nonlinear Dyn.* **2002**, *29*, 57–98. [CrossRef]
21. Abdeljawad, T.; Mert, R.; Torres, D.F.M. Variable order Mittag-Leffler fractional operators on isolated time scales and application to the calculus of variations. In *Fractional Derivatives with Mittag-Leffler Kernel*; Springer: Cham, Switzerland, 2019; Volume 194, pp. 35–47.
22. Hassani, H.; Naraghirad, E. A new computational method based on optimization scheme for solving variable-order time fractional Burgers' equation. *Math. Comput. Simul.* **2019**, *162*, 1–17. [CrossRef]
23. Odzijewicz, T.; Malinowska, A.B.; Torres, D.F.M. Fractional variational calculus of variable order. In *Advances in Harmonic Analysis and Operator Theory*; Birkhäuser/Springer Basel AG: Basel, Switzerland, 2013; Volume 229, pp. 291–301. [CrossRef]
24. Yan, R.; Han, M.; Ma, Q.; Ding, X. A spectral collocation method for nonlinear fractional initial value problems with a variable-order fractional derivative. *Comput. Appl. Math.* **2019**, *38*, 66. [CrossRef]
25. Almeida, R.; Tavares, D.; Torres, D.F.M. The variable-order fractional calculus of variations. In *Springer Briefs in Applied Sciences and Technology*; Springer: Cham, Switzerland, 2019. [CrossRef]
26. Heydari, M.H.; Avazzadeh, Z. A new wavelet method for variable-order fractional optimal control problems. *Asian J. Control* **2018**, *20*, 1804–1817. [CrossRef]
27. Mohammadi, F.; Hassani, H. Numerical solution of two-dimensional variable-order fractional optimal control problem by generalized polynomial basis. *J. Optim. Theory Appl.* **2019**, *180*, 536–555. [CrossRef]
28. Costabile, F.; Dell'Accio, F.; Gualtieri, M.I. A new approach to Bernoulli polynomials. *Rend. Mat. Appl.* **2006**, *26*, 1–12.
29. Arfken, G. *Mathematical Methods for Physicists*; Academic Press: New York, NY, USA; London, UK, 1966.
30. Shen, J.; Tang, T.; Wang, L.L. *Spectral Methods*; Springer Series in Computational Mathematics; Springer: Heidelberg, Germany, 2011; Volume 41. [CrossRef]
31. Canuto, C.; Hussaini, M.Y.; Quarteroni, A.; Zang, T.A. Spectral Methods. In *Scientific Computation*; Springer: Berlin, Germany, 2006.
32. Lotfi, A.; Yousefi, S.A.; Dehghan, M. Numerical solution of a class of fractional optimal control problems via the Legendre orthonormal basis combined with the operational matrix and the Gauss quadrature rule. *J. Comput. Appl. Math.* **2013**, *250*, 143–160. [CrossRef]

© 2020 by the authors. Licensee MDPI, Basel, Switzerland. This article is an open access article distributed under the terms and conditions of the Creative Commons Attribution (CC BY) license (http://creativecommons.org/licenses/by/4.0/).

Article

Inertial Iterative Self-Adaptive Step Size Extragradient-Like Method for Solving Equilibrium Problems in Real Hilbert Space with Applications

Wiyada Kumam [1] and Kanikar Muangchoo [2],*

[1] Program in Applied Statistics, Department of Mathematics and Computer Science, Faculty of Science and Technology, Rajamangala University of Technology Thanyaburi, Thanyaburi, Pathumthani 12110, Thailand; wiyada.kum@rmutt.ac.th
[2] Faculty of Science and Technology, Rajamangala University of Technology Phra Nakhon (RMUTP), 1381 Pracharat 1 Road, Wongsawang, Bang Sue, Bangkok 10800, Thailand
* Correspondence: kanikar.m@rmutp.ac.th; Tel.: +66-2-836-3000 (ext. 4193)

Received: 2 September 2020; Accepted: 22 October 2020; Published: 31 October 2020

Abstract: A number of applications from mathematical programmings, such as minimization problems, variational inequality problems and fixed point problems, can be written as equilibrium problems. Most of the schemes being used to solve this problem involve iterative methods, and for that reason, in this paper, we introduce a modified iterative method to solve equilibrium problems in real Hilbert space. This method can be seen as a modification of the paper titled "A new two-step proximal algorithm of solving the problem of equilibrium programming" by Lyashko et al. (Optimization and its applications in control and data sciences, Springer book pp. 315–325, 2016). A weak convergence result has been proven by considering the mild conditions on the cost bifunction. We have given the application of our results to solve variational inequality problems. A detailed numerical study on the Nash–Cournot electricity equilibrium model and other test problems is considered to verify the convergence result and its performance.

Keywords: pseudomonotone bifunction; Lipschitz-type conditions; equilibrium problem; variational inequalities

1. Introduction

An equilibrium problem (EP) is a generalized concept that unifies several mathematical problems, such as the variational inequality problems, minimization problems, complementarity problems, the fixed point problems, non-cooperative games of Nash equilibrium, the saddle point problems and scalar and vector minimization problems (see e.g., [1–3]). The particular form of an equilibrium problem was firstly established in 1992 by Muu and Oettli [4] and then further elaborated by Blum and Oettli [1]. Next, we consider the concept of an equilibrium problem introduced by Blum and Oettli in [1]. Let C be a non-empty, closed and convex subset $\mathbb{H}$ of a real Hilbert space and $f : \mathbb{H} \times \mathbb{H} \to \mathbb{R}$ is bifunction with $f(v,v) = 0$, for each $v \in C$. A equilibrium problem regarding f on the set C is defined in the following way:

$$\text{Find } p \in C \text{ such that } f(p,v) \geq 0, \text{ for all } v \in C. \tag{1}$$

Many methods have been already established over the past couple of years to figure out the equilibrium problem in Hilbert spaces [5–15], the inertial methods [11,16–18] and others in [18–24]. In particular, Tran et al. introduced an iterative scheme in [8], in that a sequence $\{u_n\}$ was generated in the following way:

$$\begin{cases} u_0 \in C, \\ v_n = \arg\min\{\lambda f(u_n, y) + \tfrac{1}{2}\|u_n - y\|^2 : y \in C\}, \\ u_{n+1} = \arg\min\{\lambda f(v_n, y) + \tfrac{1}{2}\|u_n - y\|^2 : y \in C\}, \end{cases} \quad (2)$$

where $0 < \lambda < \min\{\tfrac{1}{2c_1}, \tfrac{1}{2c_2}\}$ and c_1, c_2 are Lipschitz constants. Lyashko et al. [25] in 2016 introduced an improvement of the method (2) to solve equilibrium problem and sequence $\{u_n\}$ was generated in the following way:

$$\begin{cases} u_0, v_0 \in C, \\ u_{n+1} = \arg\min\{\lambda f(v_n, y) + \tfrac{1}{2}\|u_n - y\|^2 : y \in C\}, \\ v_{n+1} = \arg\min\{\lambda f(v_n, y) + \tfrac{1}{2}\|u_{n+1} - y\|^2 : y \in C\}, \end{cases} \quad (3)$$

where $0 < \lambda < \tfrac{1}{2c_2 + 4c_1}$ and c_1, c_2 are Lipschitz constants.

In this paper, we consider the extragradient method in (3) and to provide its improvement by using the inertial scheme [26] and continue to improve the step size rule for its second step. The step size is not fixed in our case, but it is dependent on a particular formula by using prior information of the bifunction values. A weak convergence theorem dealing with the suggested technique is presented by having the specific bi-functional condition. We have also considered how our results are presented to the problems of a variational inequality. A few other formulations of the problem of variational inequalities are discussed, and many computational examples in finite and infinite dimensions spaces are also presented to demonstrate the applicability of our proposed results.

In this study, we study the equilibrium problem through the following assumptions:

(f_1) A bifunction $f : \mathbb{H} \times \mathbb{H} \to \mathbb{R}$ is said to be (see [1,27]) *pseudomonotone* on C if

$$f(v_1, v_2) \geq 0 \implies f(v_2, v_1) \leq 0, \text{ for all } v_1, v_2 \in C.$$

(f_2) A bifunction $f : \mathbb{H} \times \mathbb{H} \to \mathbb{R}$ is said to be Lipschitz-type continuous [28] on C if there exist $c_1, c_2 > 0$ such that

$$f(v_1, v_3) \leq f(v_1, v_2) + f(v_2, v_3) + c_1 \|v_1 - v_2\|^2 + c_2 \|v_2 - v_3\|^2, \text{ for all } v_1, v_2, v_3 \in C.$$

(f_3) $\limsup_{n \to +\infty} f(v_n, z) \leq f(v^*, z)$ for each $z \in C$ and $\{v_n\} \subset C$ satisfying $v_n \rightharpoonup v^*$;

(f_4) $f(u, \cdot)$ is convex and subdifferentiable on $\mathbb{H}$ for each $u \in \mathbb{H}$.

The rest of this paper will be organized as follows: In Section 2, we give a few definitions and important lemmas to be used in this paper. Section 3 includes the main algorithm involving pseudomonotone bifunction and provides a weak convergence theorem. Section 4 describes some applications in the variational inequality problems. Section 5 sets out the numerical studies to demonstrate the algorithmic efficiency.

2. Preliminaries

In this section, some important lemmas and basic definitions are provided. Moreover, $EP(f, C)$ denotes the solution set of an equilibrium problem on the set C and p is any arbitrary element of $EP(f, C)$.

A *metric projection* $P_C(u)$ of u onto a closed, convex subset C of $\mathbb{H}$ is defined by

$$P_C(u) = \arg\min_{v \in C}\{\|v - u\|\}.$$

Lemma 1. [29] *Let $P_C : \mathbb{H} \to C$ be a metric projection from $\mathbb{H}$ onto C. Then*

(i) For all $u \in C$, $v \in \mathbb{H}$ and
$$\|u - P_C(v)\|^2 + \|P_C(v) - v\|^2 \leq \|u - v\|^2.$$

(ii) $w = P_C(u)$ if and only if
$$\langle u - w, v - w \rangle \leq 0, \text{ for all } v \in C.$$

Lemma 2. [29] *For all $u, v \in \mathbb{H}$ with $\daleth \in \mathbb{R}$. Then, the following relationship is holds.*
$$\|\daleth u + (1 - \daleth)v\|^2 = \daleth \|u\|^2 + (1 - \daleth)\|v\|^2 - \daleth(1 - \daleth)\|u - v\|^2.$$

Assume that $g : C \to \mathbb{R}$ be a convex function and subdifferential of g at $u \in C$ is defined by
$$\partial g(u) = \{w \in C : g(v) - g(u) \geq \langle w, v - u \rangle, \text{ for all } v \in C\}.$$

Given that $f(u, .)$ is convex and subdifferentiable on $\mathbb{H}$ for each fixed $u \in \mathbb{H}$ and *subdifferential of* $f(u, .)$ at $x \in \mathbb{H}$ defined by
$$\partial_2 f(u, .)(x) = \partial_2 f(u, x) = \{z \in \mathbb{H} : f(u, v) - f(u, x) \geq \langle z, v - x \rangle, \text{ for all } v \in \mathbb{H}\}.$$

A normal cone of C at $u \in C$ is defined by
$$N_C(u) = \{w \in \mathbb{H} : \langle w, v - u \rangle \leq 0, \text{ for all } v \in C\}.$$

Lemma 3. [30] *Assume that C is a nonempty, closed and convex subset of a real Hilbert space $\mathbb{H}$ and $h : C \to \mathbb{R}$ be a convex, lower semi-continuous and subdifferentiable function on C. Then, $u \in C$ is a minimizer of a function h if and only if $0 \in \partial h(u) + N_C(u)$ where $\partial h(u)$ and $N_C(u)$ denotes the subdifferential of h at u and the normal cone of C at u, respectively.*

Lemma 4. [31] *Let a_n, b_n and c_n are non-negative real sequences such that*
$$a_{n+1} \leq a_n + b_n(a_n - a_{n-1}) + c_n, \text{ for all } n \geq 1, \text{ with } \sum_{n=1}^{+\infty} c_n < +\infty,$$
where $b > 0$ such that $0 \leq b_n \leq b < 1$ for all $n \in \mathbb{N}$. Then, the following relations are true.

(i) $\sum_{n=1}^{+\infty} [a_n - a_{n-1}]_+ < +\infty$, with $[s]_+ := \max\{s, 0\}$;
(ii) $\lim_{n \to +\infty} a_n = a^* \in [0, +\infty)$.

Lemma 5. [32] *Let a sequence $\{a_n\}$ in $\mathbb{H}$ and $C \subset \mathbb{H}$ and the following conditions have been met:*

(i) *for each $a \in C$, $\lim_{n \to +\infty} \|a_n - a\|$ exists;*
(ii) *each weak sequentially limit point of $\{a_n\}$ belongs to set C.*

Then, $\{a_n\}$ weakly converges to an element in C.

3. Main Results

In this section, we present our main algorithm and provide a weak convergence theorem for our proposed method. The detailed method is given below.

Remark 1. *By Expression* (5), *we obtain*
$$\lambda_{n+1} \left[f(v_{n-1}, u_{n+1}) - f(v_{n-1}, v_n) - c_1 \|v_{n-1} - v_n\|^2 - c_2 \|v_n - u_{n+1}\|^2 \right] \leq \mu f(v_n, u_{n+1}). \quad (4)$$

Lemma 6. *Let $\{u_n\}$ be a sequence generated by Algorithm 1. Then, the following inequality holds.*

$$\mu\lambda_n f(v_n,y) - \mu\lambda_n f(v_n,u_{n+1}) \geq \langle \rho_n - u_{n+1}, y - u_{n+1}\rangle, \text{ for all } y \in C.$$

Algorithm 1 Modified Popov's subgradient extragradient-like iterative scheme.

Step 1: Choose $u_{-1}, v_{-1}, u_0, v_0 \in \mathbb{H}$ and a sequence $\wp_n$ is non-decreasing such that $0 \leq \wp_n \leq \wp < \frac{1}{3}$, $\lambda_0 > 0$ and $0 < \sigma < \min\left\{\frac{1-3\wp}{(1-\wp)^2+4c_1(\wp+\wp^2)}, \frac{1}{2c_2+4c_1(1+\wp)}\right\}$ and $\mu \in (0,\sigma)$.

Step 2: Evaluate
$$u_{n+1} = \arg\min\{\mu\lambda_n f(v_n,y) + \tfrac{1}{2}\|\rho_n - y\|^2 : y \in C\},$$
where $\rho_n = u_n + \wp_n(u_n - u_{n-1})$.

Step 3: Updated the step size in the following order:

$$\lambda_{n+1} = \begin{cases} \min\left\{\sigma, \frac{\mu f(v_n,u_{n+1})}{f(v_{n-1},u_{n+1})-f(v_{n-1},v_n)-c_1\|v_{n-1}-v_n\|^2-c_2\|u_{n+1}-v_n\|^2+1}\right\}, \\ \quad \text{if } \frac{\mu f(v_n,u_{n+1})}{f(v_{n-1},u_{n+1})-f(v_{n-1},v_n)-c_1\|v_{n-1}-v_n\|^2-c_2\|u_{n+1}-v_n\|^2+1} > 0, \\ \lambda_0 \qquad \text{else.} \end{cases} \quad (5)$$

Step 4: Evaluate
$$v_{n+1} = \arg\min\{\lambda_{n+1} f(v_n,y) + \tfrac{1}{2}\|\rho_{n+1} - y\|^2 : y \in C\},$$
where $\rho_{n+1} = u_{n+1} + \wp_{n+1}(u_{n+1} - u_n)$. If $u_{n+1} = v_n = \rho_n$ or $\rho_{n+1} = v_{n+1} = v_n$ then Stop. Else, take $n := n+1$ and go back to **Step 2**.

Proof. By the use of Lemma 3, we get

$$0 \in \partial_2\left\{\mu\lambda_n f(v_n,y) + \tfrac{1}{2}\|\rho_n - y\|^2\right\}(u_{n+1}) + N_C(u_{n+1}).$$

From above there is a $\omega \in \partial_2 f(v_n,u_{n+1})$ and $\overline{\omega} \in N_C(u_{n+1})$ such that

$$\mu\lambda_n \omega + u_{n+1} - \rho_n + \overline{\omega} = 0.$$

Therefore, we obtain

$$\langle \rho_n - u_{n+1}, y - u_{n+1}\rangle = \mu\lambda_n\langle \omega, y - u_{n+1}\rangle + \langle \overline{\omega}, y - u_{n+1}\rangle, \text{ for all } y \in C.$$

Due to $\overline{\omega} \in N_C(u_{n+1})$ then $\langle \overline{\omega}, y - u_{n+1}\rangle \leq 0$, for each $y \in C$. It implies that

$$\mu\lambda_n\langle \omega, y - u_{n+1}\rangle \geq \langle \rho_n - u_{n+1}, y - u_{n+1}\rangle, \text{ for all } y \in C. \quad (6)$$

Given that $\omega \in \partial_2 f(v_n,u_{n+1})$, we have

$$f(v_n,y) - f(v_n,u_{n+1}) \geq \langle \omega, y - u_{n+1}\rangle, \text{ for all } y \in \mathbb{H}. \quad (7)$$

By combining Expressions (6) and (7), we obtain

$$\mu\lambda_n f(v_n,y) - \mu\lambda_n f(v_n,u_{n+1}) \geq \langle \rho_n - u_{n+1}, y - u_{n+1}\rangle, \text{ for all } y \in C.$$

□

Lemma 7. Let $\{v_n\}$ be a sequence generated by Algorithm 1. Then, the following inequality holds.

$$\lambda_{n+1} f(v_n, y) - \lambda_{n+1} f(v_n, v_{n+1}) \geq \langle \rho_{n+1} - v_{n+1}, y - v_{n+1}\rangle, \text{ for all } y \in C.$$

Proof. The proof is same as the proof of Lemma 6. □

Lemma 8. If $u_{n+1} = v_n = \rho_n$ and $\rho_{n+1} = v_{n+1} = v_n$ in Algorithm 1, then $v_n \in EP(f, C)$.

Proof. The proof of this can easily be seen from Lemmas 6 and 7. □

Lemma 9. Let $f : \mathbb{H} \times \mathbb{H} \to \mathbb{R}$ be a bifunction and satisfies the conditions (f_1)–(f_4). Then, for each $p \in EP(f, C) \neq \emptyset$, we have

$$\|u_{n+1} - p\|^2$$
$$\leq \|\rho_n - p\|^2 - (1 - \lambda_{n+1})\|u_{n+1} - \rho_n\|^2 + 4c_1\lambda_{n+1}\lambda_n\|\rho_n - v_{n-1}\|^2$$
$$- \lambda_{n+1}(1 - 4c_1\lambda_n)\|\rho_n - v_n\|^2 - \lambda_{n+1}(1 - 2c_2\lambda_n)\|u_{n+1} - v_n\|^2.$$

Proof. By Lemma 6, we obtain

$$\mu \lambda_n f(v_n, p) - \mu \lambda_n f(v_n, u_{n+1}) \geq \langle \rho_n - u_{n+1}, p - u_{n+1}\rangle. \quad (8)$$

Thus, $p \in EP(f, C)$ and the condition (f_1) implies that $f(v_n, p) \leq 0$. From (8), we have

$$\langle \rho_n - u_{n+1}, u_{n+1} - p\rangle \geq \mu \lambda_n f(v_n, u_{n+1}). \quad (9)$$

From Expression (4), we obtain

$$\mu f(v_n, u_{n+1}) \geq \lambda_{n+1}\big(f(v_{n-1}, u_{n+1}) - f(v_{n-1}, v_n) - c_1\|v_{n-1} - v_n\|^2 - c_2\|v_n - u_{n+1}\|^2\big). \quad (10)$$

Combining expression (9) and (10), implies that

$$\langle \rho_n - u_{n+1}, u_{n+1} - p\rangle \geq \lambda_{n+1}\Big[\lambda_n\big\{f(v_{n-1}, u_{n+1}) - f(v_{n-1}, v_n)\big\}$$
$$- c_1\lambda_n\|v_{n-1} - v_n\|^2 - c_2\lambda_n\|u_{n+1} - v_n\|^2\Big]. \quad (11)$$

By Lemma 7, we have

$$\lambda_n\big\{f(v_{n-1}, u_{n+1}) - f(v_{n-1}, v_n)\big\} \geq \langle \rho_n - v_n, u_{n+1} - v_n\rangle. \quad (12)$$

Thus, combining (11) and (12) we get

$$\langle \rho_n - u_{n+1}, u_{n+1} - p\rangle \geq \lambda_{n+1}\Big[\langle \rho_n - v_n, u_{n+1} - v_n\rangle$$
$$- c_1\lambda_n\|v_{n-1} - v_n\|^2 - c_2\lambda_n\|u_{n+1} - v_n\|^2\Big]. \quad (13)$$

We have the following mathematical expressions:

$$2\langle \rho_n - u_{n+1}, u_{n+1} - p\rangle = \|\rho_n - p\|^2 - \|u_{n+1} - \rho_n\|^2 - \|u_{n+1} - p\|^2.$$

$$2\langle \rho_n - v_n, u_{n+1} - v_n\rangle = \|\rho_n - v_n\|^2 + \|u_{n+1} - v_n\|^2 - \|\rho_n - u_{n+1}\|^2.$$

From the above equation and (13), we have

$$\|u_{n+1} - p\|^2$$
$$\leq \|\rho_n - p\|^2 - (1 - \lambda_{n+1})\|u_{n+1} - \rho_n\|^2 - \lambda_{n+1}(1 - 2c_2\lambda_n)\|u_{n+1} - v_n\|^2$$
$$- \lambda_{n+1}\|\rho_n - v_n\|^2 + \lambda_{n+1}(2c_1\lambda_n)\|v_{n-1} - v_n\|^2$$

We also have

$$\|v_{n-1} - v_n\|^2 \leq (\|v_{n-1} - \rho_n\| + \|\rho_n - v_n\|)^2 \leq 2\|v_{n-1} - \rho_n\|^2 + 2\|\rho_n - v_n\|^2.$$

Finally, we get

$$\|u_{n+1} - p\|^2$$
$$\leq \|\rho_n - p\|^2 - (1 - \lambda_{n+1})\|u_{n+1} - \rho_n\|^2 + 4c_1\lambda_n\lambda_{n+1}\|\rho_n - v_{n-1}\|^2$$
$$- \lambda_{n+1}(1 - 4c_1\lambda_n)\|\rho_n - v_n\|^2 - \lambda_{n+1}(1 - 2c_2\lambda_n)\|u_{n+1} - v_n\|^2.$$

□

Theorem 1. *Assume that $f : \mathbb{H} \times \mathbb{H} \to \mathbb{R}$ satisfies the conditions (f_1)–(f_4). Then, for some $p \in EP(f, C) \neq \emptyset$, the sequence $\{\rho_n\}$, $\{u_n\}$ and $\{v_n\}$ generated by Algorithm 1, weakly converge to $p \in EP(f, C)$.*

Proof. By Lemma 9, we obtain

$$\|u_{n+1} - p\|^2$$
$$\leq \|\rho_n - p\|^2 - (1 - \lambda_{n+1})\|u_{n+1} - \rho_n\|^2 + 4c_1\lambda_n\lambda_{n+1}\|\rho_n - v_{n-1}\|^2$$
$$- \lambda_{n+1}(1 - 4c_1\lambda_n)\|\rho_n - v_n\|^2 - \lambda_{n+1}(1 - 2c_2\lambda_n)\|u_{n+1} - v_n\|^2. \tag{14}$$

By definition of ρ_n in the Algorithm 1, we have

$$\|\rho_n - v_{n-1}\|^2 = \|u_n + \wp_n(u_n - u_{n-1}) - v_{n-1}\|^2$$
$$= \|(1 + \wp_n)(u_n - v_{n-1}) - \wp_n(u_{n-1} - v_{n-1})\|^2$$
$$= (1 + \wp_n)\|u_n - v_{n-1}\|^2 - \wp_n\|u_{n-1} - v_{n-1}\|^2 + \wp_n(1 + \wp_n)\|u_n - u_{n-1}\|^2$$
$$\leq (1 + \wp)\|u_n - v_{n-1}\|^2 + \wp(1 + \wp)\|u_n - u_{n-1}\|^2. \tag{15}$$

Adding the term $4c_1\sigma\lambda_{n+1}(1+\wp)\|u_{n+1}-v_n\|^2$ on both sides in (14) with (15) for $n\geq 1$, we have

$$\|u_{n+1}-p\|^2 + 4c_1\sigma\lambda_{n+1}(1+\wp)\|u_{n+1}-v_n\|^2$$
$$\leq \|\rho_n-p\|^2 - (1-\sigma)\|u_{n+1}-\rho_n\|^2 + 4c_1\sigma\lambda_{n+1}(1+\wp)\|u_{n+1}-v_n\|^2$$
$$+ 4c_1\sigma\lambda_n\left[(1+\wp)\|u_n-v_{n-1}\|^2 + \wp(1+\wp)\|u_n-u_{n-1}\|^2\right]$$
$$- \lambda_{n+1}(1-4c_1\sigma)\|\rho_n-v_n\|^2 - \lambda_{n+1}(1-2c_2\sigma)\|u_{n+1}-v_n\|^2 \tag{16}$$
$$\leq \|\rho_n-p\|^2 - (1-\sigma)\|u_{n+1}-\rho_n\|^2 + 4c_1\sigma\lambda_n(1+\wp)\|u_n-v_{n-1}\|^2$$
$$+ 4c_1\sigma(\wp+\wp^2)\|u_n-u_{n-1}\|^2 - \lambda_{n+1}(1-4c_1\sigma)\|\rho_n-v_n\|^2$$
$$- \lambda_{n+1}(1-2c_2\sigma-4c_1\sigma(1+\wp))\|u_{n+1}-v_n\|^2 \tag{17}$$
$$\leq \|\rho_n-p\|^2 - (1-\sigma)\|u_{n+1}-\rho_n\|^2 + 4c_1\sigma\lambda_n(1+\wp)\|u_n-v_{n-1}\|^2$$
$$+ 4c_1\sigma(\wp+\wp^2)\|u_n-u_{n-1}\|^2$$
$$- \frac{\lambda_{n+1}}{2}(1-2c_2\sigma-4c_1\sigma(1+\wp))\left[2\|u_{n+1}-v_n\|^2 + 2\|\rho_n-v_n\|^2\right] \tag{18}$$
$$\leq \|\rho_n-p\|^2 - (1-\sigma)\|u_{n+1}-\rho_n\|^2 + 4c_1\sigma\lambda_n(1+\wp)\|u_n-v_{n-1}\|^2$$
$$+ 4c_1\sigma(\wp+\wp^2)\|u_n-u_{n-1}\|^2$$
$$- \frac{\lambda_{n+1}}{2}(1-2c_2\sigma-4c_1\sigma(1+\wp))\|u_{n+1}-\rho_n\|^2. \tag{19}$$

Given that $0<\lambda_n\leq \sigma<\frac{1}{2c_2+4c_1(1+\wp)}$, then the last inequality turns into

$$\|u_{n+1}-p\|^2 + 4c_1\sigma\lambda_{n+1}(1+\wp)\|u_{n+1}-v_n\|^2$$
$$\leq \|\rho_n-p\|^2 - (1-\sigma)\|u_{n+1}-\rho_n\|^2 + 4c_1\sigma\lambda_n(1+\wp)\|u_n-v_{n-1}\|^2$$
$$+ 4c_1\sigma(\wp+\wp^2)\|u_n-u_{n-1}\|^2. \tag{20}$$

From the definition of ρ_n, we have

$$\|\rho_n-p\|^2 = \|u_n + \wp_n(u_n-u_{n-1}) - p\|^2$$
$$= \|(1+\wp_n)(u_n-p) - \wp_n(u_{n-1}-p)\|^2$$
$$= (1+\wp_n)\|u_n-p\|^2 - \wp_n\|u_{n-1}-p\|^2 + \wp_n(1+\wp_n)\|u_n-u_{n-1}\|^2. \tag{21}$$

From ρ_{n+1}, we obtain

$$\|u_{n+1}-\rho_n\|^2 = \|u_{n+1}-u_n - \wp_n(u_n-u_{n-1})\|^2$$
$$= \|u_{n+1}-u_n\|^2 + \wp_n^2\|u_n-u_{n-1}\|^2 - 2\wp_n\langle u_{n+1}-u_n, u_n-u_{n-1}\rangle \tag{22}$$
$$\geq \|u_{n+1}-u_n\|^2 + \wp_n^2\|u_n-u_{n-1}\|^2 - 2\wp_n\|u_{n+1}-u_n\|\|u_n-u_{n-1}\|$$
$$\geq \|u_{n+1}-u_n\|^2 + \wp_n^2\|u_n-u_{n-1}\|^2 - \wp_n\|u_{n+1}-u_n\|^2 - \wp_n\|u_n-u_{n-1}\|^2$$
$$= (1-\wp_n)\|u_{n+1}-u_n\|^2 + (\wp_n^2-\wp_n)\|u_n-u_{n-1}\|^2. \tag{23}$$

Combining the Expressions (20), (21) and (23) we have

$$\|u_{n+1} - p\|^2 + 4c_1\sigma\lambda_{n+1}(1+\wp)\|u_{n+1} - v_n\|^2$$
$$\leq (1+\wp_n)\|u_n - p\|^2 - \wp_n\|u_{n-1} - p\|^2 + \wp_n(1+\wp_n)\|u_n - u_{n-1}\|^2$$
$$- (1-\sigma)\left[(1-\wp_n)\|u_{n+1} - u_n\|^2 + (\wp_n^2 - \wp_n)\|u_n - u_{n-1}\|^2\right]$$
$$+ 4c_1\sigma\lambda_n(1+\wp)\|u_n - v_{n-1}\|^2 + 4c_1\sigma(\wp + \wp^2)\|u_n - u_{n-1}\|^2 \qquad (24)$$
$$\leq (1+\wp_n)\|u_n - p\|^2 - \wp_n\|u_{n-1} - p\|^2 + 4c_1\sigma\lambda_n(1+\wp)\|u_n - v_{n-1}\|^2$$
$$+ \left[\wp(1+\wp) - (1-\sigma)(\wp_n^2 - \wp_n) + 4c_1\sigma(\wp + \wp^2)\right]\|u_n - u_{n-1}\|^2$$
$$- (1-\sigma)(1-\wp_n)\|u_{n+1} - u_n\|^2 \qquad (25)$$
$$\leq (1+\wp_n)\|u_n - p\|^2 - \wp_n\|u_{n-1} - p\|^2 + 4c_1\sigma\lambda_n(1+\wp)\|u_n - v_{n-1}\|^2$$
$$+ r_n\|u_n - u_{n-1}\|^2 - q_n\|u_{n+1} - u_n\|^2, \qquad (26)$$

where
$$r_n = \left[\wp(1+\wp) - (1-\sigma)(\wp_n^2 - \wp_n) + 4c_1\sigma(\wp + \wp^2)\right];$$
$$q_n = (1-\sigma)(1-\wp_n).$$

Assume that
$$\Gamma_n = \Psi_n + r_n\|u_n - u_{n-1}\|^2,$$

where $\Psi_n = \|u_n - p\|^2 - \wp_n\|u_{n-1} - p\|^2 + 4c_1\sigma\lambda_n(1+\wp)\|u_n - v_{n-1}\|^2$. Next, (26) implies that

$$\Gamma_{n+1} - \Gamma_n$$
$$= \|u_{n+1} - p\|^2 - \wp_{n+1}\|u_n - p\|^2 + 4c_1\sigma\lambda_{n+1}(1+\wp)\|u_{n+1} - v_n\|^2 + r_{n+1}\|u_{n+1} - u_n\|^2$$
$$- \|u_n - p\|^2 + \wp_n\|u_{n-1} - p\|^2 - 4c_1\sigma\lambda_n(1+\wp)\|u_n - v_{n-1}\|^2 - r_n\|u_n - u_{n-1}\|^2$$
$$\leq \|u_{n+1} - p\|^2 - (1+\wp_n)\|u_n - p\|^2 + \wp_n\|u_{n-1} - p\|^2 + 4c_1\sigma\lambda_{n+1}(1+\wp)\|u_{n+1} - v_n\|^2$$
$$+ r_{n+1}\|u_{n+1} - u_n\|^2 - 4c_1\sigma\lambda_n(1+\wp)\|u_n - v_{n-1}\|^2 - r_n\|u_n - u_{n-1}\|^2$$
$$\leq -(q_n - r_{n+1})\|u_{n+1} - u_n\|^2. \qquad (27)$$

Next, we have to compute

$$(q_n - r_{n+1}) = (1-\sigma)(1-\wp_n) - \wp(1+\wp) + (1-\sigma)(\wp_n^2 - \wp_n) - 4c_1\sigma(\wp + \wp^2)$$
$$\geq (1-\sigma)(1-\wp)^2 - \wp(1+\wp) - 4c_1\sigma(\wp + \wp^2)$$
$$= (1-\wp)^2 - \wp(1+\wp) - \sigma(1-\wp)^2 - 4c_1\sigma(\wp + \wp^2)$$
$$= 1 - 3\wp - \sigma\left((1-\wp)^2 + 4c_1(\wp + \wp^2)\right)$$
$$\geq 0. \qquad (28)$$

By the use of (27) and (28) for some $\delta \geq 0$ implies that

$$\Gamma_{n+1} - \Gamma_n \leq -(q_n - r_{n+1})\|u_{n+1} - u_n\|^2 \leq -\delta\|u_{n+1} - u_n\|^2 \leq 0. \qquad (29)$$

The relation (29) implies that $\{\Gamma_n\}$ is non-increasing. From Γ_{n+1} we have

$$\Gamma_{n+1} = \|u_{n+1} - p\|^2 - \wp_{n+1}\|u_n - p\|^2 + r_{n+1}\|u_{n+1} - u_n\|^2 + 4c_1\sigma\lambda_{n+1}(1+\wp)\|u_{n+1} - v_n\|^2$$
$$\geq -\wp_{n+1}\|u_n - p\|^2. \qquad (30)$$

By definition Γ_n, we have

$$\begin{aligned}
\|u_n - p\|^2 &\leq \Gamma_n + \wp_n \|u_{n-1} - p\|^2 \\
&\leq \Gamma_1 + \wp \|u_{n-1} - p\|^2 \\
&\leq \cdots \leq \Gamma_1(\wp^{n-1} + \cdots + 1) + \wp^n \|u_0 - p\|^2 \\
&\leq \frac{\Gamma_1}{1 - \wp} + \wp^n \|u_0 - p\|^2.
\end{aligned} \quad (31)$$

From Equations (30) and (31), we obtain

$$\begin{aligned}
-\Gamma_{n+1} &\leq \wp_{n+1} \|u_n - p\|^2 \\
&\leq \wp \|u_n - p\|^2 \\
&\leq \wp \frac{\Gamma_1}{1 - \wp} + \wp^{n+1} \|u_0 - p\|^2.
\end{aligned} \quad (32)$$

It follows (29) and (32) that

$$\begin{aligned}
\delta \sum_{n=1}^{k} \|u_{n+1} - u_n\|^2 &\leq \Gamma_1 - \Gamma_{k+1} \\
&\leq \Gamma_1 + \wp \frac{\Gamma_1}{1 - \wp} + \wp^{k+1} \|u_0 - p\|^2 \\
&\leq \frac{\Gamma_1}{1 - \wp} + \|u_0 - p\|^2.
\end{aligned} \quad (33)$$

By letting $k \to +\infty$ in (33), we obtain

$$\sum_{n=1}^{+\infty} \|u_{n+1} - u_n\|^2 < +\infty \quad \text{implies that} \quad \lim_{n \to +\infty} \|u_{n+1} - u_n\| = 0. \quad (34)$$

From Expressions (22) with (34), we obtain

$$\|u_{n+1} - \rho_n\| \to 0 \quad \text{as} \quad n \to +\infty. \quad (35)$$

From (32), we have

$$-\Psi_{n+1} \leq \wp \frac{\Gamma_1}{1 - \wp} + \wp^{n+1} \|u_0 - p\|^2 + r_{n+1} \|u_{n+1} - u_n\|^2. \quad (36)$$

From Expression (18) and using (21), we have

$$\begin{aligned}
&\lambda_{n+1}(1 - 2c_2\sigma - 4c_1\sigma(1 + \wp))\left[\|u_{n+1} - v_n\|^2 + \|\rho_n - v_n\|^2\right] \\
&\leq \Psi_n - \Psi_{n+1} + \wp(1 + \wp)\|u_n - u_{n-1}\|^2 + 4c_1\sigma\wp(1 + \wp)\|u_n - u_{n-1}\|^2.
\end{aligned} \quad (37)$$

Fix $k \in \mathbb{N}$ and using above expression for $n = 1, 2, \cdots, k$. Summing them up, we obtain

$$\lambda_{n+1}(1 - 2c_2\sigma - 4c_1\sigma(1+\wp)) \sum_{n=1}^{k} \left[\|u_{n+1} - v_n\|^2 + \|\rho_n - v_n\|^2 \right]$$

$$\leq \Psi_0 - \Psi_{k+1} + \wp(1+\wp) \sum_{n=1}^{k} \|u_n - u_{n-1}\|^2 + 4c_1\sigma\wp(1+\wp) \sum_{n=1}^{k} \|u_n - u_{n-1}\|^2$$

$$\leq \Psi_0 + \wp \frac{\Gamma_1}{1-\wp} + \wp^{k+1} \|u_0 - p\|^2 + r_{k+1} \|u_{k+1} - u_k\|^2$$

$$+ \wp(1+\wp) \sum_{n=1}^{k} \|u_n - u_{n-1}\|^2 + 4c_1\sigma\wp(1+\wp) \sum_{n=1}^{k} \|u_n - u_{n-1}\|^2, \tag{38}$$

letting $k \to +\infty$, and due to sum of the positive terms series, we obtain

$$\sum_{n=1}^{+\infty} \|u_{n+1} - v_n\|^2 < +\infty \quad \text{and} \quad \sum_{n=1}^{+\infty} \|\rho_n - v_n\|^2 < +\infty. \tag{39}$$

Moreover, we obtain

$$\lim_{n \to +\infty} \|u_{n+1} - v_n\| = \lim_{n \to +\infty} \|\rho_n - v_n\| = 0. \tag{40}$$

By using the triangular inequality, we get

$$\lim_{n \to +\infty} \|u_n - v_n\| = \lim_{n \to +\infty} \|u_n - \rho_n\| = \lim_{n \to +\infty} \|v_{n-1} - v_n\| = 0. \tag{41}$$

It is follow from the relation (24), we obtain

$$\|u_{n+1} - p\|^2$$
$$\leq (1 + \wp_n) \|u_n - p\|^2 - \wp_n \|u_{n-1} - p\|^2 + \wp(1+\wp) \|u_n - u_{n-1}\|^2$$
$$+ 4c_1\sigma(1+\wp) \|u_n - v_{n-1}\|^2 + 4c_1\sigma(\wp + \wp^2) \|u_n - u_{n-1}\|^2, \tag{42}$$

with (34), (39) and Lemma 4 imply that the sequences $\|u_n - p\|$, $\|\rho_n - p\|$ and $\|v_n - p\|$ limits exist for every $p \in EP(f, C)$. It means that $\{u_n\}$, $\{\rho_n\}$ and $\{v_n\}$ are bounded sequences. Take z an arbitrary sequential cluster point of the sequence $\{u_n\}$. Now our aim to prove that $z \in EP(f, C)$. By Lemma 6 with Expressions (10) and (12), we write

$$\mu \lambda_{n_k} f(v_{n_k}, y) \geq \mu \lambda_{n_k} f(v_{n_k}, u_{n_k+1}) + \langle \rho_{n_k} - u_{n_k+1}, y - u_{n_k+1} \rangle$$
$$\geq \lambda_{n_k} \lambda_{n_k+1} f(v_{n_k-1}, u_{n_k+1}) - \lambda_{n_k} \lambda_{n_k+1} f(v_{n_k-1}, v_{n_k}) - c_1 \lambda_{n_k} \lambda_{n_k+1} \|v_{n_k-1} - v_{n_k}\|^2$$
$$- c_2 \lambda_{n_k} \lambda_{n_k+1} \|v_{n_k} - u_{n_k+1}\|^2 + \langle \rho_{n_k} - u_{n_k+1}, y - u_{n_k+1} \rangle$$
$$\geq \lambda_{n_k+1} \langle \rho_{n_k} - v_{n_k}, u_{n_k+1} - v_{n_k} \rangle - c_1 \lambda_{n_k} \lambda_{n_k+1} \|v_{n_k-1} - v_{n_k}\|^2$$
$$- c_2 \lambda_{n_k} \lambda_{n_k+1} \|v_{n_k} - u_{n_k+1}\|^2 + \langle \rho_{n_k} - u_{n_k+1}, y - u_{n_k+1} \rangle \tag{43}$$

where y in C. Next, from (35), (40), (41) and due to boundedness of $\{u_n\}$ gives that the right hand side reaches to zero. Due to μ, $\lambda_{n_k} > 0$ and $v_{n_k} \rightharpoonup z$, we have

$$0 \leq \limsup_{k \to +\infty} f(v_{n_k}, y) \leq f(z, y), \quad \text{for all } y \in C. \tag{44}$$

Thus, $z \in C$ implies that $f(z, y) \geq 0$, for all $y \in C$. It proves that $z \in EP(f, C)$. By Lemma 5, the sequence $\{u_n\}$ converges weakly to $p \in EP(f, C)$. □

If $\wp_n = 0$ in Algorithm 1, we have a better version of Lyashko et al. [25] extragradient method in terms of step size improvement.

Corollary 1. Let $f : \mathbb{H} \times \mathbb{H} \to \mathbb{R}$ satisfy the conditions (f_1)-(f_4). For some $p \in EP(f,C) \neq \emptyset$, the sequence $\{u_n\}$ and $\{v_n\}$ generated in the following way:

(i) Given $u_0, v_{-1}, v_0 \in \mathbb{H}$, $0 < \sigma < \min\{1, \frac{1}{2c_2+4c_1}\}$, $\mu \in (0, \sigma)$ and $\lambda_0 > 0$.

(ii) Compute
$$\begin{cases} u_{n+1} = \underset{y \in C}{\arg\min}\{\mu\lambda_n f(v_n, y) + \frac{1}{2}\|u_n - y\|^2\}, \\ v_{n+1} = \underset{y \in C}{\arg\min}\{\lambda_{n+1} f(v_n, y) + \frac{1}{2}\|u_{n+1} - y\|^2\}, \end{cases}$$

where
$$\lambda_{n+1} = \min\left\{\sigma, \frac{\mu f(v_n, u_{n+1})}{f(v_{n-1}, u_{n+1}) - f(v_{n-1}, v_n) - c_1\|v_{n-1} - v_n\|^2 - c_2\|u_{n+1} - v_n\|^2 + 1}\right\}.$$

Then, the sequences $\{u_n\}$ and $\{v_n\}$ converge weakly to $p \in EP(f,C)$.

4. Applications

Now, we consider the applications of Theorem 1 to solve the variational inequality problems involving pseudomonotone and Lipschitz continuous operator. A variational inequality problem is defined in the following way:

Find $p^* \in C$ such that $\langle F(p^*), v - p^* \rangle \geq 0$, for all $v \in C$.

We consider that F meets the following conditions.

(F_1) Solution set $VI(F,C)$ is non-empty and F is pseudomonotone on C, i.e.,

$$\langle F(u), v - u \rangle \geq 0 \quad \text{implies that} \quad \langle F(v), u - v \rangle \leq 0, \text{ for all } u, v \in C;$$

(F_2) F is L-Lipschitz continuous on C if there exists a positive constants $L > 0$ such that

$$\|F(u) - F(v)\| \leq L\|u - v\|, \text{ for all } u, v \in C.$$

(F_3) $\underset{n \to +\infty}{\limsup} \langle F(u_n), v - u_n \rangle \leq \langle F(p^*), v - p^* \rangle$ for every $v \in C$ and $\{u_n\} \subset C$ satisfying $u_n \rightharpoonup p^*$.

Corollary 2. Assume that $F : C \to \mathbb{H}$ meet the conditions (F_1)–(F_3). Let $\{\rho_n\}$, $\{u_n\}$ and $\{v_n\}$ be the sequences are generated in the following way:

(i) Choose $u_{-1}, v_{-1}, u_0, v_0 \in \mathbb{H}$ and a sequence $\wp_n$ is non-decreasing such that $0 \leq \wp_n \leq \wp < \frac{1}{3}$, $\lambda_0 > 0$, $0 < \sigma < \min\left\{\frac{1-3\wp}{(1-\wp)^2+2L(\wp+\wp^2)}, \frac{1}{3L+2\wp L}\right\}$ and $\mu \in (0, \sigma)$.

(ii) Compute
$$\begin{cases} u_{n+1} = P_C(\rho_n - \mu\lambda_n F(v_n)), & \text{where} \quad \rho_n = u_n + \wp_n(u_n - u_{n-1}), \\ v_{n+1} = P_C(\rho_{n+1} - \lambda_{n+1} F(v_n)), & \text{where} \quad \rho_{n+1} = u_{n+1} + \wp_{n+1}(u_{n+1} - u_n), \end{cases}$$

while
$$\lambda_{n+1} = \min\left\{\sigma, \frac{\mu\langle Fv_n, u_{n+1} - v_n\rangle}{\langle Fv_{n-1}, u_{n+1} - v_n\rangle - \frac{L}{2}\|v_{n-1} - v_n\|^2 - \frac{L}{2}\|u_{n+1} - v_n\|^2 + 1}\right\}.$$

Then, the sequence $\{\rho_n\}$, $\{u_n\}$ and $\{v_n\}$ converge weakly to p.

Corollary 3. *Assume that $F : C \to \mathbb{H}$ meets the conditions (F_1)-(F_3). Let $\{u_n\}$ and $\{v_n\}$ be the sequences are generated in the following way:*

(i) *Choose $v_{-1}, u_0, v_0 \in \mathbb{H}$, $0 < \sigma < \min\{1, \frac{1}{3L}\}$ and $\lambda_0 > 0$.*
(ii) *Compute*

$$\begin{cases} u_{n+1} = P_C(u_n - \mu\lambda_n F(v_n)), \\ v_{n+1} = P_C(u_{n+1} - \lambda_{n+1} F(v_n)), \end{cases}$$

while

$$\lambda_{n+1} = \min\left\{\sigma, \frac{\mu\langle Fv_n, u_{n+1} - v_n\rangle}{\langle Fv_{n-1}, u_{n+1} - v_n\rangle - \frac{L}{2}\|v_{n-1} - v_n\|^2 - \frac{L}{2}\|u_{n+1} - v_n\|^2 + 1}\right\}.$$

Then, the sequence $\{u_n\}$ and $\{v_n\}$ converge weakly to p.

5. Computational Illustration

Numerical findings are discussed in this section to show the efficiency of our suggested method. Moreover, for Lyashko et al.'s [25] method (L.EgA), our proposed algorithm (Algo.1) and we use error term $D_n = \|u_{n+1} - u_n\|$.

Example 1. *Consider the Nash–Cournot equilibrium of electricity markets as in [7]. In this problem, there are total three electricity producing firms: i ($i = 1, 2, 3$). The firm's 1,2,3 have generating units named as $I_1 = \{1\}$, $I_2 = \{2, 3\}$ and $I_3 = \{4, 5, 6\}$, respectively. Assume that u_j denote the producing power of the unit for $i = \{1, 2, 3, 4, 5, 6\}$. Suppose that the value p of electricity can be taken as $p = 378.4 - 2\sum_{j=1}^{6} u_j$. The cost of the manufacture j unit follows:*

$$c_j(u_j) := \max\{\mathring{c}_j(u_j), \dot{c}_j(u_j)\},$$

where $\mathring{c}_j(u_j) := \frac{\mathring{\alpha}_j}{2}u_j^2 + \mathring{\beta}_j u_j + \mathring{\gamma}_j$ and $\dot{c}_j(u_j) := \dot{\alpha}_j u_j + \frac{\dot{\beta}_j}{\dot{\beta}_j + 1}\dot{\gamma}_j^{\frac{-1}{\dot{\beta}_j}}(u_j)^{(\dot{\beta}_j+1)/\dot{\beta}_j}$. The values are provided in $\mathring{\alpha}_j, \mathring{\beta}_j, \mathring{\gamma}_j, \dot{\alpha}_j, \dot{\beta}_j$ and $\dot{\gamma}_j$ in Table 1. Profit of the firm i is

$$f_i(u) := p\sum_{j \in I_i} u_j - \sum_{j \in I_i} c_j(u_j) = \left(378.4 - 2\sum_{l=1}^{6} u_l\right)\sum_{j \in I_i} u_j - \sum_{j \in I_i} c_j(u_j),$$

where $u = (u_1, \cdots, u_6)^T$ with reference to set $u \in C := \{u \in \mathbb{R}^6 : u_j^{\min} \leq u_j \leq u_j^{\max}\}$, with $u_j^{\min}$ and $u_j^{\max}$ give in Table 2. Define the equilibrium bifunction f in the following way:

$$f(u,v) := \sum_{i=1}^{3}(\phi_i(u,u) - \phi_i(u,v)),$$

where

$$\phi_i(u,v) := \left[378.4 - 2\left(\sum_{j \notin I_i} u_j + \sum_{j \in I_i} v_j\right)\right]\sum_{j \in I_i} v_j - \sum_{j \in I_i} c_j(v_j).$$

This model of electricity markets can be viewed as an equilibrium problem

Find $u^* \in C$ such that $f(u^*, v) \geq 0$, for all $v \in C$.

Numerical conclusions have shown in Figures 1–4 and Table 3. For these numerical experiments we take $u_{-1} = v_{-1} = u_0 = v_0 = (48, 48, 30, 27, 18, 24)^T$ and $\lambda = 0.01$, $\sigma = 0.026$, $\mu = 0.024$, $\wp_n = 0.20$, $\lambda_0 = 0.1$.

Table 1. Parameters for cost bi-function.

Unit j	$\mathring{\alpha}_j$	$\mathring{\beta}_j$	$\mathring{\gamma}_j$	$\dot{\alpha}_j$	$\dot{\beta}_j$	$\dot{\gamma}_j$
1	0.04	2	0	2	1	25
2	0.035	1.75	0	1.75	1	28.5714
3	0.125	1	0	1	1	8
4	0.0116	3.25	0	3.25	1	86.2069
5	0.05	3	0	3	1	20
6	0.05	3	0	3	1	20

Table 2. Values used for constraint set.

j	u_j^{min}	u_j^{max}
1	0	80
2	0	80
3	0	50
4	0	55
5	0	30
6	0	40

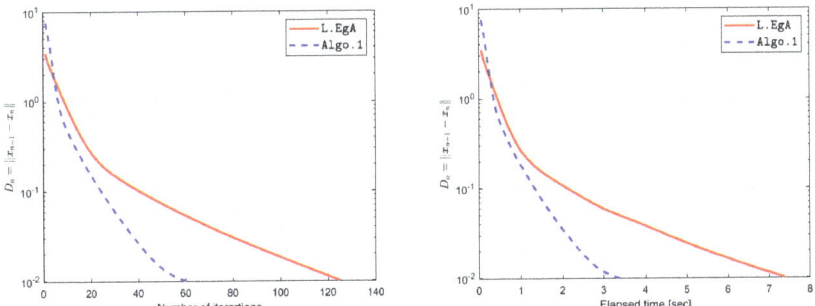

Figure 1. Example 1 while tolerance is 0.01.

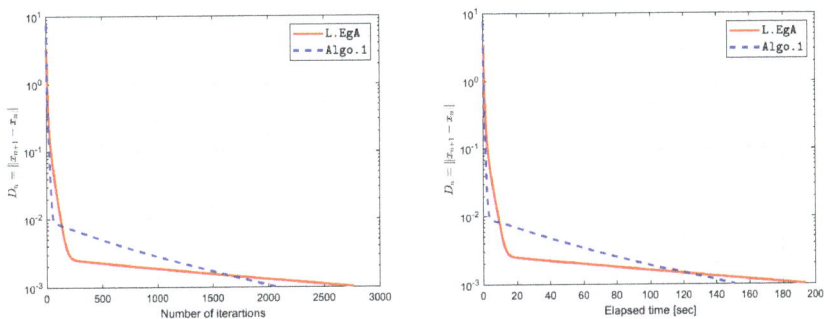

Figure 2. Example 1 while tolerance is 0.001.

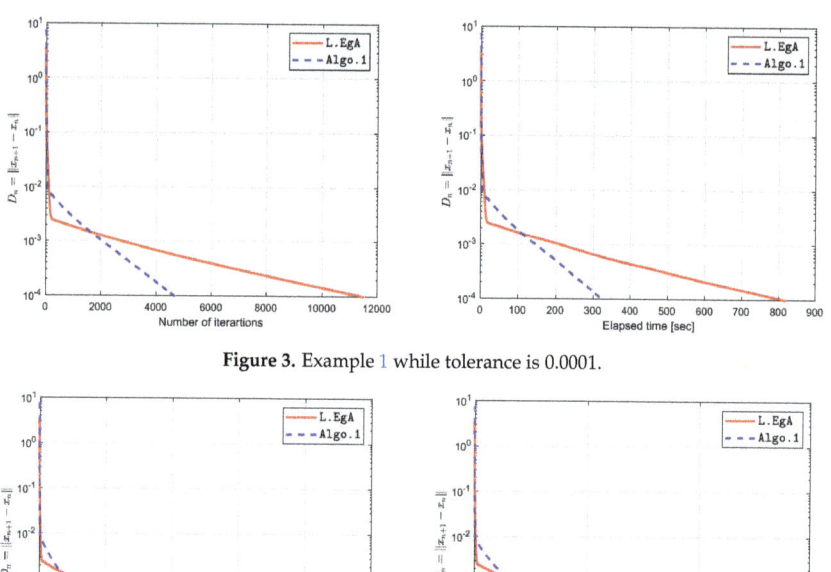

Figure 3. Example 1 while tolerance is 0.0001.

Figure 4. Example 1 while tolerance is 0.00001.

Table 3. Figures 1–4 numerical values.

TOL		L.EgA		Algo.1	
		Iter.	time (s)	Iter.	time (s)
0.01		125	7.3692	61	3.4055
$u^*_{\text{L.EgA}} = (47.3245, 47.3245, 47.3245, 47.3245, 47.3245, 47.3245)$					
$u^*_{\text{Algo.1}} = (47.3245, 47.3245, 47.3245, 47.3245, 47.3245, 47.3245)$					
0.001		2761	193.3939	2063	150.6757
$u^*_{\text{L.EgA}} = (47.3245, 47.3245, 47.3245, 47.3245, 47.3245, 47.3245)$					
$u^*_{\text{Algo.1}} = (47.3245, 47.3245, 47.3245, 47.3245, 47.3245, 47.3245)$					
0.0001		11,526	818.7184	4687	324.3571
$u^*_{\text{L.EgA}} = (47.3245, 47.3245, 47.3245, 47.3245, 47.3245, 47.3245)$					
$u^*_{\text{Algo.1}} = (47.3245, 47.3245, 47.3245, 47.3245, 47.3245, 47.3245)$					
0.00001		20,946	1449.3959	7307	526.9766
$u^*_{\text{L.EgA}} = (47.3245, 47.3245, 47.3245, 47.3245, 47.3245, 47.3245)$					
$u^*_{\text{Algo.1}} = (47.3245, 47.3245, 47.3245, 47.3245, 47.3245, 47.3245)$					

Example 2. *Assume that the following cost bifunction f defined by*

$$f(u,v) = \langle (AA^T + B + C)u, v - u \rangle,$$

on the convex set $C = \{u \in \mathbb{R}^n : Du \leq d\}$ while D is an $100 \times n$ matrix and d is a non-negative vector. In the above bifunction definition we take A is an $n \times n$ matrix, B is an $n \times n$ skew-symmetric matrix, C is an $n \times n$ diagonal matrix having diagonal entries are non-negative. The matrices are generated as; A = rand(n), K = rand(n), $B = 0.5K - 0.5K^T$ and C = diag(rand(n,1)). The bifunction f is monotone and having Lipschitz-type constants are $c_1 = c_2 = \frac{1}{2}\|AA^T + B + C\|$. Numerical results are presented in the

Figures 5–8 and Table 4. For these numerical experiments we take $u_{-1} = v_{-1} = u_0 = v_0 = (1, 1, \cdots, 1)^T$ and $\lambda = \frac{1}{10c_1}, \sigma = \frac{1}{8c_1}, \mu = \frac{1}{8.2c_1}, \wp_n = \frac{1}{5}, \lambda_0 = 1/4c_1$.

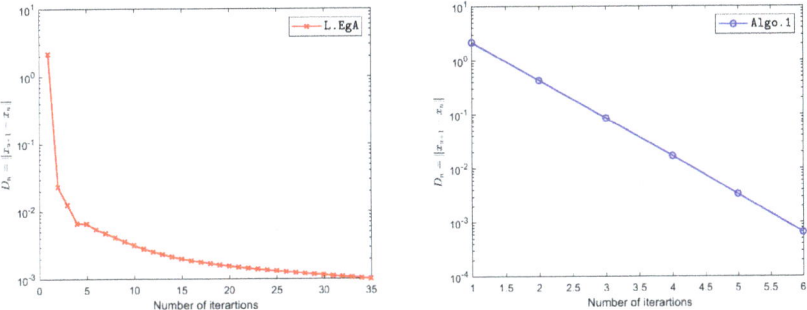

Figure 5. Example 2 for average number of iterations while $n = 5$.

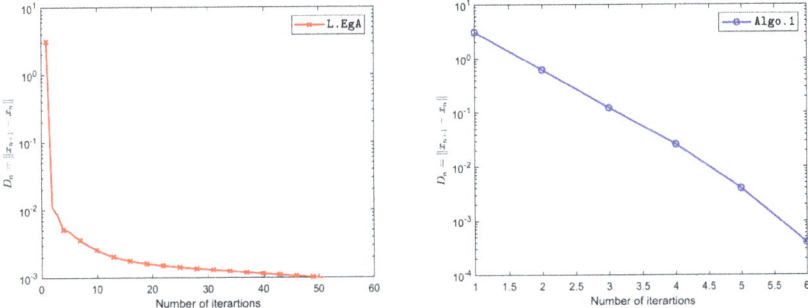

Figure 6. Example 2 for average number of iterations while $n = 10$.

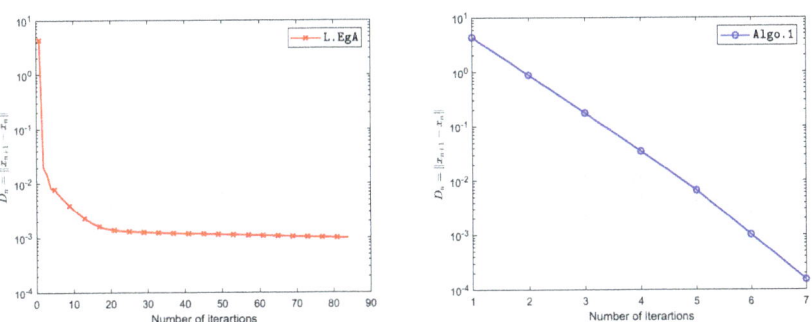

Figure 7. Example 2 for average number of iterations while $n = 20$.

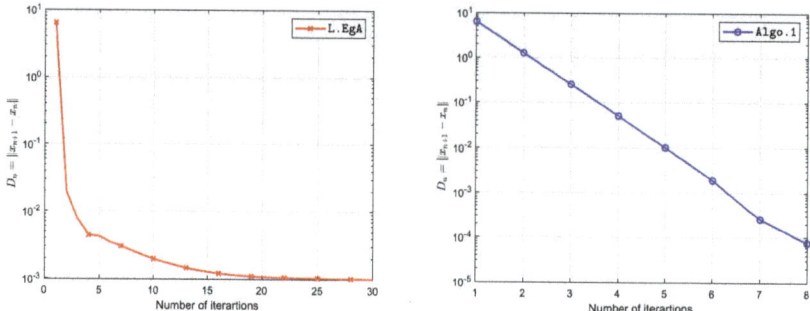

Figure 8. Example 2 for average number of iterations while $n = 40$.

Table 4. Numerical results for Figures 5–8.

		L.EgA		Algo.1	
n	T. Samples	Avg Iter.	Avg time(s)	Avg Iter.	Avg time(s)
5	10	35	0.8066	6	0.1438
10	10	51	1.1779	6	0.1302
20	10	84	1.7441	7	0.1801
40	10	30	0.6859	8	0.1999

Example 3. Assume that $F : \mathbb{R}^2 \to \mathbb{R}^2$ is defined by

$$F(u) = \begin{pmatrix} 0.5u_1u_2 - 2u_2 - 10^7 \\ -4u_1 - 0.1u_2^2 - 10^7 \end{pmatrix}$$

with $C = \{u \in \mathbb{R}^2 : (u_1 - 2)^2 + (u_2 - 2)^2 \leq 1\}$. It is not hard to check that F is Lipschitz continuous with $L = 5$ and pseudomonotone. The step size $\lambda = 10^{-6}$ for Lyashko et al. [25] and $\lambda_0 = 0.1$, $\sigma = 0.129$, $\wp_n = 0.20$ and $\mu = 0.119$. Computational results are shown in the Table 5 and in Figures 9–12.

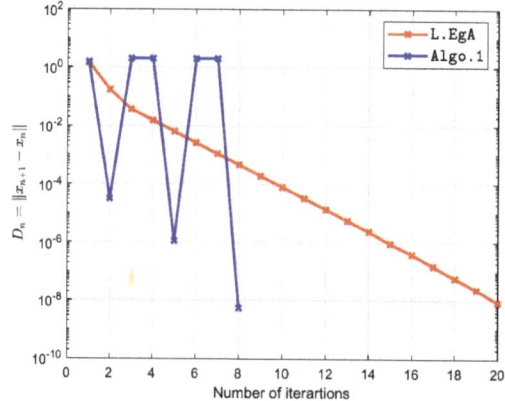

Figure 9. Example 3 while $u_0 = (1.5, 1.7)$.

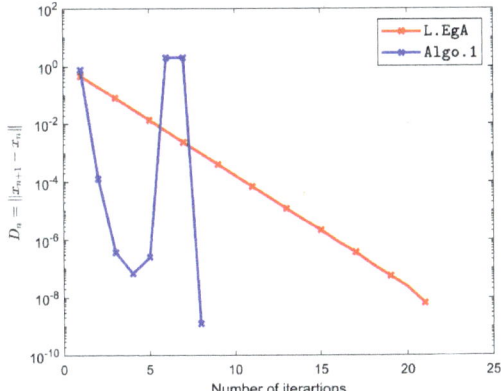

Figure 10. Example 3 while $u_0 = (2, 3)$.

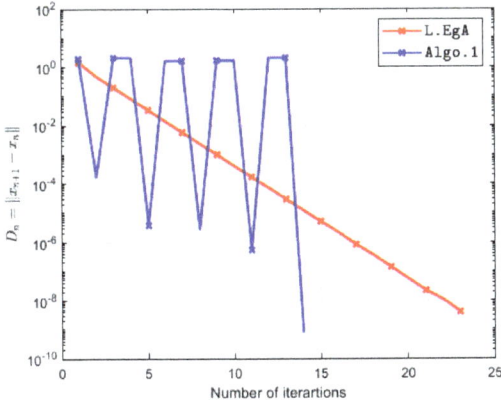

Figure 11. Example 3 while $u_0 = (1, 2)$.

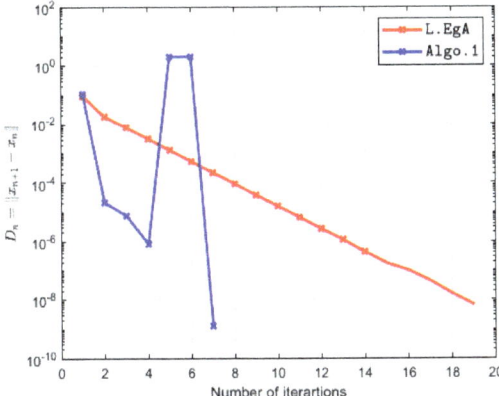

Figure 12. Example 3 while $u_0 = (2.7, 2.6)$.

Table 5. Numerical results for Figures 9–12.

u_0	L.EgA		Algo.1	
	Iter.	time(s)	Iter.	time(s)
(1.5, 1.7)	20	0.7506	8	0.5316
(2.0, 3.0)	21	0.7879	8	0.6484
(1.0, 2.0)	23	1.1450	14	0.9730
(2.7, 2.6)	19	0.7254	7	0.5835

Example 4. *Let $F : \mathbb{R}^2 \to \mathbb{R}^2$ is defined by*

$$F(u) = \begin{pmatrix} (u_1^2 + (u_2-1)^2)(1+u_2) \\ -u_1^3 - u_1(u_2-1)^2 \end{pmatrix}$$

and $C = \{u \in \mathbb{R}^2 : (u_1 - 2)^2 + (u_2 - 2)^2 \leq 1\}$. Here, F is not monotone but pseudomonotone on C and L-Lipschitz continuous through $L = 5$ (see, e.g., [33]). The stepsize $\lambda = 10^{-2}$ for Lyashko et al. [25] and $\lambda_0 = 0.01$, $\sigma = 0.129$, $\wp_n = 0.15$ and $\mu = 0.119$. The computational experimental findings are written in Table 6 and in Figures 13–15.

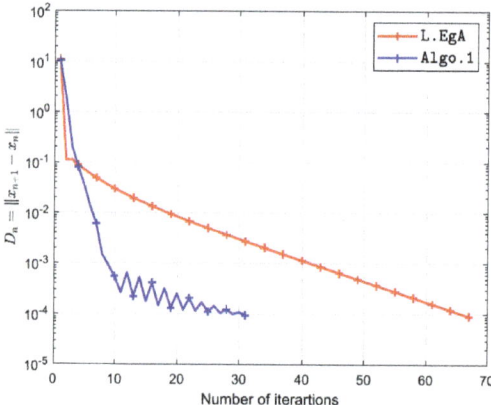

Figure 13. Example 4 while $u_0 = (10, 10)$.

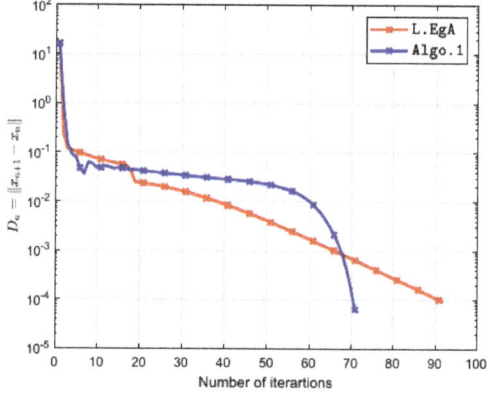

Figure 14. Example 4 while $u_0 = (-10, -10)$.

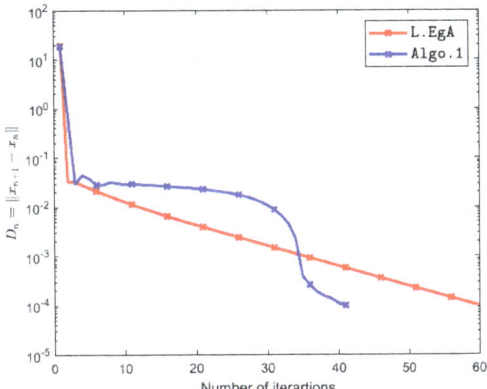

Figure 15. Example 4 while $u_0 = (10, 20)$.

Table 6. Figures 13–15 numerical values.

	L.EgA		Algo.1	
u_0	Iter.	time(s)	Iter.	time(s)
(10, 10)	67	1.9151	31	1.0752
(−10, −10)	92	2.5721	71	2.0469
(10, 20)	60	1.7689	41	1.1864

6. Conclusions

We have established an extragradient-like method to solve pseudomonotone equilibrium problems in real Hilbert space. The main advantage of the proposed method is that an iterative sequence has been incorporated with a certain step size evaluation formula. The step size formula is updated for each iteration based on the previous iterations. Numerical findings were presented to show our algorithm's numerical efficiency compared with other methods. Such numerical investigations indicate that inertial effects often generally improve the effectiveness of the iterative sequence in this context.

Author Contributions: Conceptualization, W.K. and K.M.; methodology, W.K. and K.M.; writing–original draft preparation, W.K. and K.M.; writing–review and editing, W.K. and K.M.; software, W.K. and K.M.; supervision, W.K.; project administration and funding acquisition, K.M. All authors have read and agreed to the published version of the manuscript.

Funding: This project was supported by Rajamangala University of Technology Phra Nakhon (RMUTP).

Acknowledgments: The first author would like to thanks the Rajamangala University of Technology Thanyaburi (RMUTT) (Grant No. NSF62D0604). The second author would like to thanks the Rajamangala University of Technology Phra Nakhon (RMUTP).

Conflicts of Interest: The authors declare that they have no conflict of interest.

References

1. Blum, E. From optimization and variational inequalities to equilibrium problems. *Math. Stud.* **1994**, *63*, 123–145.
2. Facchinei, F.; Pang, J.S. *Finite-Dimensional Variational Inequalities and Complementarity Problems*; Springer Science & Business Media: Berlin/Heidelberg, Germany, 2007.
3. Konnov, I. *Equilibrium Models and Variational Inequalities*; Elsevier: Amsterdam, The Netherlands, 2007; Volume 210.
4. Muu, L.D.; Oettli, W. Convergence of an adaptive penalty scheme for finding constrained equilibria. *Nonlinear Anal. Theory Methods Appl.* **1992**, *18*, 1159–1166.

5. Combettes, P.L.; Hirstoaga, S.A. Equilibrium programming in Hilbert spaces. *J. Nonlinear Convex Anal.* **2005**, *6*, 117–136.
6. Flåm, S.D.; Antipin, A.S. Equilibrium programming using proximal-like algorithms. *Math. Program.* **1996**, *78*, 29–41.
7. Quoc, T.D.; Anh, P.N.; Muu, L.D. Dual extragradient algorithms extended to equilibrium problems. *J. Glob. Optim.* **2012**, *52*, 139–159.
8. Quoc Tran, D.; Le Dung, M.; Nguyen, V.H. Extragradient algorithms extended to equilibrium problems. *Optimization* **2008**, *57*, 749–776.
9. Santos, P.; Scheimberg, S. An inexact subgradient algorithm for equilibrium problems. *Comput. Appl. Math.* **2011**, *30*, 91–107.
10. Takahashi, S.; Takahashi, W. Viscosity approximation methods for equilibrium problems and fixed point problems in Hilbert spaces. *J. Math. Anal. Appl.* **2007**, *331*, 506–515.
11. Ur Rehman, H.; Kumam, P.; Kumam, W.; Shutaywi, M.; Jirakitpuwapat, W. The Inertial Sub-Gradient Extra-Gradient Method for a Class of Pseudo-Monotone Equilibrium Problems. *Symmetry* **2020**, *12*, 463. [CrossRef]
12. Ur Rehman, H.; Kumam, P.; Argyros, I.K.; Alreshidi, N.A.; Kumam, W.; Jirakitpuwapat, W. A Self-Adaptive Extra-Gradient Methods for a Family of Pseudomonotone Equilibrium Programming with Application in Different Classes of Variational Inequality Problems. *Symmetry* **2020**, *12*, 523. [CrossRef]
13. Ur Rehman, H.; Kumam, P.; Shutaywi, M.; Alreshidi, N.A.; Kumam, W. Inertial Optimization Based Two-Step Methods for Solving Equilibrium Problems with Applications in Variational Inequality Problems and Growth Control Equilibrium Models. *Energies* **2020**, *13*, 3293. [CrossRef]
14. Hieu, D.V. New extragradient method for a class of equilibrium problems in Hilbert spaces. *Appl. Anal.* **2017**, *97*, 811–824. [CrossRef]
15. Hammad, H.A.; ur Rehman, H.; De la Sen, M. Advanced Algorithms and Common Solutions to Variational Inequalities. *Symmetry* **2020**, *12*, 1198. [CrossRef]
16. Ur Rehman, H.; Kumam, P.; Cho, Y.J.; Yordsorn, P. Weak convergence of explicit extragradient algorithms for solving equilibirum problems. *J. Inequalities Appl.* **2019**, *2019*. [CrossRef]
17. Ur Rehman, H.; Kumam, P.; Abubakar, A.B.; Cho, Y.J. The extragradient algorithm with inertial effects extended to equilibrium problems. *Comput. Appl. Math.* **2020**, *39*. [CrossRef]
18. Ur Rehman, H.; Kumam, P.; Argyros, I.K.; Deebani, W.; Kumam, W. Inertial Extra-Gradient Method for Solving a Family of Strongly Pseudomonotone Equilibrium Problems in Real Hilbert Spaces with Application in Variational Inequality Problem. *Symmetry* **2020**, *12*, 503. [CrossRef]
19. Koskela, P.; Manojlović, V. Quasi-Nearly Subharmonic Functions and Quasiconformal Mappings. *Potential Anal.* **2012**, *37*, 187–196. [CrossRef]
20. Ur Rehman, H.; Kumam, P.; Argyros, I.K.; Shutaywi, M.; Shah, Z. Optimization Based Methods for Solving the Equilibrium Problems with Applications in Variational Inequality Problems and Solution of Nash Equilibrium Models. *Mathematics* **2020**, *8*, 822. [CrossRef]
21. Rehman, H.U.; Kumam, P.; Dong, Q.L.; Peng, Y.; Deebani, W. A new Popov's subgradient extragradient method for two classes of equilibrium programming in a real Hilbert space. *Optimization* **2020**, 1–36. [CrossRef]
22. Yordsorn, P.; Kumam, P.; ur Rehman, H.; Ibrahim, A.H. A Weak Convergence Self-Adaptive Method for Solving Pseudomonotone Equilibrium Problems in a Real Hilbert Space. *Mathematics* **2020**, *8*, 1165. [CrossRef]
23. Todorčević, V. *Harmonic Quasiconformal Mappings and Hyperbolic Type Metrics*; Springer International Publishing: Berlin/Heidelberg, Germany, 2019. [CrossRef]
24. Yordsorn, P.; Kumam, P.; Rehman, H.U. Modified two-step extragradient method for solving the pseudomonotone equilibrium programming in a real Hilbert space. *Carpathian J. Math.* **2020**, *36*, 313–330.
25. Lyashko, S.I.; Semenov, V.V. A new two-step proximal algorithm of solving the problem of equilibrium programming. In *Optimization and Its Applications in Control and Data Sciences*; Springer: Berlin/Heidelberg, Germany, 2016; pp. 315–325.
26. Polyak, B.T. Some methods of speeding up the convergence of iteration methods. *USSR Comput. Math. Math. Phys.* **1964**, *4*, 1–17. [CrossRef]
27. Bianchi, M.; Schaible, S. Generalized monotone bifunctions and equilibrium problems. *J. Optim. Theory Appl.* **1996**, *90*, 31–43. [CrossRef]

28. Mastroeni, G. On Auxiliary Principle for Equilibrium Problems. In *Nonconvex Optimization and Its Applications*; Springer: New York, NY, USA; 2003; pp. 289–298._15. [CrossRef]
29. Bauschke, H.H.; Combettes, P.L. *Convex Analysis and Monotone Operator Theory in Hilbert Spaces*; Springer: Berlin/Heidelberg, Germany, 2011; Volume 408.
30. Tiel, J.V. *Convex Analysis*; John Wiley: Hoboken, NJ, USA, 1984.
31. Alvarez, F.; Attouch, H. An inertial proximal method for maximal monotone operators via discretization of a nonlinear oscillator with damping. *Set Valued Anal.* **2001**, *9*, 3–11. [CrossRef]
32. Opial, Z. Weak convergence of the sequence of successive approximations for nonexpansive mappings. *Bull. Am. Math. Soc.* **1967**, *73*, 591–597. [CrossRef]
33. Shehu, Y.; Dong, Q.L.; Jiang, D. Single projection method for pseudo-monotone variational inequality in Hilbert spaces. *Optimization* **2019**, *68*, 385–409. [CrossRef]

Publisher's Note: MDPI stays neutral with regard to jurisdictional claims in published maps and institutional affiliations.

© 2020 by the authors. Licensee MDPI, Basel, Switzerland. This article is an open access article distributed under the terms and conditions of the Creative Commons Attribution (CC BY) license (http://creativecommons.org/licenses/by/4.0/).

Article

Approximation Results for Equilibrium Problems Involving Strongly Pseudomonotone Bifunction in Real Hilbert Spaces

Wiyada Kumam [1] and Kanikar Muangchoo [2],*

[1] Program in Applied Statistics, Department of Mathematics and Computer Science, Faculty of Science and Technology, Rajamangala University of Technology Thanyaburi, Thanyaburi, Pathumthani 12110, Thailand; wiyada.kum@rmutt.ac.th

[2] Faculty of Science and Technology, Rajamangala University of Technology Phra Nakhon (RMUTP), 1381 Pracharat 1 Road, Wongsawang, Bang Sue, Bangkok 10800, Thailand

* Correspondence: kanikar.m@rmutp.ac.th; Tel.: +66-0-2836-3000 (ext. 4193)

Received: 26 September 2020; Accepted: 21 November 2020; Published: 26 November 2020

Abstract: A plethora of applications in non-linear analysis, including minimax problems, mathematical programming, the fixed-point problems, saddle-point problems, penalization and complementary problems, may be framed as a problem of equilibrium. Most of the methods used to solve equilibrium problems involve iterative methods, which is why the aim of this article is to establish a new iterative method by incorporating an inertial term with a subgradient extragradient method to solve the problem of equilibrium, which includes a bifunction that is strongly pseudomonotone and meets the Lipschitz-type condition in a real Hilbert space. Under certain mild conditions, a strong convergence theorem is proved, and a required sequence is generated without the information of the Lipschitz-type cost bifunction constants. Thus, the method operates with the help of a slow-converging step size sequence. In numerical analysis, we consider various equilibrium test problems to validate our proposed results.

Keywords: equilibrium problem; variational inequalities; strongly pseudomonotone bifunction; Lipschitz-type conditions

1. Background

Assume that a bifunction $f : \mathcal{H} \times \mathcal{H} \to \mathbb{R}$ satisfying the conditions $f(v,v) = 0$ for each $v \in \mathcal{K}$. A *equilibrium problem* [1,2] for f on $\mathcal{K}$ is said to be:

$$\text{Find } v^* \in \mathcal{K} \text{ such that } f(v^*, v) \geq 0, \ \forall v \in \mathcal{K}. \tag{1}$$

where $\mathcal{K}$ is a non-empty closed and convex subset of a Hilbert space $\mathcal{H}$. Next, we present the definitions of the important classification of the problems of equilibrium [1,3]. A function $f : \mathcal{H} \times \mathcal{H} \to \mathbb{R}$ on $\mathcal{K}$ for $\gamma > 0$ is said to be

(i) strongly monotone if

$$f(v_1, v_2) + f(v_2, v_1) \leq -\gamma \|v_1 - v_2\|^2, \ \forall v_1, v_2 \in \mathcal{K};$$

(ii) monotone if

$$f(v_1, v_2) + f(v_2, v_1) \leq 0, \ \forall v_1, v_2 \in \mathcal{K};$$

(iii) γ-strongly pseudo-monotone if

$$f(v_1, v_2) \geq 0 \implies f(v_2, v_1) \leq -\gamma \|v_1 - v_2\|^2, \ \forall v_1, v_2 \in \mathcal{K};$$

(iv) pseudo-monotone if
$$f(v_1, v_2) \geq 0 \Longrightarrow f(v_2, v_1) \leq 0, \forall v_1, v_2 \in \mathcal{K};$$

and

(v) satisfy the Lipschitz-type conditions on $\mathcal{K}$ for $L_1, L_2 > 0$, such that
$$f(v_1, v_3) - L_1\|v_1 - v_2\|^2 - L_2\|v_2 - v_3\|^2 \leq f(v_1, v_2) + f(v_2, v_3), \forall v_1, v_2, v_3 \in \mathcal{K}.$$

The above well-defined simple mathematical problem (1) includes many mathematical and applied sciences problems as a special case, consisting of the fixed point problems, vector and scalar minimization problems, problems of variational inequalities (VIP), the complementarity problems, the Nash equilibrium problems in non-cooperative games, and inverse optimization problems [1,4,5]. This problem is also seen as a problem of Ky Fan inequality based on his initial contribution [2]. Several researchers have developed and generalized numerous findings on the nature of a solution to an equilibrium problem. (e.g., see [2,4,6,7]). Due to the basic formulation of a problem (1) and its application in both the theoretical and applied sciences, it has been extensively studied in recent times by several authors [8,9] (see also [10–16]).

Many methods have been previously established and considered their convergence investigation to deal with the problem (1). There is an impressive number of numerical methods have been designed along with their well-defined convergence analysis and theoretical properties to solve the problem (1) in different dimensional spaces [17–22]. Regularization is one of the most significant methods to figure out various ill-posed problems in the many fields of pure and applied mathematics. The prominent aspect of the regularization method is to employ it on monotone equilibrium problems and the initial problem converts into strongly monotone equilibrium sub-problem. Therefore, each computationally efficient sub-problem is strongly monotone and a unique solution exists.

A proximal method is another approach to deal with equilibrium problems that rely on numerical minimization problems [23]. This method has also been identified as the extragradient method [24] based on the initial contribution of the Korpelevich [25] method to solve the saddle point problems. Hieu [26] established an algorithmic sequence $\{u_n\}$ as follows:

$$\begin{cases} u_0 \in \mathcal{K} \\ v_n = \arg\min_{v \in \mathcal{K}}\{\zeta_n f(u_n, v) + \frac{1}{2}\|u_n - v\|^2\}, \\ u_{n+1} = \arg\min_{v \in \mathcal{K}}\{\zeta_n f(v_n, v) + \frac{1}{2}\|u_n - v\|^2\}, \end{cases} \quad (2)$$

while $\{\zeta_n\}$ meet the following conditions:

$$\mathcal{C}_1: \lim_{n \to +\infty} \zeta_n = 0 \text{ and } \mathcal{C}_2: \sum_{n=1}^{+\infty} \zeta_n = +\infty. \quad (3)$$

Inertial-like methods are two-step iterative methods, where the next iteration is carried out by employing the previous two iterations [27,28]. The inertial interpolation term is required to boost the sequence and help to improve the convergence rate of the iterative sequence. Such inertial methods are essentially used to speed up the iterative sequence to the appropriate solution and to improve the convergence rate. Numerical descriptions demonstrate that inertial effects also enhance the numerical performance. Such impressive attributes increase the curiosity of researchers in creating inertial methods. Recently, various inertial methods have also been established for specific types of equilibrium problems [29–32].

In this paper, we use the projection method that is simple to carry out due to its low cost and efficient numerical computations. Inspired by the works of Fan et al. [33], Thong and Hieu [34], and Censor et al. [35], we set up an accelerated extragradient-like algorithm to solve the problem (1) and other special class of equilibrium problem, such as variational inequalities. We prove a strong convergence theorem corresponding to the sequence generated to solve the problem of equilibrium under certain mild conditions. At the end, the computational tests show that the algorithm is more efficient than the current ones [26,29,36–38].

The rest of the article has been organized as follows. Section 2 consists of some basic results which are used throughout the article. Section 3 includes our proposed method and its convergence analysis. Section 4 includes numerical experiments that demonstrate practical effectiveness.

2. Preliminaries

Assume that a convex function $g : \mathcal{K} \to \mathbb{R}$ and subdifferential of g on $v_1 \in \mathcal{K}$ is defined as follows:

$$\partial g(v_1) = \{v_3 \in \mathcal{H} : g(v_2) - g(v_1) \geq \langle v_3, v_2 - v_1 \rangle, \forall v_2 \in \mathcal{K}\}.$$

A *normal cone for* $\mathcal{K}$ on $v_1 \in \mathcal{K}$ is defined as follows:

$$N_{\mathcal{K}}(v_1) = \{v_3 \in \mathcal{H} : \langle v_3, v_2 - v_1 \rangle \leq 0, \forall v_2 \in \mathcal{K}\}.$$

Lemma 1 ([39]). *Assume the three sequences* α_n, β_n *and* γ_n *are in* $[0, +\infty)$ *such that*

$$\alpha_{n+1} \leq \alpha_n + \beta_n(\alpha_n - \alpha_{n-1}) + \gamma_n, \text{for all} n \geq 1, \text{ having } \sum_{n=1}^{+\infty} \gamma_n < +\infty,$$

where $0 < \beta$ *with* $0 \leq \beta_n \leq \beta < 1$ *for each* $n \in \mathbb{N}$. *Thus, we have*

(i) $\sum_{n=1}^{+\infty} [\alpha_n - \alpha_{n-1}]_+ < +\infty$, *with* $[q]_+ := \max\{q, 0\}$;

(ii) $\lim_{n \to +\infty} \alpha_n = \alpha^* \in [0, +\infty)$.

Lemma 2 ([40]). *For each* $v_1, v_2 \in \mathcal{H}$ *and* $r \in \mathbb{R}$, *the following equality holds*

$$\|rv_1 + (1-r)v_2\|^2 = r\|v_1\|^2 + (1-r)\|v_2\|^2 - r(1-r)\|v_1 - v_2\|^2.$$

Lemma 3 ([41]). *Let* $\{p_n\}$ *and* $\{q_n\} \subset [0, +\infty)$ *be two sequences such that*

$$\sum_{n=1}^{+\infty} p_n = +\infty \quad \text{and} \quad \sum_{n=1}^{+\infty} p_n q_n < +\infty.$$

Then, $\liminf_{n \to +\infty} q_n = 0$.

Lemma 4 ([42]). *Assume that a function* $h : \mathcal{K} \to \mathbb{R}$ *is subdifferentiable, convex, and lower semi-continuous on* $\mathcal{K}$. *Then,* $v_1 \in \mathcal{K}$ *is a function h minimizer if and only if* $0 \in \partial h(v_1) + N_{\mathcal{K}}(v_1)$ *while* $\partial h(v_1)$ *and* $N_{\mathcal{K}}(v_1)$ *stand for the subdifferential of h on* $v_1 \in \mathcal{K}$ *and a normal cone of* $\mathcal{K}$ *at* v_1, *respectively.*

Suppose that $f : \mathcal{H} \times \mathcal{H} \to \mathbb{R}$ satisfies the following conditions:

(C1) $f(v_1, v_1) = 0$, for all $v_1 \in \mathcal{K}$ and f is strongly pseudomonotone on $\mathcal{K}$;
(C2) f meet the Lipschitz-type condition with two constants L_1 and L_2; and
(C3) $f(v_1, .)$ is convex and sub-differentiable on $\mathcal{H}$ for fixed each $v_1 \in \mathcal{H}$.

3. Main Results

The following is the main method (Algorithm 1) in more detail.

Algorithm 1. Modified subgradient extragradient method for equilibrium problems.

Step 0: Choose $u_{-1}, u_0 \in \mathcal{H}$ arbitrarily. Let ζ_n satisfy the conditions (3). $\{\theta_n\}$ and $\{\vartheta_n\}$ are control parameter sequences.
Step 1: Compute
$$v_n = \arg\min_{v \in \mathcal{K}}\{\zeta_n f(w_n, v) + \frac{1}{2}\|w_n - v\|^2\},$$
where $w_n = u_n + \theta_n(u_n - u_{n-1})$. If $v_n = w_n$, then STOP and $w_n \in EP(f, \mathcal{K})$.
Step 2: Compute a set
$$\mathcal{H}_n = \{z \in \mathcal{H} : \langle w_n - \zeta_n t_n - v_n, z - v_n \rangle \leq 0\},$$
where $t_n \in \partial_2 f(w_n, v_n)$.
Step 3: Compute
$$\eta_n = \arg\min_{v \in \mathcal{H}_n}\{\zeta_n f(v_n, v) + \frac{1}{2}\|w_n - v\|^2\}.$$
Step 4: Compute
$$u_{n+1} = (1 - \vartheta_n)w_n + \vartheta_n \eta_n,$$
where $\{\vartheta_n\}$ and $\{\theta_n\}$ are real sequences meet the conditions:

(i) $\{\theta_n\}$ sequence is non-decreasing and $0 \leq \theta_n \leq \theta < 1$ for each $n \geq 1$;
(ii) there exists $\vartheta, \delta, \sigma > 0$ such that
$$\delta > \frac{4\theta[\theta(1+\theta) + \sigma]}{1 - \theta^2}, \tag{4}$$
and
$$0 < \vartheta \leq \vartheta_n \leq \frac{\delta - 4\theta[\theta(1+\theta) + \sigma + \frac{1}{4}\theta\delta]}{4\delta[\theta(1+\theta) + \sigma + \frac{1}{4}\theta\delta]}. \tag{5}$$

Set $n := n + 1$ and switch to **Step 1**.

Lemma 5. *Suppose that $f : \mathcal{H} \times \mathcal{H} \to \mathbb{R}$ satisfies the conditions (C1)-(C3). For $v^* \in EP(f, \mathcal{K}) \neq \emptyset$, we have*
$$\|\eta_n - v^*\|^2 \leq \|w_n - v^*\|^2 - (1 - 2L_1\zeta_n)\|w_n - v_n\|^2 - (1 - 2L_2\zeta_n)\|\eta_n - v_n\|^2$$
$$- 2\gamma\zeta_n\|v_n - v^*\|^2.$$

Proof. By value of η_n and Lemma 4, we have
$$0 \in \partial_2\left\{\zeta_n f(v_n, v) + \frac{1}{2}\|w_n - v\|^2\right\}(\eta_n) + N_{\mathcal{H}_n}(\eta_n).$$
Thus, there exists $\omega \in \partial f(v_n, \eta_n)$ and $\overline{\omega} \in N_{\mathcal{H}_n}(\eta_n)$ such that
$$\zeta_n \omega + \eta_n - w_n + \overline{\omega} = 0.$$
Thus, the above implies that
$$\langle w_n - \eta_n, v - \eta_n \rangle = \zeta_n \langle \omega, v - \eta_n \rangle + \langle \overline{\omega}, v - \eta_n \rangle, \ \forall v \in \mathcal{H}_n.$$
Since $\overline{\omega} \in N_{\mathcal{H}_n}(\eta_n)$, it implies that $\langle \overline{\omega}, v - \eta_n \rangle \leq 0$, for all $v \in \mathcal{H}_n$. This gives that
$$\zeta_n \langle \omega, v - \eta_n \rangle \geq \langle w_n - \eta_n, v - \eta_n \rangle, \ \forall v \in \mathcal{H}_n. \tag{6}$$

By $\omega \in \partial f(v_n, \eta_n)$, we have

$$f(v_n, v) - f(v_n, \eta_n) \geq \langle \omega, v - \eta_n \rangle, \ \forall v \in \mathcal{H}. \tag{7}$$

From (6) and (7), we obtain

$$\zeta_n f(v_n, v) - \zeta_n f(v_n, \eta_n) \geq \langle w_n - \eta_n, v - \eta_n \rangle, \ \forall v \in \mathcal{H}_n. \tag{8}$$

By the use of $v = v^*$, we get

$$\zeta_n f(v_n, v^*) - \zeta_n f(v_n, \eta_n) \geq \langle w_n - \eta_n, v^* - \eta_n \rangle. \tag{9}$$

By given $v^* \in EP(f, \mathcal{K})$, $f(v^*, v_n) \geq 0$, which implies that $f(v_n, v^*) \leq -\gamma \|v_n - v^*\|^2$. From the expression (9), we obtain

$$\langle w_n - \eta_n, \eta_n - v^* \rangle \geq \zeta_n f(v_n, \eta_n) + \gamma \zeta_n \|v_n - v^*\|^2. \tag{10}$$

Due to the Lipschitz-type continuity of a bifunction f,

$$f(w_n, \eta_n) \leq f(w_n, v_n) + f(v_n, \eta_n) + L_1 \|w_n - v_n\|^2 + L_2 \|v_n - \eta_n\|^2. \tag{11}$$

Expressions (10) and (11) gives that

$$\langle w_n - \eta_n, \eta_n - v^* \rangle \geq \zeta_n \{f(w_n, \eta_n) - f(w_n, v_n)\} \\ - L_1 \zeta_n \|w_n - v_n\|^2 - L_2 \zeta_n \|v_n - \eta_n\|^2 + \gamma \zeta_n \|v_n - v^*\|^2. \tag{12}$$

By value $\eta_n \in \mathcal{H}_n$,

$$\langle w_n - \zeta_n t_n - v_n, \eta_n - v_n \rangle \leq 0.$$

The above implies that

$$\langle w_n - v_n, \eta_n - v_n \rangle \leq \zeta_n \langle t_n, \eta_n - v_n \rangle. \tag{13}$$

$t_n \in \partial_2 f(w_n, v_n)$ gives that

$$f(w_n, v) - f(w_n, v_n) \geq \langle t_n, v - v_n \rangle, \forall v \in \mathcal{H}.$$

Substituting $v = \eta_n$ into the above expression,

$$f(w_n, \eta_n) - f(w_n, v_n) \geq \langle t_n, \eta_n - v_n \rangle. \tag{14}$$

Expressions (13) and (14) imply that

$$\zeta_n \{f(w_n, \eta_n) - f(w_n, v_n)\} \geq \langle w_n - v_n, \eta_n - v_n \rangle. \tag{15}$$

Combining expressions (12) and (15) implies that

$$\langle w_n - \eta_n, \eta_n - v^* \rangle \geq \langle w_n - v_n, \eta_n - v_n \rangle \\ - L_1 \zeta_n \|w_n - v_n\|^2 - L_2 \zeta_n \|v_n - \eta_n\|^2 + \gamma \zeta_n \|v_n - v^*\|^2. \tag{16}$$

We have the following facts:

$$2\langle w_n - \eta_n, \eta_n - v^* \rangle = \|w_n - v^*\|^2 - \|\eta_n - w_n\|^2 - \|\eta_n - v^*\|^2.$$

$$2\langle v_n - w_n, v_n - \eta_n \rangle = \|w_n - v_n\|^2 + \|\eta_n - v_n\|^2 - \|w_n - \eta_n\|^2.$$

Thus, we finally obtain

$$\|\eta_n - v^*\|^2 \leq \|w_n - v^*\|^2 - (1 - 2L_1\zeta_n)\|w_n - v_n\|^2 - (1 - 2L_2\zeta_n)\|\eta_n - v_n\|^2$$
$$- 2\gamma\zeta_n\|v_n - v^*\|^2.$$

□

Theorem 1. *The sequences $\{w_n\}$, $\{v_n\}$, $\{\eta_n\}$ and $\{u_n\}$ generated by Algorithm 1 strongly converge to v^*.*

Proof. By the value of u_{n+1}, we have

$$\begin{aligned}\|u_{n+1} - v^*\|^2 &= \|(1 - \vartheta_n)w_n + \vartheta_n\eta_n - v^*\|^2 \\ &= \|(1 - \vartheta_n)(w_n - v^*) + \vartheta_n(\eta_n - v^*)\|^2 \\ &= (1 - \vartheta_n)\|w_n - v^*\|^2 + \vartheta_n\|\eta_n - v^*\|^2 - \vartheta_n(1 - \vartheta_n)\|w_n - \eta_n\|^2 \\ &\leq (1 - \vartheta_n)\|w_n - v^*\|^2 + \vartheta_n\|\eta_n - v^*\|^2.\end{aligned} \quad (17)$$

From Lemma 5, we obtain

$$\|\eta_n - v^*\|^2 \leq \|w_n - v^*\|^2 - (1 - 2L_1\zeta_n)\|w_n - v_n\|^2 - (1 - 2L_2\zeta_n)\|\eta_n - v_n\|^2$$
$$- 2\gamma\zeta_n\|v_n - v^*\|^2. \quad (18)$$

By combining expressions (17) and (18), we get

$$\begin{aligned}\|u_{n+1} - v^*\|^2 &\leq (1 - \vartheta_n)\|w_n - v^*\|^2 + \vartheta_n\|w_n - v^*\|^2 - 2\gamma\vartheta_n\zeta_n\|v_n - v^*\|^2 \\ &\quad - \vartheta_n(1 - 2L_1\zeta_n)\|w_n - v_n\|^2 - \vartheta_n(1 - 2L_2\zeta_n)\|\eta_n - v_n\|^2 \quad (19)\\ &= \|w_n - v^*\|^2 - \vartheta_n(1 - b\zeta_n)[\|w_n - v_n\|^2 + \|\eta_n - v_n\|^2] \quad (20)\\ &= \|w_n - v^*\|^2 - \frac{\vartheta_n(1 - b\zeta_n)}{2}[2\|w_n - v_n\|^2 + 2\|\eta_n - v_n\|^2] \\ &\leq \|w_n - v^*\|^2 - \frac{\vartheta_n(1 - b\zeta_n)}{2}[\|w_n - v_n\| + \|\eta_n - v_n\|]^2 \\ &\leq \|w_n - v^*\|^2 - \frac{\vartheta_n(1 - b\zeta_n)}{2}\|\eta_n - w_n\|^2, \quad (21)\end{aligned}$$

where $b = \max\{2L_1, 2L_2\}$. It continues from u_{n+1} such that

$$\|u_{n+1} - w_n\| = \|(1 - \vartheta_n)w_n + \vartheta_n\eta_n - w_n\| = \|\vartheta_n(\eta_n - w_n)\|. \quad (22)$$

Combining (21) and (22), we have

$$\|u_{n+1} - v^*\|^2 \leq \|w_n - v^*\|^2 - \frac{(1 - b\zeta_n)}{2\vartheta_n}\|u_{n+1} - w_n\|^2. \quad (23)$$

Since $\zeta_n \to 0$, thus there is $n_0 > 0$ in order that $\zeta_n \leq \frac{1}{2b}$ for each $n \geq n_0$. This implies $\frac{1-b\zeta_n}{2} \geq \frac{1}{4}$ for every $n \geq n_0$. The expression (23) for $n \geq n_0$, turn as

$$\|u_{n+1} - v^*\|^2 \leq \|w_n - v^*\|^2 - \frac{1}{4\vartheta_n}\|u_{n+1} - w_n\|^2. \quad (24)$$

By description of w_n, we have

$$\begin{aligned}\|w_n - v^*\|^2 &= \|u_n + \theta_n(u_n - u_{n-1}) - v^*\|^2 \\ &= \|(1+\theta_n)(u_n - v^*) - \theta_n(u_{n-1} - v^*)\|^2 \\ &= (1+\theta_n)\|u_n - v^*\|^2 - \theta_n\|u_{n-1} - v^*\|^2 + \theta_n(1+\theta_n)\|u_n - u_{n-1}\|^2.\end{aligned} \quad (25)$$

By value of w_n, we have

$$\begin{aligned}\|u_{n+1} - w_n\|^2 &= \|u_{n+1} - u_n - \theta_n(u_n - u_{n-1})\|^2 \\ &= \|u_{n+1} - u_n\|^2 + \theta_n^2\|u_n - u_{n-1}\|^2 + 2\theta_n\langle u_n - u_{n+1}, u_n - u_{n-1}\rangle \\ &\geq \|u_{n+1} - u_n\|^2 + \theta_n^2\|u_n - u_{n-1}\|^2 - \rho_n\theta_n\|u_{n+1} - u_n\|^2 - \frac{\theta_n}{\rho_n}\|u_n - u_{n-1}\|^2 \\ &\geq (1 - \rho_n\theta_n)\|u_{n+1} - u_n\|^2 + \left(\theta_n^2 - \frac{\theta_n}{\rho_n}\right)\|u_n - u_{n-1}\|^2,\end{aligned} \quad (26)(27)$$

where $\rho_n = \frac{1}{\delta\vartheta_n + \theta_n}$. Combining (24), (25), and (27) gives that

$$\begin{aligned}\|u_{n+1} - v^*\|^2 &\leq (1+\theta_n)\|u_n - v^*\|^2 - \theta_n\|u_{n-1} - v^*\|^2 + \theta_n(1+\theta_n)\|u_n - u_{n-1}\|^2 \\ &\quad - \frac{1}{4\vartheta_n}\left[(1-\rho_n\theta_n)\|u_{n+1} - u_n\|^2 + \left(\theta_n^2 - \frac{\theta_n}{\rho_n}\right)\|u_n - u_{n-1}\|^2\right] \\ &= (1+\theta_n)\|u_n - v^*\|^2 - \theta_n\|u_{n-1} - v^*\|^2 - \frac{1}{4\vartheta_n}(1-\rho_n\theta_n)\|u_{n+1} - u_n\|^2 \\ &\quad + \left[\theta_n(1+\theta_n) - \frac{1}{4\vartheta_n}\left(\theta_n^2 - \frac{\theta_n}{\rho_n}\right)\right]\|u_n - u_{n-1}\|^2 \\ &= (1+\theta_n)\|u_n - v^*\|^2 - \theta_n\|u_{n-1} - v^*\|^2 - \frac{1}{4\vartheta_n}(1-\rho_n\theta_n)\|u_{n+1} - u_n\|^2 \\ &\quad + \gamma_n\|u_n - u_{n-1}\|^2,\end{aligned} \quad (28)(29)$$

where

$$\gamma_n = \theta_n(1+\theta_n) - \frac{1}{4\vartheta_n}\left(\theta_n^2 - \frac{\theta_n}{\rho_n}\right) = \theta_n(1+\theta_n) + \frac{1}{4\vartheta_n}\left(\frac{\theta_n}{\rho_n} - \theta_n^2\right) > 0. \quad (30)$$

By the above expression and the choice of $\{\rho_n\}$, we have

$$\gamma_n = \theta_n(1+\theta_n) + \frac{1}{4\vartheta_n}\left(\frac{\theta_n}{\rho_n} - \theta_n^2\right) \leq \theta(1+\theta) + \frac{1}{4}\theta\delta. \quad (31)$$

We substitute

$$\Psi_n = \|u_n - p\|^2 - \theta_n\|u_{n-1} - p\|^2 + \gamma_n\|u_n - u_{n-1}\|^2.$$

It follows (29) such that

$$\begin{aligned}\Psi_{n+1} - \Psi_n &= \|u_{n+1} - p\|^2 - \theta_{n+1}\|u_n - p\|^2 + \gamma_{n+1}\|u_{n+1} - u_n\|^2 \\ &\quad - \|u_n - p\|^2 + \theta_n\|u_{n-1} - p\|^2 - \gamma_n\|u_n - u_{n-1}\|^2 \\ &\leq \|u_{n+1} - p\|^2 - (1+\theta_n)\|u_n - p\|^2 + \theta_n\|u_{n-1} - p\|^2 \\ &\quad + \gamma_{n+1}\|u_{n+1} - u_n\|^2 - \gamma_n\|u_n - u_{n-1}\|^2 \\ &= -\left(\frac{1}{4\vartheta_n}(1 - \rho_n\theta_n) - \gamma_{n+1}\right)\|u_{n+1} - u_n\|^2.\end{aligned} \quad (32)$$

We claim that

$$\frac{1}{4\vartheta_n}(1 - \rho_n\theta_n) - \gamma_{n+1} \geq \sigma.$$

The above inequality implies that

$$\frac{1}{4\vartheta_n}(1-\rho_n\theta_n) - \gamma_{n+1} \geq \sigma$$
$$\text{iff} \quad (1-\rho_n\theta_n) - 4\vartheta_n\gamma_{n+1} \geq 4\vartheta_n\sigma$$
$$\text{iff} \quad (1-\rho_n\theta_n) - 4\vartheta_n(\gamma_{n+1}+\sigma) \geq 0$$
$$\text{iff} \quad \frac{\delta\vartheta_n}{\delta\vartheta_n+\theta_n} - 4\vartheta_n(\gamma_{n+1}+\sigma) \geq 0$$
$$\text{iff} \quad -4(\gamma_{n+1}+\sigma)(\delta\vartheta_n+\theta_n) \geq -\delta \tag{33}$$

(31) and (5) give that

$$-4(\gamma_{n+1}+\sigma)(\delta\vartheta_n+\theta_n) \geq -4\left[\theta(1+\theta)+\frac{1}{4}\theta\delta+\sigma\right](\delta\vartheta_n+\theta_n) \geq -\delta. \tag{34}$$

Expression (32) implies that

$$\Psi_{n+1} - \Psi_n \leq -\sigma\|u_{n+1}-u_n\|^2 \leq 0, \quad \text{for all } n \geq n_0. \tag{35}$$

Thus, we obtain a non-increasing sequence $\{\Psi_n\}$ for $n \geq n_0$. By the value of Ψ_{n+1}, we have

$$\begin{aligned}\Psi_{n+1} &= \|u_{n+1}-p\|^2 - \theta_{n+1}\|u_n-p\|^2 + \gamma_{n+1}\|u_{n+1}-u_n\|^2 \\ &\geq -\theta_{n+1}\|u_n-p\|^2.\end{aligned} \tag{36}$$

By the value of Ψ_n, we have

$$\begin{aligned}\Psi_n &= \|u_n-p\|^2 - \theta_n\|u_{n-1}-p\|^2 + \gamma_n\|u_n-u_{n-1}\|^2 \\ &\geq \|u_n-p\|^2 - \theta_n\|u_{n-1}-p\|^2.\end{aligned} \tag{37}$$

Thus, expression (37) for $n \geq n_0$ is such that

$$\begin{aligned}\|u_n-p\|^2 &\leq \Psi_n + \theta_n\|u_{n-1}-p\|^2 \\ &\leq \Psi_{n_0} + \theta\|u_{n-1}-p\|^2 \\ &\leq \cdots \leq \Psi_{n_0}(\theta^{n-n_0}+\cdots+1) + \theta^{n-n_0}\|u_{n_0}-p\|^2 \\ &\leq \frac{\Psi_{n_0}}{1-\theta} + \theta^{n-n_0}\|u_{n_0}-p\|^2.\end{aligned} \tag{38}$$

By (36) and (38) for all $n \geq n_0$, we get

$$\begin{aligned}-\Psi_{n+1} &\leq \theta_{n+1}\|u_n-p\|^2 \\ &\leq \theta\|u_n-p\|^2 \\ &\leq \theta\frac{\Psi_{n_0}}{1-\theta} + \theta^{n-n_0+1}\|u_{n_0}-p\|^2.\end{aligned} \tag{39}$$

It follows from (35) and (39) that

$$\begin{aligned}\sigma\sum_{n=n_0}^{k}\|u_{n+1}-u_n\|^2 &\leq \Psi_{n_0} - \Psi_{k+1} \\ &\leq \Psi_{n_0} + \theta\frac{\Psi_{n_0}}{1-\theta} + \theta^{n-n_0+1}\|u_{n_0}-p\|^2 \\ &\leq \frac{\Psi_{n_0}}{1-\theta} + \|u_{n_0}-p\|^2.\end{aligned} \tag{40}$$

Sending $k \to +\infty$ implies that

$$\sum_{n=1}^{+\infty} \|u_{n+1} - u_n\|^2 < +\infty. \qquad (41)$$

It continues from that

$$\lim_{n \to +\infty} \|u_{n+1} - u_n\| = 0. \qquad (42)$$

Equations (26) and (42) provide that

$$\lim_{n \to +\infty} \|u_{n+1} - w_n\| = 0. \qquad (43)$$

By the value of u_{n+1}, we have

$$\|u_{n+1} - w_n\| = \|(1 - \vartheta_n)w_n + \vartheta_n \eta_n - w_n\| = \vartheta_n \|\eta_n - w_n\|. \qquad (44)$$

By Equations (43) and (44), we obtain

$$\lim_{n \to +\infty} \|\eta_n - w_n\| = 0. \qquad (45)$$

By the use of triangular inequality and (42) with (43), we obtain

$$\lim_{n \to +\infty} \|u_n - w_n\| \leq \lim_{n \to +\infty} \|u_n - u_{n+1}\| + \lim_{n \to +\infty} \|u_{n+1} - w_n\| = 0 \qquad (46)$$

and

$$\lim_{n \to +\infty} \|u_n - \eta_n\| \leq \lim_{n \to +\infty} \|u_n - w_n\| + \lim_{n \to +\infty} \|w_n - \eta_n\| = 0. \qquad (47)$$

Expressions (28) and (41) with Lemma 1 imply that

$$\lim_{n \to +\infty} \|u_n - v^*\|^2 = b \quad \text{for some} \quad b \geq 0. \qquad (48)$$

Expressions (46) and (47) imply that

$$\lim_{n \to +\infty} \|w_n - v^*\|^2 = \lim_{n \to +\infty} \|\eta_n - v^*\|^2 = b. \qquad (49)$$

Thus, Lemma 5 implies that

$$(1 - 2L_2\zeta)\|w_n - v_n\|^2 \leq \|w_n - v^*\|^2 - \|\eta_n - v^*\|^2. \qquad (50)$$

The above expression with (48) and (49) gives that

$$\lim_{n \to +\infty} \|w_n - v_n\| = 0 \quad \text{and} \quad \lim_{n \to +\infty} \|v_n - v^*\|^2 = b. \qquad (51)$$

The argument referred to above concludes that the sequences $\{w_n\}$, $\{v_n\}$, $\{\eta_n\}$, and $\{\eta_n\}$ are bounded for each $v^* \in EP(f, \mathcal{K})$ the $\lim_{n \to +\infty} \|u_n - v^*\|^2$ exists. It follows from (19) and (25) that we have

$$\begin{aligned}
2\gamma \vartheta_n \zeta_n \|v_n - v^*\|^2 &\leq -\|u_{n+1} - v^*\|^2 + (1 + \theta_n)\|u_n - v^*\|^2 - \theta_n \|u_{n-1} - v^*\|^2 \\
&\quad + \theta_n(1 + \theta_n)\|u_n - u_{n-1}\|^2 \\
&\leq (\|u_n - v^*\|^2 - \|u_{n+1} - v^*\|^2) + 2\theta \|u_n - u_{n-1}\|^2 \\
&\quad + (\theta_n \|u_n - v^*\|^2 - \theta_{n-1}\|u_{n-1} - v^*\|^2).
\end{aligned} \qquad (52)$$

The above expression for $k \geq n_0$ gives that

$$\sum_{n=n_0}^{k} 2\gamma\vartheta_n\zeta_n\|v_n - v^*\|^2 \leq (\|u_{n_0} - v^*\|^2 - \|u_{k+1} - v^*\|^2) + 2\theta \sum_{n=n_0}^{k} \|u_n - u_{n-1}\|^2$$
$$+ (\theta_k\|u_k - v^*\|^2 - \theta_0\|u_{n_0} - v^*\|^2)$$
$$\leq \|u_{n_0} - v^*\|^2 + \theta\|u_k - v^*\|^2 + 2\theta \sum_{n=n_0}^{k} \|u_n - u_{n-1}\|^2, \tag{53}$$

letting $k \to +\infty$ in (53), we obtain

$$\sum_{n=n_0}^{k} 2\gamma\vartheta_n\zeta_n\|v_n - v^*\|^2 < +\infty. \tag{54}$$

From Lemma 3 and (54),

$$\liminf \|v_n - p\| = 0. \tag{55}$$

By expressions (46), (47), (49), (51) and (55),

$$\lim_{n \to +\infty} \|v_n - p\| = \lim_{n \to +\infty} \|w_n - p\| = \lim_{n \to +\infty} \|\eta_n - p\| = \lim_{n \to +\infty} \|u_n - p\| = 0. \tag{56}$$

This completes the proof. □

Next, we consider the application of our results to solve variational inequality problems. A function $G : \mathcal{H} \to \mathcal{H}$ is said to be

(G1) *strongly pseudo-monotone* over $\mathcal{K}$ for $\gamma > 0$ if

$$\langle G(v_1), v_2 - v_1 \rangle \geq 0 \quad \text{implies that} \quad \langle G(v_2), v_1 - v_2 \rangle \leq -\gamma\|v_1 - v_2\|^2, \forall v_1, v_2 \in \mathcal{K};$$

and

(G2) *L-Lipschitz continuity* on C if

$$\|G(v_1) - G(v_2)\| \leq L\|v_1 - v_2\|, \forall v_1, v_2 \in \mathcal{K}.$$

Let a bifunction $f(v_1, v_2) := \langle G(v_1), v_2 - v_1 \rangle$ for all $v_1, v_2 \in \mathcal{K}$ then equilibrium problem turns into problem of variational inequality with $L = 2L_1 = 2L_2$. By the value of v_n,

$$v_n = \arg\min_{v \in \mathcal{K}} \left\{ \zeta_n f(w_n, v) + \frac{1}{2}\|w_n - v\|^2 \right\}$$
$$= \arg\min_{v \in \mathcal{K}} \left\{ \zeta_n \langle G(w_n), v - w_n \rangle + \frac{1}{2}\|w_n - v\|^2 \right\}$$
$$= \arg\min_{v \in \mathcal{K}} \left\{ \zeta_n \langle G(w_n), v - w_n \rangle + \frac{1}{2}\|w_n - v\|^2 + \frac{\zeta_n^2}{2}\|G(w_n)\|^2 - \frac{\zeta_n^2}{2}\|G(w_n)\|^2 \right\}$$
$$= \arg\min_{v \in \mathcal{K}} \left\{ \frac{1}{2}\|v - (w_n - \zeta_n G(w_n))\|^2 \right\} - \frac{\zeta_n^2}{2}\|G(w_n)\|^2$$
$$= P_{\mathcal{K}}(w_n - \zeta_n G(w_n)). \tag{57}$$

Similar to above, the value of η_n turns into

$$\eta_n = P_{\mathcal{H}_n}(w_n - \zeta_n G(v_n)).$$

Corollary 1. *Assume that an operator $G : \mathcal{K} \to \mathcal{H}$ satisfies Conditions (G1)–(G2). Let $\{w_n\}, \{v_n\}, \{\eta_n\}$, and $\{u_n\}$ be the sequences generated as follows:*

(S1) Let $u_{-1}, u_0 \in \mathcal{H}$ arbitrarily.
(S2) Choose ζ_n satisfying condition (3) and $\{\theta_n\}, \{\vartheta_n\}$ are control parameters.
(S3) Compute
$$v_n = P_\mathcal{K}(w_n - \zeta_n G(w_n)),$$
where $w_n = u_n + \theta_n(u_n - u_{n-1})$. If $v_n = w_n$, then STOP.
(S4) Determine a half space first $\mathcal{H}_n = \{z \in \mathcal{H} : \langle w_n - \zeta_n G(w_n) - v_n, z - v_n \rangle \le 0\}$ and evaluate
$$\eta_n = P_{\mathcal{H}_n}(w_n - \zeta_n G(v_n)).$$

(S5) Compute
$$u_{n+1} = (1 - \vartheta_n)w_n + \vartheta_n \eta_n,$$
where $\{\theta_n\}$ and $\{\vartheta_n\}$ satisfies the following conditions:

(i) non-decreasing sequence $\{\theta_n\}$ through $0 \le \theta_n \le \theta < 1$, for each $n \ge 1$; and
(ii) there exists $\vartheta, \delta, \sigma > 0$, thus that

$$\delta > \frac{4\theta[\theta(1+\theta) + \sigma]}{1 - \theta^2} \tag{58}$$

and

$$0 < \vartheta \le \vartheta_n \le \frac{\delta - 4\theta[\theta(1+\theta) + \sigma + \frac{1}{4}\theta\delta]}{4\delta[\theta(1+\theta) + \sigma + \frac{1}{4}\theta\delta]}. \tag{59}$$

Then, $\{w_n\}, \{v_n\}, \{\eta_n\}$, and $\{u_n\}$ strongly converge to $v^* \in VI(G, \mathcal{K})$.

4. Numerical Illustration

Numerical findings are summarized in this section to demonstrate the effectiveness of the proposed methods. The following control parameters are used in this section.

(1) For Hieu et al. [26] (Hieu-EgA), we use $D_n = \|u_n - v_n\|^2$.

(2) For Hieu et al. [29] (Hieu-mEgA), we use $\theta = 0.5$ and $D_n = \max\{\|u_{n+1} - v_n\|^2, \|u_{n+1} - w_n\|^2\}$.

(3) For Algorithm 1 (iEgA), we use $\alpha_n = 0.50$, $\beta_n = 0.80$, and $D_n = \|w_n - v_n\|^2$.

Example 1. *Let bifunction f have the following form*
$$f(u, v) = \langle Au + Bv + c, v - u \rangle$$

where $c \in \mathbb{R}^5$ and A and B are

$$A = \begin{pmatrix} 3.1 & 2 & 0 & 0 & 0 \\ 2 & 3.6 & 0 & 0 & 0 \\ 0 & 0 & 3.5 & 2 & 0 \\ 0 & 0 & 2 & 3.3 & 0 \\ 0 & 0 & 0 & 0 & 3 \end{pmatrix} \quad B = \begin{pmatrix} 1.6 & 1 & 0 & 0 & 0 \\ 1 & 1.6 & 0 & 0 & 0 \\ 0 & 0 & 1.5 & 1 & 0 \\ 0 & 0 & 1 & 1.5 & 0 \\ 0 & 0 & 0 & 0 & 2 \end{pmatrix}$$

and
$$c = \begin{pmatrix} 1 \\ -2 \\ -1 \\ 2 \\ -1 \end{pmatrix}$$

where Lipschitz parameters $L_1 = L_2 = \frac{1}{2}\|A - B\|$ [26]. The feasible set $\mathcal{K} \subset \mathbb{R}^5$ is

$$\mathcal{K} := \{u \in \mathbb{R}^5 : -5 \leq u_i \leq 5\}.$$

Table 1 and Figures 1–3 show the numerical results by $u_{-1} = u_0 = v_0 = (1, \cdots, 1)$, and $TOL = 10^{-12}$.

Table 1. Example 1: Numerical values for Figures 1–3.

			Hieu-EgA [26]		Hieu-mEgA [29]		iEgA	Algorithm 1
n	TOL	ζ_n	Iter.	Time	Iter.	Time	Iter.	Time
5	10^{-12}	$\frac{1}{\log(n+3)(n+1)}$	320	5.8584	59	0.5979	64	0.2830
5	10^{-12}	$\frac{1}{n+1}$	222	3.1116	43	0.4158	39	0.1696
5	10^{-12}	$\frac{\log(n+3)}{n+1}$	122	1.5466	40	0.3732	33	0.1581

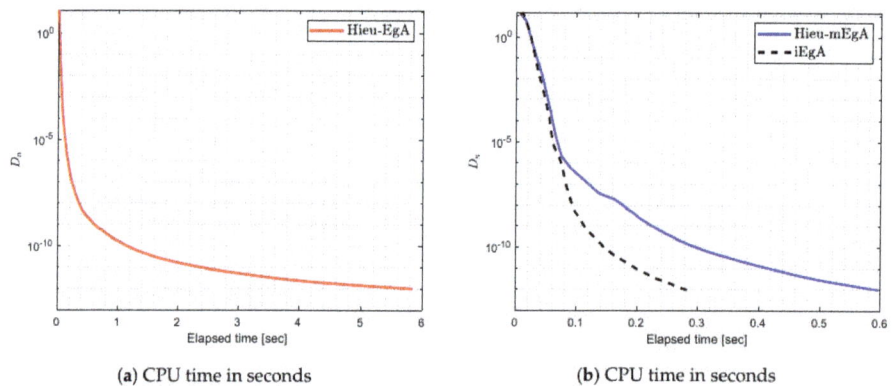

Figure 1. Example 1: Numerical comparison for Algorithm 1 while $\zeta_n = \frac{1}{(n+1)\log(n+3)}$.

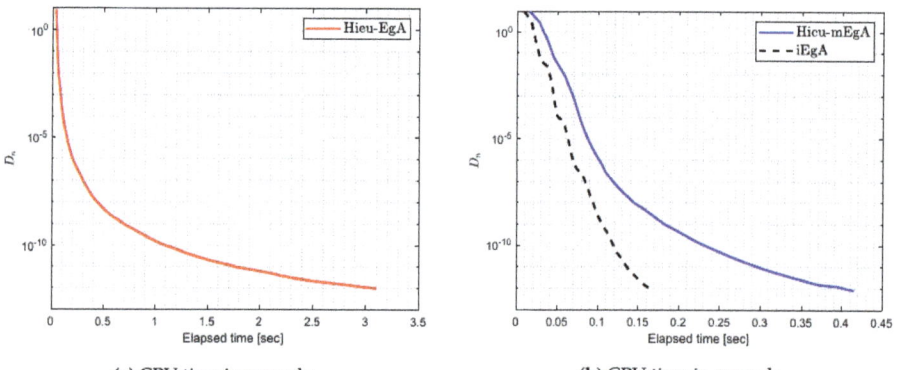

Figure 2. Example 1: Numerical comparison for Algorithm 1 while $\zeta_n = \frac{1}{n+1}$.

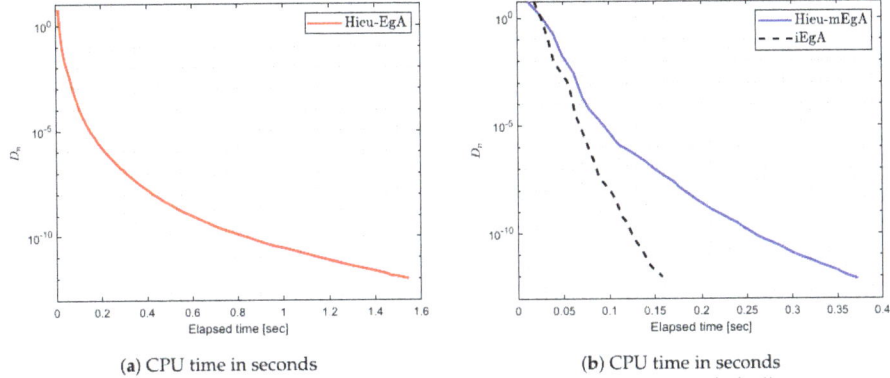

(a) CPU time in seconds

(b) CPU time in seconds

Figure 3. Example 1: Numerical comparison for Algorithm 1 while $\zeta_n = \frac{\log(n+3)}{n+1}$.

Example 2. *Let a bifunction f be defined on the convex set $\mathcal{K}$ as*

$$f(u,v) = \langle (BB^T + S + D)u, v - u \rangle,$$

where B is a 50×50 matrix, S is a 50×50 skew-symmetric matrix, and D is a 50×50 diagonal matrix. The set $\mathcal{K} \subset \mathbb{R}^{50}$ is defined by

$$\mathcal{K} := \{u \in \mathbb{R}^{50} : Au \leq b\}$$

with matrix A as 100×50 and vector b as a non-negative vector. Observe that f is monotone and Lipschitz-type constants are $c_1 = c_2 = \frac{\|BB^T + S + D\|}{2}$. We generate random matrices in our case [$B = rand(n)$, $C = rand(n)$, $S = 0.5C - 0.5C^T$, $D = diag(rand(n,1))$] and the numerical findings regarding Example 2 are shown in Figures 4–7 with $u_{-1} = u_0 = v_0 = (1, \cdots, 1)$ and TOL $= 10^{-12}$.

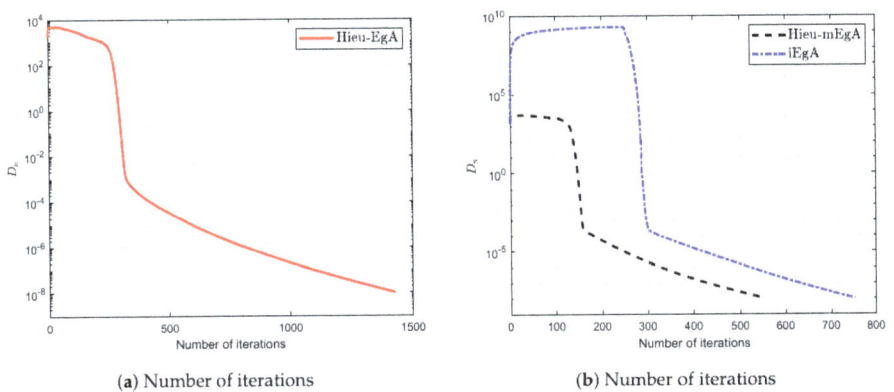

(a) Number of iterations

(b) Number of iterations

Figure 4. Example 2: Numerical comparison for Algorithm 1 while $\zeta_n = \frac{1}{n+1}$.

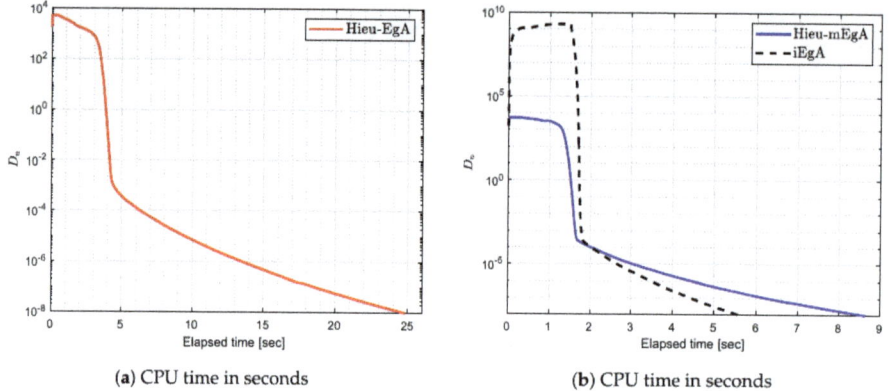

(a) CPU time in seconds

(b) CPU time in seconds

Figure 5. Example 2: Numerical comparison for Algorithm 1 while $\zeta_n = \frac{1}{n+1}$.

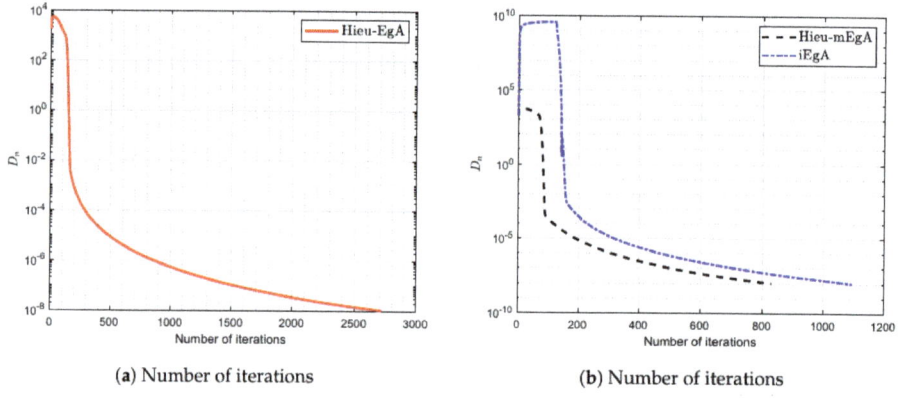

(a) Number of iterations

(b) Number of iterations

Figure 6. Example 2: Numerical comparison for Algorithm 1 while $\zeta_n = \frac{\log(n+3)}{n+1}$.

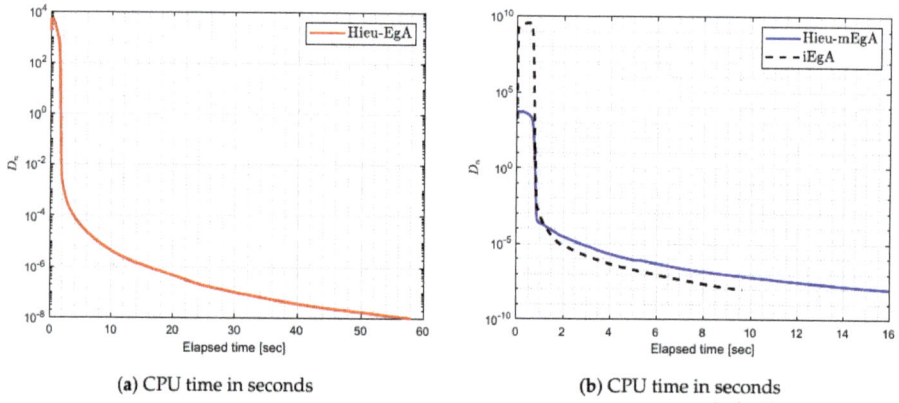

(a) CPU time in seconds

(b) CPU time in seconds

Figure 7. Example 2: Numerical comparison for Algorithm 1 while $\zeta_n = \frac{\log(n+3)}{n+1}$.

Example 3. *Let $G : \mathbb{R}^5 \to \mathbb{R}^5$ be defined by*

$$G(u) = Au + B(u) + c,$$

where $n \times n$ symmetric semi-definite matrix A and $B(u)$ is the function depends on the proximal operator [43] through $h(u) = \frac{1}{4}\|u\|^4$ such that

$$B(u) = \arg\min_{v \in \mathbb{R}^n}\left\{\frac{\|u\|^4}{4} + \frac{1}{2}\|v - u\|^2\right\}.$$

The feasible set $\mathcal{K}$ is considered as

$$\mathcal{K} := \{u \in \mathbb{R}^5 : -2 \leq u_i \leq 5\}.$$

The entries of A and c are taken as follows:

$$A = \begin{pmatrix} 3 & 1 & 0 & 1 & 2 \\ 1 & 5 & -1 & 0 & 1 \\ 0 & 1 & -4 & 2 & -2 \\ 1 & 0 & 2 & 6 & -1 \\ 2 & 1 & -2 & -1 & 4 \end{pmatrix} \quad c = \begin{pmatrix} 1 \\ -2 \\ -1 \\ 2 \\ -1 \end{pmatrix}$$

Figures 8–11 and Table 2 show the numerical results by using $u_{-1} = u_0 = v_0 = (1, \cdots, 1)$ and $TOL = 10^{-12}$.

Table 2. Example 3: Numerical results for Figures 8–11.

			Hieu-EgA [26]		Hieu-mEgA [29]		iEgA Algorithm 1	
n	TOL	ζ_n	Iter.	Time	Iter.	Time	Iter.	Time
5	10^{-10}	$\frac{1}{(n+1)\log(n+3)}$	440	29.7625	190	16.2712	247	10.8531
5	10^{-10}	$\frac{1}{n+1}$	198	13.8482	104	11.8096	145	5.8483
5	10^{-10}	$\frac{\log(n+3)}{n+1}$	178	12.2979	98	7.8478	120	5.2870
5	10^{-10}	$\frac{1}{\sqrt{n+1}}$	251	16.7337	110	9.6097	148	6.0004

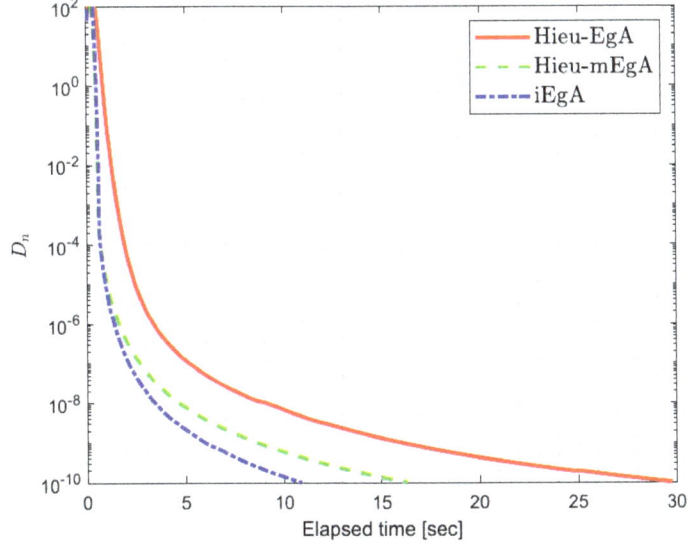

Figure 8. Example 3: Numerical comparison for Algorithm 1 while $\zeta_n = \frac{1}{(n+1)\log(n+3)}$.

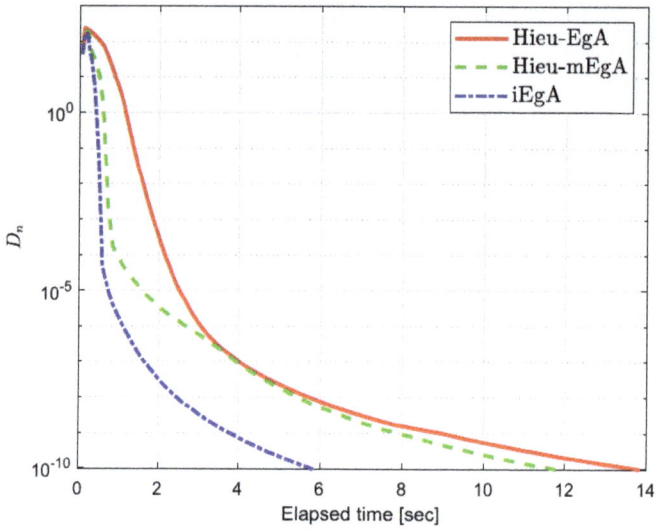

Figure 9. Example 3: Numerical comparison for Algorithm 1 while $\zeta_n = \frac{1}{n+1}$.

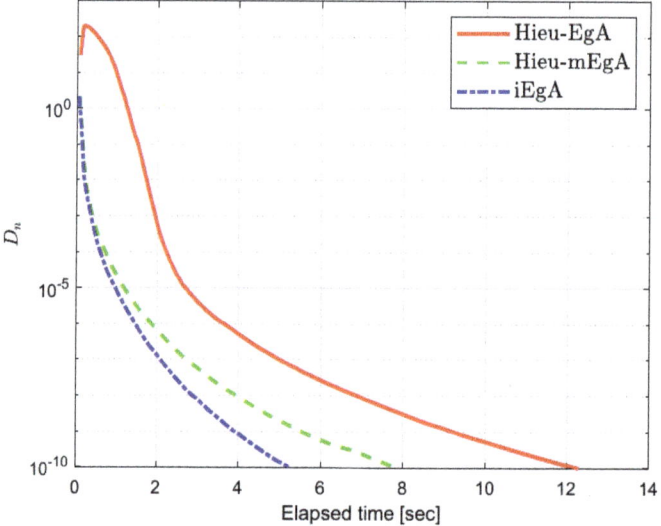

Figure 10. Example 3: Numerical comparison for Algorithm 1 while $\zeta_n = \frac{\log(n+3)}{n+1}$.

Example 4. *Suppose that* $\mathcal{K} \subset G : \mathbb{R}^2 \to \mathbb{R}^2$ *is defined by*

$$G\begin{pmatrix} v_1 \\ v_2 \end{pmatrix} = \begin{pmatrix} v_1 + v_2 + sin(v_1) \\ -v_1 + v_2 + sin(v_2) \end{pmatrix}, \text{ for all } (v_1, v_2) \in \mathbb{R}^2.$$

where $\mathcal{K} = [-5, 5] \times [-5, 5]$. *It is easy that G is Lipschitz continuous and strongly pseudomonotone operator. Figures 12–15 show the numerical results with* $u_{-1} = u_0 = v_0$ *and* $TOL = 10^{-10}$.

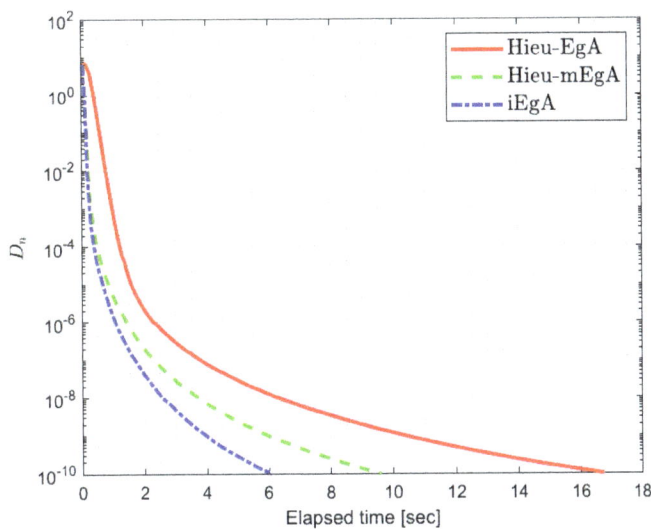

Figure 11. Example 3: Numerical comparison for Algorithm 1 while $\zeta_n = \frac{1}{\sqrt{n+1}}$.

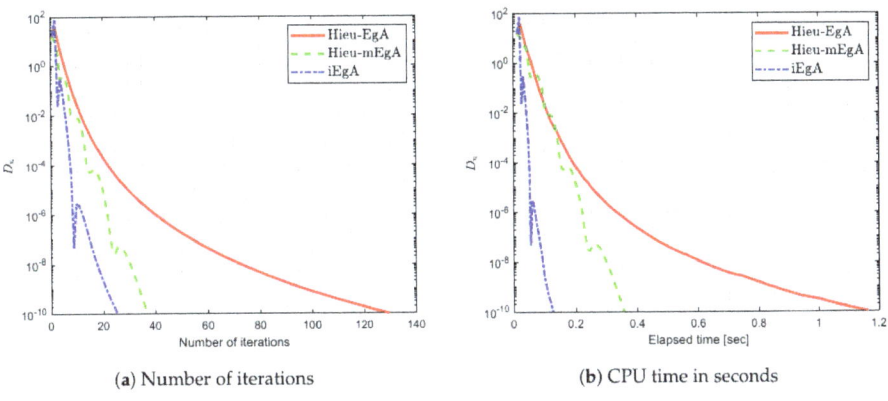

(**a**) Number of iterations

(**b**) CPU time in seconds

Figure 12. Example 4: Numerical comparison for Algorithm 1 while $u_0 = (1,1)$ and $\zeta_n = \frac{1}{n+1}$.

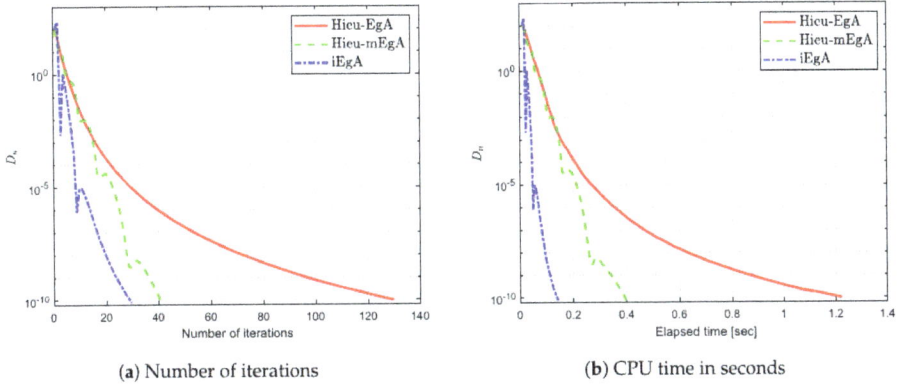

(**a**) Number of iterations

(**b**) CPU time in seconds

Figure 13. Example 4: Numerical comparison for Algorithm 1 while $u_0 = (4,4)$ and $\zeta_n = \frac{1}{n+1}$.

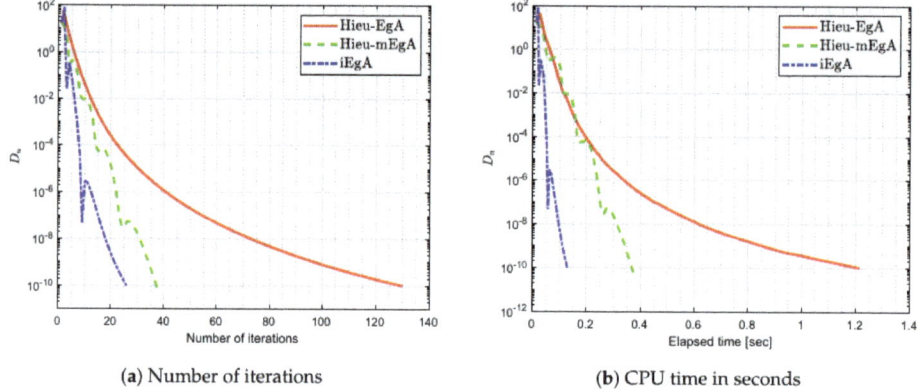

(a) Number of iterations (b) CPU time in seconds

Figure 14. Example 4: Numerical comparison for Algorithm 1 while $u_0 = (-1, -1)$ and $\zeta_n = \frac{1}{n+1}$.

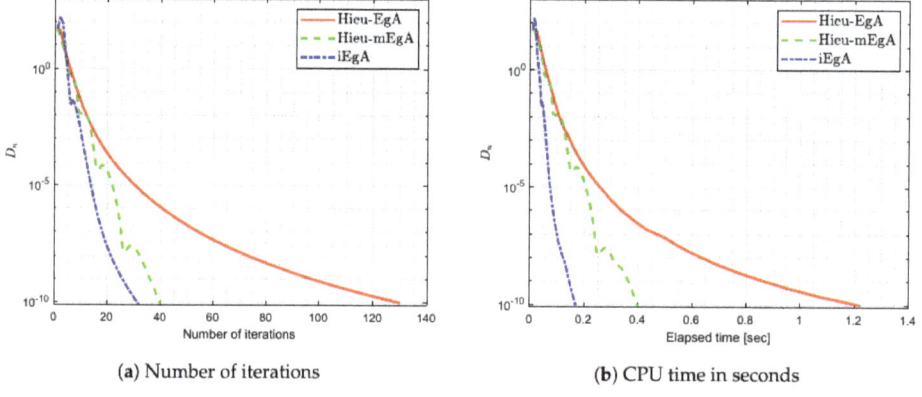

(a) Number of iterations (b) CPU time in seconds

Figure 15. Example 4: Numerical comparison for Algorithm 1 while $u_0 = (-2, -2)$ and $\zeta_n = \frac{1}{n+1}$.

5. Conclusions

In this paper, we set up a new method by combining an inertial term with an extragradient method for solving a family of strongly pseudomonotone equilibrium problems. The introduced method involves a sequence of diminishing and non-summable step size rule and the method operates without previous information of the Lipschitz-type constants. Four numerical examples are described to show the computational performance of the proposed method in relation to other existing methods. Numerical experiments clearly point out that the method with an inertial term performs better than those without an inertial term.

Author Contributions: Conceptualization, W.K. and K.M.; methodology, W.K. and K.M.; writing—original draft preparation, W.K. and K.M.; writing—review and editing, W.K. and K.M.; software, W.K. and K.M.; supervision, W.K.; and project administration and funding acquisition, K.M. All authors have read and agreed to the published version of the manuscript.

Funding: This project was supported by Rajamangala University of Technology Phra Nakhon (RMUTP).

Acknowledgments: The first author thanks the Rajamangala University of Technology Thanyaburi (RMUTT) (Grant No. NSF62D0604). The second author thanks the Rajamangala University of Technology Phra Nakhon (RMUTP).

Conflicts of Interest: The authors declare that they have no conflict of interest.

References

1. Blum, E. From optimization and variational inequalities to equilibrium problems. *Math. Stud.* **1994**, *63*, 123–145.
2. Fan, K. *A Minimax Inequality and Applications, Inequalities III*; Shisha, O., Ed.; Academic Press: New York, NY, USA, 1972.
3. Bianchi, M.; Schaible, S. Generalized monotone bifunctions and equilibrium problems. *J. Optim. Theory Appl.* **1996**, *90*, 31–43. [CrossRef]
4. Bigi, G.; Castellani, M.; Pappalardo, M.; Passacantando, M. Existence and solution methods for equilibria. *Eur. J. Oper. Res.* **2013**, *227*, 1–11. [CrossRef]
5. Muu, L.; Oettli, W. Convergence of an adaptive penalty scheme for finding constrained equilibria. *Nonlinear Anal. Theory Methods Appl.* **1992**, *18*, 1159–1166. [CrossRef]
6. Combettes, P.L.; Hirstoaga, S.A.; others. Equilibrium programming in Hilbert spaces. *J. Nonlinear Convex Anal.* **2005**, *6*, 117–136.
7. Antipin, A. Equilibrium programming: Proximal methods. *Comput. Math. Math. Phys.* **1997**, *37*, 1285–1296.
8. Giannessi, F.; Maugeri, A.; Pardalos, P.M. *Equilibrium Problems: Nonsmooth Optimization and Variational Inequality Models*; Springer Science & Business Media: Berlin/Heidelberg, Germany, 2006; Volume 58.
9. Dafermos, S. Traffic Equilibrium and Variational Inequalities. *Transp. Sci.* **1980**, *14*, 42–54. [CrossRef]
10. Todorčević, V. *Harmonic Quasiconformal Mappings and Hyperbolic Type Metrics*; Springer International Publishing: Berlin/Heidelberg, Germany, 2019. [CrossRef]
11. ur Rehman, H.; Kumam, P.; Cho, Y.J.; Yordsorn, P. Weak convergence of explicit extragradient algorithms for solving equilibirum problems. *J. Inequalities Appl.* **2019**, *2019*. [CrossRef]
12. ur Rehman, H.; Kumam, P.; Je Cho, Y.; Suleiman, Y.I.; Kumam, W. Modified Popov's explicit iterative algorithms for solving pseudomonotone equilibrium problems. *Optim. Methods Softw.* **2020**, 1–32. [CrossRef]
13. Todorčević, V. Subharmonic behavior and quasiconformal mappings. *Anal. Math. Phys.* **2019**, *9*, 1211–1225. [CrossRef]
14. Koskela, P.; Manojlović, V. Quasi-Nearly Subharmonic Functions and Quasiconformal Mappings. *Potential Anal.* **2011**, *37*, 187–196. [CrossRef]
15. ur Rehman, H.; Kumam, P.; Abubakar, A.B.; Cho, Y.J. The extragradient algorithm with inertial effects extended to equilibrium problems. *Comput. Appl. Math.* **2020**, *39*. [CrossRef]
16. Hammad, H.A.; ur Rehman, H.; la Sen, M.D. Advanced Algorithms and Common Solutions to Variational Inequalities. *Symmetry* **2020**, *12*, 1198. [CrossRef]
17. Hieu, D.V.; Quy, P.K.; Vy, L.V. Explicit iterative algorithms for solving equilibrium problems. *Calcolo* **2019**, *56*. [CrossRef]
18. Hieu, D.V. New inertial algorithm for a class of equilibrium problems. *Numer. Algorithms* **2018**, *80*, 1413–1436. [CrossRef]
19. Anh, P.K.; Hai, T.N. Novel self-adaptive algorithms for non-Lipschitz equilibrium problems with applications. *J. Glob. Optim.* **2018**, *73*, 637–657. [CrossRef]
20. Anh, P.N.; Anh, T.T.H.; Hien, N.D. Modified basic projection methods for a class of equilibrium problems. *Numer. Algorithms* **2017**, *79*, 139–152. [CrossRef]
21. ur Rehman, H.; Kumam, P.; Kumam, W.; Shutaywi, M.; Jirakitpuwapat, W. The Inertial Sub-Gradient Extra-Gradient Method for a Class of Pseudo-Monotone Equilibrium Problems. *Symmetry* **2020**, *12*, 463. [CrossRef]
22. ur Rehman, H.; Kumam, P.; Argyros, I.K.; Deebani, W.; Kumam, W. Inertial Extra-Gradient Method for Solving a Family of Strongly Pseudomonotone Equilibrium Problems in Real Hilbert Spaces with Application in Variational Inequality Problem. *Symmetry* **2020**, *12*, 503. [CrossRef]
23. Flåm, S.D.; Antipin, A.S. Equilibrium programming using proximal-like algorithms. *Math. Program.* **1996**, *78*, 29–41. [CrossRef]
24. Tran, D.Q.; Dung, M.L.; Nguyen, V.H. Extragradient algorithms extended to equilibrium problems. *Optimization* **2008**, *57*, 749–776. [CrossRef]
25. Korpelevich, G. The extragradient method for finding saddle points and other problems. *Matecon* **1976**, *12*, 747–756.

26. Hieu, D.V. New extragradient method for a class of equilibrium problems in Hilbert spaces. *Appl. Anal.* **2017**, *97*, 811–824. [CrossRef]
27. Polyak, B. Some methods of speeding up the convergence of iteration methods. *USSR Comput. Math. Math. Phys.* **1964**, *4*, 1–17. [CrossRef]
28. Beck, A.; Teboulle, M. A Fast Iterative Shrinkage-Thresholding Algorithm for Linear Inverse Problems. *SIAM J. Imaging Sci.* **2009**, *2*, 183–202. [CrossRef]
29. Hieu, D.V.; Cho, Y.J.; bin Xiao, Y. Modified extragradient algorithms for solving equilibrium problems. *Optimization* **2018**, *67*, 2003–2029. [CrossRef]
30. Rehman, H.U.; Kumam, P.; Dong, Q.L.; Peng, Y.; Deebani, W. A new Popov's subgradient extragradient method for two classes of equilibrium programming in a real Hilbert space. *Optimization* **2020**, 1–36. [CrossRef]
31. Yordsorn, P.; Kumam, P.; ur Rehman, H.; Ibrahim, A.H. A Weak Convergence Self-Adaptive Method for Solving Pseudomonotone Equilibrium Problems in a Real Hilbert Space. *Mathematics* **2020**, *8*, 1165. [CrossRef]
32. Yordsorn, P.; Kumam, P.; Rehman, H.U. Modified two-step extragradient method for solving the pseudomonotone equilibrium programming in a real Hilbert space. *Carpathian J. Math.* **2020**, *36*, 313–330.
33. Fan, J.; Liu, L.; Qin, X. A subgradient extragradient algorithm with inertial effects for solving strongly pseudomonotone variational inequalities. *Optimization* **2019**, 1–17. [CrossRef]
34. Thong, D.V.; Hieu, D.V. Inertial extragradient algorithms for strongly pseudomonotone variational inequalities. *J. Comput. Appl. Math.* **2018**, *341*, 80–98. [CrossRef]
35. Censor, Y.; Gibali, A.; Reich, S. The Subgradient Extragradient Method for Solving Variational Inequalities in Hilbert Space. *J. Optim. Theory Appl.* **2010**, *148*, 318–335. [CrossRef] [PubMed]
36. ur Rehman, H.; Kumam, P.; Argyros, I.K.; Alreshidi, N.A.; Kumam, W.; Jirakitpuwapat, W. A Self-Adaptive Extra-Gradient Methods for a Family of Pseudomonotone Equilibrium Programming with Application in Different Classes of Variational Inequality Problems. *Symmetry* **2020**, *12*, 523. [CrossRef]
37. ur Rehman, H.; Kumam, P.; Argyros, I.K.; Shutaywi, M.; Shah, Z. Optimization Based Methods for Solving the Equilibrium Problems with Applications in Variational Inequality Problems and Solution of Nash Equilibrium Models. *Mathematics* **2020**, *8*, 822. [CrossRef]
38. ur Rehman, H.; Kumam, P.; Shutaywi, M.; Alreshidi, N.A.; Kumam, W. Inertial Optimization Based Two-Step Methods for Solving Equilibrium Problems with Applications in Variational Inequality Problems and Growth Control Equilibrium Models. *Energies* **2020**, *13*, 3292. [CrossRef]
39. Attouch, F.A.H. An Inertial Proximal Method for Maximal Monotone Operators via Discretization of a Nonlinear Oscillator with Damping. *Set-Valued Var. Anal.* **2001**, *9*, 3–11. [CrossRef]
40. Heinz, H.; Bauschke, P.L. *Convex Analysis and Monotone Operator Theory in Hilbert Spaces*, 2nd ed.; CMS Books in Mathematics; Springer International Publishing: Berlin/Heidelberg, Germany, 2017.
41. Ofoedu, E. Strong convergence theorem for uniformly L-Lipschitzian asymptotically pseudocontractive mapping in real Banach space. *J. Math. Anal. Appl.* **2006**, *321*, 722–728. [CrossRef]
42. Tiel, J.V. *Convex Analysis: An Introductory Text*, 1st ed.; Wiley: New York, NY, USA, 1984.
43. Kreyszig, E. *Introductory Functional Analysis with Applications*, 1st ed.; Wiley: New York, NY, USA, 1978.

Publisher's Note: MDPI stays neutral with regard to jurisdictional claims in published maps and institutional affiliations.

© 2020 by the authors. Licensee MDPI, Basel, Switzerland. This article is an open access article distributed under the terms and conditions of the Creative Commons Attribution (CC BY) license (http://creativecommons.org/licenses/by/4.0/).

Article

Towards the Dependence on Parameters for the Solution of the Thermostatted Kinetic Framework

Bruno Carbonaro [†] and Marco Menale [*,†]

Dipartimento di Matematica e Fisica, Università degli Studi della Campania "L. Vanvitelli", Viale Lincoln 5, I-81100 Caserta, Italy; bruno.carbonaro@unicampania.it
* Correspondence: marco.menale@unicampania.it
† These authors contributed equally to this work.

Abstract: A complex system is a system involving particles whose pairwise interactions cannot be composed in the same way as in classical Mechanics, i.e., the result of interaction of each particle with all the remaining ones cannot be expressed as a sum of its interactions with each of them (we cannot even know the functional dependence of the total interaction on the single interactions). Moreover, in view of the wide range of its applications to biologic, social, and economic problems, the variables describing the state of the system (i.e., the states of all of its particles) are not always (only) the usual mechanical variables (position and velocity), but (also) many additional variables describing e.g., health, wealth, social condition, social rôle ..., and so on. Thus, in order to achieve a mathematical description of the problems of everyday's life of any human society, either at a microscopic or at a macroscpoic scale, a new mathematical theory (or, more precisely, a scheme of mathematical models), called KTAP, has been devised, which provides an equation which is a generalized version of the Boltzmann equation, to describe in terms of probability distributions the evolution of a non-mechanical complex system. In connection with applications, the classical problems about existence, uniqueness, continuous dependence, and stability of its solutions turn out to be particularly relevant. As far as we are aware, however, the problem of continuous dependence and stability of solutions with respect to perturbations of the parameters expressing the interaction rates of particles and the transition probability densities (see Section The Basic Equations has not been tackled yet). Accordingly, the present paper aims to give some initial results concerning these two basic problems. In particular, Theorem 2 reveals to be stable with respect to small perturbations of parameters, and, as far as instability of solutions with respect to perturbations of parameters is concerned, Theorem 3 shows that solutions are unstable with respect to "large" perturbations of interaction rates; these hints are illustrated by numerical simulations that point out how much solutions corresponding to different values of parameters stay away from each other as $t \to +\infty$.

Keywords: kinetic theory; complex systems; stability; parameters; differential equations

MSC: 82B40; 37F05; 45M10; 35B30; 34A12

Citation: Carbonaro, B.; Menale, M. Towards the Dependence on Parameters for the Solution of the Thermostatted Kinetic Framework. *Axioms* **2021**, *10*, 59. https://doi.org/10.3390/axioms10020059

Academic Editors: Jesús Martín Vaquero, Deolinda M. L. Dias Rasteiro, Araceli Queiruga-Dios and Fatih Yilmaz

Received: 13 February 2021
Accepted: 31 March 2021
Published: 12 April 2021

Publisher's Note: MDPI stays neutral with regard to jurisdictional claims in published maps and institutional affiliations.

Copyright: © 2021 by the authors. Licensee MDPI, Basel, Switzerland. This article is an open access article distributed under the terms and conditions of the Creative Commons Attribution (CC BY) license (https://creativecommons.org/licenses/by/4.0/).

1. Introduction

The present paper deals with the system of equations governing the behavior of so-called *complex systems* (see Section 2 for details). Roughly speaking, a *complex system* is a set of a large number of individuals (particles) whose behavior is strongly influenced by their mutual interactions, in addition to external forces and possibly to a *thermostat* [1–4], so that the evolution of the system cannot be by no means *deterministic*, but must be described in terms of the *probability distribution fuction* on the set of possible values of a suitable variable describing the state of each individual. In this connection, it must be carefully noted that—though the notion of a complex system was originated in a purely mechanical framework and could be traced back to Boltzmann's Kinetic Theory of Gases [5–8]—a complex system is not nowadays considered as simply consisting of material particles, whose state is completely described by the two variables *position* and *velocity*. The notion of complex system has been exported in several different contexts (biology [9,10], medicine [11–15], economy [16,17], psychology [18,19], social dynamics [20–23] ...), in which the state variables are non-mechanical and, in at least one case, vectorial [24].

Though the behavior of each particle of the system is of course deterministic, i.e., it is uniquely determined by its interactions with other particles; nevertheless, the number of particles and interactions is so large as to prevent us from following the evolution of the state of each particle. Accordingly, the model is based on the choice to describe the evolution of the system as a whole, turning the attention to the probability distribution on the states of particles; thus, the evolution equation takes the form (1) (see Section 2). As usual for nonlinear differential and integral equations, also in this case we have to tackle the classical problems about existence, uniqueness, continuous dependence, and stability with respect to initial values (and boundary values, when required). These problems have been tackled in [25–30].

In many cases of interest, the solutions to the problems about stability and continuous dependence of solutions depend on the coefficients of the equations, especially in the cases in which they are not constants but functions of the independent variables. In this last case, the question of whether two solutions, corresponding to the same assigned data but to two different systems of coefficients, are close when such are the coefficients spontaneously arises. This question seems to be of special relevance for Equation (1). As we shall see in more detail in Section 2, in Equation (1), denoting by u the state variable and by D_u the state space, that is the set of all possible values of u, we find two kinds of coefficients:

1. the coefficient $\eta(u_*, u^*)$, a function defined on D_u^2, expressing the interaction rate of the particles whose state is u_* with the particles whose state is u^*, i.e.,— roughly speaking—the number of their interactions per unit time;
2. the coefficient $\mathcal{A}(u_*, u^*, u)$, a function defined on D_u^3, expressing the transition probability, i.e., the probability (density) that any individual in the state u_*, when interacting with a particle in the state u^*, *falls* in the state u.

In any context, it is obvious that, to different prescriptions on the form of function $\eta(u_*, u^*)$ or of function $\mathcal{A}(u_*, u^*, u)$, there will correspond different probability distributions on D_u (or, in a strictly statistical interpretation, different distributions of relative frequencies on D_u over the system). However, what should we expect about the dependence of the difference of distributions on the difference between prescribed coefficients? Should an accordingly small difference between the corresponding distributions correspond to *small* perturbations to the interaction rate or to the transition distribution?

These questions are quite similar to those posed in all the classical problems associated with differential and integro-differential equations, but—as far as we are aware—have not been tackled yet for Equation (1). Nevertheless, in view of the large number of applications of complex systems (and of Equation (1), which describes their evolution) to so many basic problems of collective life of the whole mankind (for instance, let us mention the prediction of the evolution of epidemic diseases, or of the emergence of unsustainable economic inequalities), these questions are of special relevance in the framework of KTAP. In addition, the present paper is the first attempt to tackle them and to give some initial results about *both* the continuous dependence of solutions of Equation (1) on the coefficients *and* their instability. In connection with this last topic, the paper also offers some numerical simulations that show the separation between solutions corresponding to different values of parameters.

The contents of the paper are distributed as follows: in Section 2, we recall the structure of KTAP theory and Equation (1), and report the Cauchy problem associated with it, in the case in which the activity variable is assumed to be continuous (Section 2.1), as well as in the case in which it is assumed to be discrete (Section 2.2); Section 3 will be devoted to draw the notion of dependence of solutions on the parameters, and we state and prove a result concerning the continuous dependence of solutions on parameters, again in both the continuous case (Section 3.1) and the discrete case (Section 3.2), and, in Section 4, we give first results about instability of solutions (in both cases); in this connection, special attention should be paid to the numerical simulations presented in Section 4.3 for the discrete case that offers a clear perception of the fact that the solutions to Equation (4) depend continuously on the parameters, but stay apart from each other when the perturbation of the parameters is greater than a well-defined threshold value. Finally, in Section 5, we outline some research perspectives based on some general and meaningful conclusions that can be drawn from the results found in the previous sections.

2. The Basic Equations
2.1. The Continuous Activity Framework

Let $D_u \subseteq \mathbb{R}$ and $F > 0$. According to what has been laid out in the Introduction, this paper is devoted to the analysis of properties of solutions $f(t,u) : [0,+\infty[\times D_u \to \mathbb{R}^+$ of the following nonlinear integro-differential equation, with quadratic nonlinearity:

$$\partial_t f(t,u) + F\partial_u((1 - u\,\mathbb{E}_1[f](t))f(t,u)) = J[f,f](t,u), \tag{1}$$

where the operator $J[f,f](t,u)$ is defined as follows:

$$\begin{aligned} J[f,f](t,u) &= G[f,f](t,u) - L[f,f](t,u) \\ &= \int_{D_u \times D_u} \eta(u_*,u^*)\,\mathcal{A}(u_*,u^*,u)\,f(t,u_*)f(t,u^*)\,du_*\,du^* + \\ &\quad - f(t,u)\int_{D_u} \eta(u,u^*)\,f(t,u^*)\,du^*, \end{aligned} \tag{2}$$

and
- $\eta(u_*,u^*) : D_u \times D_u \to \mathbb{R}^+$;
- $\mathcal{A}(u_*,u^*,u) : D_u \times D_u \times D_u \to \mathbb{R}^+$ with the property:

$$\int_{D_u} \mathcal{A}(u_*,u^*,u)\,du = 1, \qquad \forall u_*,u^* \in D_u;$$

- $\mathbb{E}_1[f](t) = \int_{D_u} u\,f(t,u)\,du$.

The *Cauchy problem* associated with Equation (1) reads

$$\begin{cases} \partial_t f(t,u) + F\partial_u((1 - u\,\mathbb{E}_1[f](t))f(t,u)) = \\ J[f,f](t,u) & (t,u) \in [0,+\infty[\times D_u \\ f(0,u) = f^0(u) & u \in D_u. \end{cases} \tag{3}$$

Let
$$\mathbb{E}_0[f](t) = \int_{D_u} f(t,u)\,du$$

and
$$\mathbb{E}_2[f](t) = \int_{D_u} u^2\,f(t,u)\,du.$$

Consider the *function space* $\mathcal{K}(D_u)$ defined as

$$\mathcal{K}(D_u) := \{f(t,u) \in [0,+\infty[\times D_u \to \mathbb{R}^+ : \mathbb{E}_0[f](t) = \mathbb{E}_2[f](t) = 1\}.$$

The existence and uniqueness of solutions

$$f(t,u) \in C\Big((0,+\infty) \times D_u; L^1(D_u)\Big) \cap \mathcal{K}(D_u)$$

of the Cauchy problem (3) are proved in [25], under the condition

$$f(t, u) = 0, \quad u \in \partial D_u.$$

The existence of solutions of the *nonequilibrium stationary problem* related to (1) is proved in [31]. A proof of the convergence of the solution of (3) to the nonequilibrium stationary solution as time goes to infinity is given in [32].

In many cases of interest, as for example in the description of the diffusion of epidemics $\eta(u_*, u^*)$ can be supposed to be constant, i.e., there exists $\eta > 0$ such that $\eta(u_*, u^*) = \eta$, for all $u_*, u^* \in D_u$.

Remark 1. *If C is a complex system, homogeneous with respect to the mechanical variables, i.e., space and velocity, (1) describes the evolution of the distribution function $f(t, u)$ of C, and is called thermostatted kinetic framework [2].*

The microscopic state is described by a scalar variable u, called activity, which attains its values in a real continuous subset D_u. In this frame:

- $\eta(u_*, u^*)$ is the interaction rate between the particles that are in the state u_* and the particles in the state u^*;
- $\mathcal{A}(u_*, u^*, u)$ is the transition probability density i.e., the probability (density) that a particle in the state u_* falls into the state u after interacting with a particle in the state u^*;
- $F > 0$ is the value of the external force field acting on the system C;
- $\mathbb{E}_0[f](t)$ is the density, $\mathbb{E}_1[f](t)$ is the linear momentum and $\mathbb{E}_2[f](t)$ is the global energy;
- $G[f, f](t, u)$ is the gain-term operator and $L[f, f](t, u)$ is the loss-term operator.

Equation (1) and the related problem (3) describe the evolution of a system C such that the global activation energy, $\mathbb{E}_2[f](t)$, is kept constant by means of a thermostat [33].

2.2. The Discrete Activity Framework

Let $I_u = \{u_1, u_2, \ldots, u_n\}$ be a discrete subset of $\mathbb{R}$. The operator $J_i[\mathbf{f}](t)$, for $i \in \{1, 2, \ldots, n\}$ is defined as:

$$J_i[\mathbf{f}](t) = G_i[\mathbf{f}](t) - L_i[\mathbf{f}](t)$$
$$= \sum_{h=1}^{n} \sum_{k=1}^{n} \eta_{hk} B_{hk}^i f_h(t) f_k(t) - f_i(t) \sum_{k=1}^{n} \eta_{ik} f_k(t),$$

where $\eta_{hk} : I_u \times I_u \to \mathbb{R}^+$, for $h, k \in \{1, 2, \ldots, n\}$, and the functions $B_{hk}^i : I_u \times I_u \times I_u \to \mathbb{R}^+$ (where $i, h, k \in \{1, 2, \ldots, n\}$) obey the condition

$$\sum_{i=1}^{n} B_{hk}^i = 1, \quad h, k \in \{1, 2, \ldots, n\}.$$

Let $\mathbf{f}(t) = (f_1(t), f_2(t), \ldots, f_n(t))$, where, for any $i \in \{1, 2, \ldots n\}$,

$$f_i(t) := f(t, u_i) : [0, +\infty[\times I_u \to \mathbb{R}^+$$

is a solution of the *nonlinear ordinary differential equation*

$$\frac{df_i}{dt}(t) = J_i[\mathbf{f}](t) + F_i(t) - \sum_{i=1}^{n} \left(\frac{u_i^2 (J_i[\mathbf{f}] + F_i)}{\mathbb{E}_2[\mathbf{f}]} \right) f_i(t), \quad (4)$$

for $\mathbf{F}(t) = (F_1(t), F_2(t), \ldots, F_n(t))$ with $F_i(t) > 0$. The 2-nd order moment function $\mathbb{E}_2[\mathbf{f}](t)$ of $\mathbf{f}$ takes now the form

$$\mathbb{E}_2[\mathbf{f}](t) = \sum_{i=1}^{n} u_i^2 f_i(t).$$

Consider the *function space*:

$$\mathcal{R}_\mathbf{f}^2 = \mathcal{R}_\mathbf{f}^2(\mathbb{R}^+; \mathbb{E}_2) = \left\{ \mathbf{f} \in C\left([0, +\infty]; (\mathbb{R}^+)^n\right) : \mathbb{E}_2[\mathbf{f}] = \mathbb{E}_2 \right\}$$

where $\mathbb{E}_2 \in \mathbb{R}^+$. The existence and uniqueness of solutions to the Cauchy problem associated with Equation (4), with initial data $\mathbf{f}^0$ such that $\sum_{i=1}^n u_i^2 f^0 = 1$, has been proved in [34] under the following assumption:

H1 There exist $\eta, F > 0$, such that $F_i(t) \leq F$, for $t > 0$, and $\eta_{hk} \leq \eta$, for $h, k \in \{1, 2, \ldots, n\}$.

A *nonequilibrium stationary solution* of Equation (4), for $i \in \{1, 2, \ldots, n\}$, is a function f_i satisfying the equation

$$J_i[\mathbf{f}] + F_i - \sum_{i=1}^n \left(\frac{u_i^2(J_i[\mathbf{f}] + F_i)}{\mathbb{E}_2} \right) f_i = 0. \quad (5)$$

Let $\tilde{\mathcal{R}}_\mathbf{f}^2$ denote the *function space*:

$$\tilde{\mathcal{R}}_\mathbf{f}^2(\mathbb{R}^+; \mathbb{E}_2) = \left\{ \mathbf{f} \in (\mathbb{R}^+)^n : \mathbb{E}_2[\mathbf{f}] = \mathbb{E}_2 \right\}.$$

The existence of *nonequilibrium stationary solutions* $g(u) \in \tilde{\mathcal{R}}_\mathbf{f}^2$ has been proved in [28], under the assumption **H1**.

In particular, under the further assumptions:

H2 $\sum_{i=1}^n u_i B_{hk}^i = 0$, for all $h, k \in \{1, 2, \ldots, n\}$,

H3 $\sum_{i=1}^n u_i^2 B_{hk}^i = u_h^2$, for all $h, k \in \{1, 2, \ldots, n\}$,

it has been proved in [28] that any nonequilibrium stationary solution is unique if the force field verifies the constraint

$$F > 2\eta \mathbb{E}_2^2 \left(1 + \frac{1}{\|u\|_2^2} \right).$$

Proposition 1 ([28]). *If assumptions* **H1–H3** *are met, together with the assumption*

H4
$$\mathbb{E}_0[\mathbf{f}] = \mathbb{E}_2[\mathbf{f}] = 1$$

then

1. *The evolution equation of* $\mathbb{E}_1[\mathbf{f}](t) = \sum_{i=1}^n u_i f_i(t)$ *takes the form*

$$\mathbb{E}_1'[\mathbf{f}](t) + \left(\eta + \sum_{i=1}^n u_i^2 f_i \right) \mathbb{E}_1[\mathbf{f}](t) - \sum_{i=1}^n u_i F_i = 0; \quad (6)$$

2. *as* $t \to +\infty$,

$$\mathbb{E}_1[\mathbf{f}](t) \to K := \frac{\sum_{i=1}^n u_i F_i}{\eta + \sum_{i=1}^n u_i^2 F_i}; \quad (7)$$

3. *Denoting by* $\mathbf{f}_0$ *the initial data of the Cauchy problem related to* (4), *one has*

$$|\mathbb{E}_1[\mathbf{f}](t) - K| \leq c \exp\left[-\left(\eta + \sum_{i=1}^n u_i^2 F_i \right) t \right], \quad (8)$$

where c is a constant depending on the system.

Remark 2. *In [35], the existence of solutions of Equation (4) and of the related nonequilibrium stationary problem has been proved for more general values of the real discrete variable u_i.*

Remark 3. *Equation (4) has been proposed in [34] in order to model a complex system which partitioned in n subsystems, called* functional subsystems. *In particular:*

- *the function $f_i(t)$, for $i \in \{1, 2, \ldots, n\}$, denotes the* distribution function *of the i-th functional subsystem;*
- *the function $\mathbf{F}(t) = (F_1(t), F_2(t), \ldots, F_n(t))$ is the* external force field *acting on the whole system;*
- *The term*

$$\alpha := \sum_{i=1}^{n} \left(\frac{u_i^2(J_i[\mathbf{f}] + F_i)}{\mathbb{E}_2} \right)$$

represents the thermostat term, *which allows for keeping constant the quantity $\mathbb{E}_2[\mathbf{f}](t)$;*
- *the term η_{hk} is the* interaction rate *related to the encounters between the functional subsystem h and the functional subsystem k, for $h, k \in \{1, 2, \ldots, n\}$;*
- *the function B_{hk}^i denotes the* transition probability density *that the functional subsystem h falls into the i after interacting with the functional subsystem k, for $i, h, k \in \{1, 2, \ldots, n\}$;*
- *the operator $J_i[\mathbf{f}](t)$, for $i \in \{1, 2, \ldots, n\}$, models the net flux to the i-th functional subsystem; $G_i[\mathbf{f}](t)$ denotes the* gain term operator *(incoming flux) and $L_i[\mathbf{f}](t)$ the* loss term operator *(outgoing flux).*

Remark 4. *Let $p \in \mathbb{N}$. Equation (4) can be further generalized as follows:*

$$\frac{df_i}{dt}(t) = J_i[\mathbf{f}](t) + F_i(t) - \sum_{i=1}^{n} \left(\frac{u_i^p(J_i[\mathbf{f}] + F_i)}{\mathbb{E}_p[\mathbf{f}]} \right) f_i(t).$$

This framework allows for keeping the p-th order moment

$$\mathbb{E}_p[\mathbf{f}](t) = \sum_{i=1}^{n} u_i^p f_i(t).$$

of the distribution **f** *constant.*

Remark 5. *The convergence of any solution of (4) to a corresponding nonequilibrium stationary state (solution to (5)), as time goes to infinity, has been proved in [27].*

3. The Continuous Dependence on the Parameters

3.1. The Continuous Activity Framework

Let $\mathcal{A}(u_*, u^*, u)$, $\tilde{\mathcal{A}}(u_*, u^*, u)$, η and $\tilde{\eta}$ two classes of parameters for Equation (1). Let $J[f, f](t, u) = G[f, f](t, u) - L[f, f](t, u)$ be the operator related to the parameters $\mathcal{A}(u_*, u^*, u)$ and η, and $\tilde{J}[f, f](t, u) = \tilde{G}[f, f](t, u) - \tilde{L}[f, f](t, u)$ the operator related to the parameters $\tilde{\mathcal{A}}(u_*, u^*, u)$ and $\tilde{\eta}$.

The related Cauchy problems, with the same initial data $f^0(u)$, are defined as follows:

$$\begin{cases} \partial_t f(t, u) + F \partial_u ((1 - u \mathbb{E}_1[f](t)) f(t, u)) = J[f, f](t, u) \\ f(0, u) = f^0(u), \end{cases} \quad (9)$$

$$\begin{cases} \partial_t f(t,u) + F\partial_u((1-u\mathbb{E}_1[f](t))f(t,u)) = \tilde{J}[f,f](t,u) \\ f(0,u) = f^0(u). \end{cases} \quad (10)$$

By [25], there exist two functions $f(t,u) \in C((0,+\infty) \times D_u; L^1(D_u)) \cap \mathcal{K}(D_u)$ and $\tilde{f}(t,u) \in C((0,+\infty) \times D_u; L^1(D_u)) \cap \mathcal{K}(D_u)$ that are solutions to problem (9) and to problem (10), respectively.

The present paper aims to give a contribution in two directions:

1. the *continuous dependence* of the solutions of Equation (1) on the *parameters* $\mathcal{A}(u_*, u^*, u)$ and η;
2. a first attempt towards the instability of the solutions of Equation (1) for certain values of the two classes of parameters $(\mathcal{A}(u_*, u^*, u), \eta)$ and $(\tilde{\mathcal{A}}(u_*, u^*, u), \tilde{\eta})$.

The first result will be a proof of the *continuous dependence* of solutions to Equation (1) with respect to the parameters.

Let $\Theta(u_*, u^*, u)$ be the function defined as:

$$\Theta(u_*, u^*, u) := |\eta \mathcal{A}(u_*, u^*, u) - \tilde{\eta} \tilde{\mathcal{A}}(u_*, u^*, u)|.$$

Theorem 1. *Let $f(t,u), \tilde{f}(t,u) \in C((0,+\infty) \times D_u; L^1(D_u)) \cap \mathcal{K}(D_u)$ the solutions to problems (9) and (10), respectively. Assume that $\Theta(u_*, u^*, u) \in L^1(D_u \times D_u \times D_u)$. If there exist $\delta, \mathring{\delta} > 0$ such that $|\eta - \tilde{\eta}| < \delta$ and $\|\Theta(u_*, u^*, u)\|_{L^1(D_u \times D_u \times D_u)} \leq \mathring{\delta}$ then, for all $T > 0$:*

$$\|f(t,u) - \tilde{f}(t,u)\|_{C((0,T) \times D_u; L^1(D_u))} \leq (\delta + \mathring{\delta}) T e^{(2\eta + \tilde{\eta})T}. \quad (11)$$

Proof. Integrating Equation (1) from 0 to t and recalling that $f(0,u) = f^0(u)$ in mind, one has

$$f(t,u) = f^0(u) + \int_0^t J[f,f](\tau,u) \, d\tau \\ - F \int_0^t \partial_u((1-u\mathbb{E}_1[f](\tau))f(\tau,u)) \, d\tau, \quad (12)$$

where

$$\int_0^t J[f,f](\tau,u) \, d\tau = \int_0^t G[f,f](\tau,u) \, d\tau - \int_0^t L[f,f](\tau,u) \, d\tau \\ = \int_0^t \int_{D_u \times D_u} \eta \, \mathcal{A}(u_*, u^*, u) f(t,u_*) f(t, u^*) \, du_* \, du^* \, d\tau \\ - \eta \int_0^t f(\tau,u) \, d\tau.$$

The integral expression (12) for the solutions $f(t,u)$ and $\tilde{f}(t,u)$ becomes

$$f(t,u) = f^0(u) + \int_0^t \int_{D_u \times D_u} \eta \, \mathcal{A}(u_*, u^*, u) f(t,u_*) f(t,u^*) \, du_* \, du^* \, d\tau \\ - \eta \int_0^t f(\tau,u) \, d\tau - F \int_0^t \partial_u((1-u\mathbb{E}_1[f](\tau))f(\tau,u)) \, d\tau, \quad (13)$$

and

$$\tilde{f}(t,u) = f^0(u) + \int_0^t \int_{D_u \times D_u} \tilde{\eta} \, \tilde{\mathcal{A}}(u_*, u^*, u) \tilde{f}(t,u_*) \tilde{f}(t,u^*) \, du_* \, du^* \, d\tau \\ - \tilde{\eta} \int_0^t \tilde{f}(\tau,u) \, d\tau - F \int_0^t \partial_u((1-u\mathbb{E}_1[\tilde{f}](\tau))\tilde{f}(\tau,u)) \, d\tau \quad (14)$$

respectively.

Subtracting side by side Equation (14) from Equation (13), we readily get

$$f(t,u) - \tilde{f}(t,u) =$$
$$= \int_0^t \int_{D_u \times D_u} \left(\eta \mathcal{A}(u_*, u^*, u) f(\tau, u_*) f(\tau, u^*) - \tilde{\eta} \tilde{\mathcal{A}}(u_*, u^*, u) \tilde{f}(\tau, u_*) \tilde{f}(\tau, u^*) \right) du_* \, du^* \, d\tau$$
$$+ \int_0^t \left[\tilde{\eta} \tilde{f}(\tau, u) - \eta f(\tau, u) \right] d\tau \qquad (15)$$
$$+ F \int_0^t \partial_u \left(\tilde{f}(\tau, u) - f(\tau, u) + f(\tau, u) \, u \, \mathbb{E}_1[f](\tau) - \tilde{f}(\tau, u) \, u \, \mathbb{E}_1[\tilde{f}](\tau) \right) d\tau$$

which leads immediately to the estimate

$$|f(t,u) - \tilde{f}(t,u)| \le$$
$$\le \left| \int_0^t \int_{D_u \times D_u} \eta \mathcal{A}(u_*, u^*, u) f(\tau, u_*) f(\tau, u^*) - \tilde{\eta} \tilde{\mathcal{A}}(u_*, u^*, u) \tilde{f}(\tau, u_*) \tilde{f}(\tau, u^*) \, du_* \, du^* \, d\tau \right|$$
$$+ \left| \tilde{\eta} \int_0^t \tilde{f}(\tau, u) \, d\tau - \eta \int_0^t f(\tau, u) \, d\tau \right| \qquad (16)$$
$$+ F \left| \int_0^t \partial_u \left(\tilde{f}(\tau, u) - f(\tau, u) + f(\tau, u) \, u \, \mathbb{E}_1[f](\tau) - \tilde{f}(\tau, u) \, u \, \mathbb{E}_1[\tilde{f}](\tau) \right) d\tau \right|.$$

Since $f(t,u) = \tilde{f}(t,u) = 0$ for $u \in \partial D_u$, the third term on the right-hand side of inequality (16) vanishes, so that we obtain the relation

$$|f(t,u) - \tilde{f}(t,u)| \le$$
$$\le \left| \int_0^t \int_{D_u \times D_u} \eta \mathcal{A}(u_*, u^*, u) f(\tau, u_*) f(\tau, u^*) - \tilde{\eta} \tilde{\mathcal{A}}(u_*, u^*, u) \tilde{f}(\tau, u_*) \tilde{f}(\tau, u^*) \, du_* \, du^* \, d\tau \right| \qquad (17)$$
$$+ \int_0^t |\tilde{\eta} \tilde{f}(\tau, u) - \eta f(\tau, u)| \, d\tau$$

and, by straightforward calculations, one finds that the first term on the right-hand side of (17) can be estimated as follows:

$$\left| \int_0^t \int_{D_u \times D_u} \eta \mathcal{A}(u_*, u^*, u) f(\tau, u_*) f(\tau, u^*) - \tilde{\eta} \tilde{\mathcal{A}}(u_*, u^*, u) \tilde{f}(\tau, u_*) \tilde{f}(\tau, u^*) \, du_* \, du^* \, d\tau \right| \le$$
$$\le \int_0^t \int_{D_u \times D_u} \eta f(\tau, u^*) \mathcal{A}(u_*, u^*, u) |f(\tau, u_*) - \tilde{f}(\tau, u_*)| \, du_* \, du^* \, d\tau$$
$$+ \int_0^t \int_{D_u \times D_u} \tilde{\eta} \tilde{f}(\tau, u_*) \tilde{\mathcal{A}}(u_*, u^*, u) |\tilde{f}(\tau, u^*) - f(\tau, u^*)| \, du^* \, du_* \, d\tau \qquad (18)$$
$$+ \int_0^t \int_{D_u \times D_u} \tilde{f}(\tau, u_*) f(\tau, u^*) [\eta \mathcal{A}(u_*, u^*, u) - \tilde{\eta} \tilde{\mathcal{A}}(u_*, u^*, u)] \, du_* \, du^* \, d\tau.$$

Now, using inequality (18) and integrating both sides of relation (17) on D_u, we get

$$\|f(t,u) - \tilde{f}(t,u)\|_{L^1(D_u)} \le \eta \int_0^t \|f(\tau, u) - \tilde{f}(\tau, u)\|_{L^1(D_u)} \, d\tau$$
$$+ \tilde{\eta} \int_0^t \|f(\tau, u) - \tilde{f}(\tau, u)\|_{L^1(D_u)} \, d\tau + \|\Theta(u_*, u^*, u)\|_{L^1(D_u \times D_u \times D_u)} \qquad (19)$$
$$+ \int_0^t \int_{D_u} |\tilde{\eta} \tilde{f}(\tau, u) - \eta f(\tau, u)| \, du \, d\tau.$$

Since:

$$|\tilde{\eta} \tilde{f}(\tau, u) - \eta f(\tau, u)| = |\tilde{\eta} \tilde{f}(\tau, u) - \eta \tilde{f}(\tau, u) + \eta \tilde{f}(\tau, u) - \eta f(\tau, u)|$$
$$\le \tilde{f}(\tau, u) |\tilde{\eta} - \eta| + \eta |\tilde{f}(\tau, u) - f(\tau, u)|,$$

relation (19), bearing in mind that $|\eta - \tilde{\eta}| < \delta$ and $\|\Theta(u_*, u^*, u)\|_{L^1(D_u \times D_u \times D_u)} \leq \hat{\delta}$, may be rewritten in the form

$$\begin{aligned}\|f(t,u) - \tilde{f}(t,u)\|_{L^1(D_u)} &\leq \int_0^t (\eta + \tilde{\eta}) \|f(\tau,u) - \tilde{f}(\tau,u)\|_{L^1(D_u)} d\tau \\ &\quad + \|\Theta(u_*, u^*, u)\|_{L^1(D_u \times D_u \times D_u)} + |\tilde{\eta} - \eta| t \\ &\quad + \eta \int_0^t \|f(\tau,u) - \tilde{f}(\tau,u)\|_{L^1(D_u)} d\tau \\ &\leq \int_0^t (2\eta + \tilde{\eta}) \|f(\tau,u) - \tilde{f}(\tau,u)\|_{L^1(D_u)} d\tau + (\delta + \hat{\delta}) t.\end{aligned} \quad (20)$$

By Grönwall's inequaility [36],

$$\|f(t,u) - \tilde{f}(t,u)\|_{L^1(D_u)} \leq (\delta + \hat{\delta}) t \, e^{(2\eta + \tilde{\eta})t}. \quad (21)$$

By (20) and (21), relation (11) is proved, i.e., for $T > 0$,

$$\|f(t,u) - \tilde{f}(t,u)\|_{C((0,T) \times D_u; L^1(D_u))} \leq (\delta + \hat{\delta}) T \, e^{(2\eta + \tilde{\eta})T}.$$

□

Remark 6. *It is worth pointing out that*
1. *the assumption $|\eta - \tilde{\eta}| < \delta$ is an estimate of the distance between the interaction rates;*
2. *the assumption $\|\Theta(u_*, u^*, u)\|_{L^1(D_u \times D_u \times D_u)} \leq \hat{\delta}$ is an estimate on the distance between the transition probability densities, "weighted" by the interaction rates.*

Remark 7. *The conclusion (11) of Theorem 1 ensures the continuous dependence of solutions on the parameters $\mathcal{A}(u_*, u^*, u)$ and η. Indeed,*

$$\|f(t,u) - \tilde{f}(t,u)\|_{C((0,+\infty) \times D_u; L^1(D_u))} \xrightarrow{\delta, \hat{\delta} \to 0} 0.$$

3.2. The Discrete Activity Framework

This section aims to prove the continuous dependence of the solutions of Equation (4) on the parameters η_{hk} and B_{hk}^i when $\mathbf{f}^0$, $\mathbf{F}(t) = \mathbf{F}$, constant in time, and $T > 0$ are fixed.

Let $\mathbf{f}(t) = (f_1(t), f_2(t), \ldots, f_n(t))$, $\hat{\mathbf{f}}(t) = \left(\hat{f}_1(t), \hat{f}_2(t), \ldots, \hat{f}_n(t)\right)$ be the solutions of the systems

$$\begin{cases} \dfrac{df_i}{dt}(t) = J_i[\mathbf{f}](t) + F_i(t) - \sum_{i=1}^n \left(\dfrac{u_i^2(J_i[\mathbf{f}] + F_i)}{\mathbb{E}_2[\mathbf{f}]}\right) f_i(t) & t \in [0,T] \\ \mathbf{f}(0) = \mathbf{f}^0, \end{cases} \quad (22)$$

and

$$\begin{cases} \dfrac{df_i}{dt}(t) = \hat{J}_i[\mathbf{f}](t) + F_i(t) - \sum_{i=1}^n \left(\dfrac{u_i^2(\hat{J}_i[\mathbf{f}] + F_i)}{\mathbb{E}_2[\mathbf{f}]}\right) f_i(t) & t \in [0,T] \\ \mathbf{f}(0) = \mathbf{f}^0, \end{cases} \quad (23)$$

respectively; the operators $\mathbf{J}[\mathbf{f}]$ and $\hat{\mathbf{J}}[\mathbf{f}]$ are defined by the parameters η_{hk}, B_{hk}^i, and $\hat{\eta}_{hk}$, $\hat{B}_{hk}^i$, respectively.

The following stability result holds.

Theorem 2. Let $\mathbf{f}(t) = (f_1(t), f_2(t), \ldots, f_n(t))$, $\hat{\mathbf{f}}(t) = (\hat{f}_1(t), \hat{f}_2(t), \ldots, \hat{f}_n(t))$ be the solutions of (12) and (13), respectively. Assume $\eta_{hk} \leq \eta$, $\hat{\eta}_{hk} \leq \hat{\eta}$, for $h, k \in \{1, 2, \ldots, n\}$, and $F_i \leq F$, for $i \in \{1, 2, \ldots, n\}$, for $\eta, \hat{\eta}, F > 0$. If $\Lambda := \sum_{i=1}^{n} \sum_{h,k=1}^{n} |\eta_{hk} B_{hk}^i - \hat{\eta}_{hk} \hat{B}_{hk}^i|$, then

$$\max_{t \in [0,T]} \left\| \mathbf{f}(t) - \hat{\mathbf{f}}(t) \right\|_1 \leq \Lambda T e^{(\eta + \hat{\eta} + \sum_{i=1}^{n} u_i^2 F_i)T} \tag{24}$$

where

$$\|\mathbf{f}(t) - \hat{\mathbf{f}}(t)\|_1 := \sum_{i=1}^{n} |f_i(t) - \hat{f}_i(t)|.$$

Proof. Bearing assumption **H4** in mind, and integrating Equations (22) and (23) on $[0, t]$, we get

$$f_i(t) = f_i^0 + \int_0^t \left(J_i[\mathbf{f}](t) + F_i - \left(\sum_{i=1}^{n} u_i^2 (J_i[\mathbf{f}](t) + F_i) \right) f_i(t) \right) dt, \tag{25}$$

and

$$\hat{f}_i(t) = \hat{f}_i^0 + \int_0^t \left(\hat{J}_i[\hat{\mathbf{f}}](t) + F_i - \left(\sum_{i=1}^{n} u_i^2 \left(\hat{J}_i[\hat{\mathbf{f}}](t) + F_i \right) \right) \hat{f}_i(t) \right) dt \tag{26}$$

for $i \in \{1, 2, \ldots, n\}$. Now, subtracting (26) from (25), we find

$$f_i(t) - \hat{f}_i(t) =$$
$$= \int_0^t J_i[\mathbf{f}](t) - \hat{J}_i[\hat{\mathbf{f}}](t) \, dt$$
$$- \int_0^t \left[\left(\sum_{i=1}^{n} u_i^2 (J_i[\mathbf{f}](t) + F_i) \right) f_i(t) - \left(\sum_{i=1}^{n} u_i^2 \left(\hat{J}_i[\hat{\mathbf{f}}](t) + F_i \right) \hat{f}_i(t) \right) \right] dt. \tag{27}$$

By taking the side-by-side sum on $i \in \{1, 2, \ldots, n\}$ of these last relations, we arrive at

$$\sum_{i=1}^{n} \left| f_i(t) - \hat{f}_i(t) \right| \leq$$

$$\leq \int_0^t \sum_{i=1}^{n} \left| J_i[\mathbf{f}](t) - \hat{J}_i[\hat{\mathbf{f}}](t) \right|$$
$$+ \int_0^t \sum_{i=1}^{n} \left| \left(\sum_{i=1}^{n} u_i^2 (J_i[\mathbf{f}](t) + F_i) \right) f_i(t) - \left(\sum_{i=1}^{n} u_i^2 \left(\hat{J}_i[\hat{\mathbf{f}}](t) + F_i \right) \hat{f}_i(t) \right) \right| dt. \tag{28}$$

Now, first of all, observe that

$$\int_0^t \sum_{i=1}^{n} |J_i[\mathbf{f}](t) - \hat{J}_i[\hat{\mathbf{f}}](t)| \leq$$

$$\leq \int_0^t \sum_{i=1}^{n} \left| \sum_{h,k=1}^{n} \left(\eta_{hk} B_{hk}^i f_h(t) f_k(t) - \hat{\eta}_{hk} \hat{B}_{hk}^i \hat{f}_h(t) \hat{f}_k(t) \right) \right| dt \tag{29}$$
$$+ \int_0^t \sum_{i=1}^{n} \left| f_i(t) \sum_{k=1}^{n} \eta_{ik} f_k(t) - \hat{f}_i(t) \sum_{k=1}^{n} \hat{\eta}_{ik} \hat{f}_k(t) \right|.$$

The integrand in the first term on the right-hand side of (29) can be estimated as follows:

$$\left| \eta_{hk} B^i_{hk} f_h(t) f_k(t) - \hat{\eta}_{hk} \hat{B}^i_{hk} \hat{f}_h(t) \hat{f}_k(t) \right| =$$
$$= \left| \eta_{hk} B^i_{hk} f_h(t) f_k(t) - B^i_{hk} \eta_{hk} f_h(t) \hat{f}_k(t) + B^i_{hk} \eta_{hk} f_h(t) \hat{f}_k(t) - \hat{\eta}_{hk} \hat{B}^i_{hk} \hat{f}_h(t) \hat{f}_k(t) \right|$$
$$\leq \left| \eta_{hk} B^i_{hk} f_h(t) \left(f_k(t) - \hat{f}_k(t) \right) + \hat{f}_k(t) \left(\eta_{hk} B^i_{hk} f_h(t) - \hat{\eta}_{hk} \hat{B}^i_{hk} \hat{f}_h(t) \right) \right|$$
$$\leq \eta_{hk} B^i_{hk} f_h(t) \left| f_k(t) - \hat{f}_k(t) \right| \qquad (30)$$
$$+ \hat{f}_k(t) \left| \eta_{hk} B^i_{hk} f_h(t) - \hat{\eta}_{hk} \hat{B}^i_{hk} f_h(t) + \hat{\eta}_{hk} \hat{B}^i_{hk} f_h(t) - \hat{\eta}_{hk} \hat{B}^i_{hk} \hat{f}_h(t) \right|$$
$$\leq \eta_{hk} B^i_{hk} f_h(t) \left| f_k(t) - \hat{f}_k(t) \right| + \hat{f}_k(t) f_h(t) \left| \eta_{hk} B^i_{hk} - \hat{\eta}_{hk} \hat{B}^i_{hk} \right|$$
$$+ \hat{\eta}_{hk} \hat{B}^i_{hk} \left| f_h(t) - \hat{f}_h(t) \right|.$$

By using this estimate, the first integral on the right-hand side of (29) turns out to be majorized as follows:

$$\int_0^t \sum_{i=1}^n \left| \sum_{h,k=1}^n \left(\eta_{hk} B^i_{hk} f_h(t) f_k(t) - \hat{\eta}_{hk} \hat{B}^i_{hk} \hat{f}_h(t) \hat{f}_k(t) \right) \right| dt$$
$$\leq \int_0^t \sum_{i=1}^n \sum_{h,k=1}^n \eta_{hk} B^i_{hk} f_h(t) \left| f_k(t) - \hat{f}_k(t) \right| dt + \int_0^t \sum_{i=1}^n \sum_{h,k=1}^n \hat{\eta}_{hk} \hat{B}^i_{hk} \hat{f}_k(t) \left| f_h(t) - \hat{f}_h(t) \right| dt \qquad (31)$$
$$+ \int_0^t \sum_{h,k=1}^n \hat{f}_k(t) f_h(t) \sum_{i=1}^n \left| \eta_{hk} B^i_{hk} - \hat{\eta}_{hk} \hat{B}^i_{hk} \right| dt$$
$$\leq \eta \int_0^t \|\mathbf{f}(t) - \hat{\mathbf{f}}(t)\|_1 \, dt + \hat{\eta} \int_0^t \|\mathbf{f}(t) - \hat{\mathbf{f}}(t)\|_1 \, dt + \Lambda t.$$

As far as the second term on the right-hand side of (28) is concerned, one has

$$\sum_{i=1}^n u_i^2 J_i[\mathbf{f}](t) = \sum_{i=1}^n u_i^2 (G_i[\mathbf{f}](t) - L_i[\mathbf{f}(t)])$$
$$= \sum_{i=1}^n u_i^2 \left(\sum_{h,k=1}^n \eta_{hk} B^i_{hk} f_h(t) f_k(t) - f_i(t) \sum_{k=1}^n \eta_{ik} f_k(t) \right) \qquad (32)$$
$$= \sum_{h,k=1}^n \left(\sum_{i=1}^n u_i^2 B^i_{hk} \right) \eta_{hk} f_h(t) f_k(t) - \sum_{i=1}^n u_i^2 f_i(t) \sum_{k=1}^n \eta_{ik} f_k(t) = 0$$

which in turn implies

$$\int_0^t \sum_{i=1}^n \left| \left(\sum_{i=1}^n u_i^2 (J_i[\mathbf{f}](t) + F_i) \right) f_i(t) - \left(\sum_{i=1}^n u_i^2 \left(\hat{J}_i[\hat{\mathbf{f}}](t) + F_i \right) \hat{f}_i(t) \right) \right| dt \leq$$
$$\leq \int_0^t \left(\sum_{i=1}^n u_i^2 F_i \right) \|\mathbf{f}(t) - \hat{\mathbf{f}}(t)\|_1 \, dt. \qquad (33)$$

Finally, by using relations (31) and (33), inequality (28) becomes

$$\|\mathbf{f}(t) - \hat{\mathbf{f}}(t)\|_1 \leq \int_0^t \left(\eta h + \hat{\eta} k + \sum_{i=1}^n u_i^2 F_i \right) \|\mathbf{f}(t) - \hat{\mathbf{f}}(t)\|_1 \, dt + \Lambda t \qquad (34)$$

and now Grönwall's inequaility [36] yields

$$\|\mathbf{f}(t) - \hat{\mathbf{f}}(t)\|_1 \leq \Lambda T e^{\left(\eta h + \hat{\eta} k + \sum_{i=1}^n u_i^2 F_i \right) t}$$

leading at once to (24). □

Remark 8. *The conclusion of Theorem 2 is the continuous dependence of solution of Equation (4) on the parameters of the system, i.e., the interaction rate η_{hk} and the transition probability density B^i_{hk}. In fact,*

$$\max_{t \in [0,T]} \|\mathbf{f}(t) - \hat{\mathbf{f}}(t)\|_1 \xrightarrow{\Lambda \to 0} 0.$$

Remark 9. *The coefficient Λ defined in Theorem 2 is a first estimate of the distance between the two classes of parameters, i.e., (η_{hk}, B^i_{hk}) and $(\hat{\eta}_{hk}, \hat{B}^i_{hk})$.*

4. A First Attempt towards the Instability with Respect to the Parameters

4.1. The Continuous Activity Framework

This section aims to give a first result about instability of solutions of Equation (1) with respect to the parameters, interaction rate η, and transition probability density $\mathcal{A}(u_*, u^*, u)$.

Theorem 3. *Let $f(t,u), \tilde{f}(t,u) \in C((0,+\infty) \times D_u; L^1(D_u)) \cap \mathcal{K}(D_u)$ be the solutions to problems (9) and (10), respectively. Assume that $\Theta(u_*, u^*, u) \in L^1(D_u \times D_u \times D_u)$. If there exist two constants M_1 and $\hat{M}_1$, with $M_1 > \hat{M}_1$ such that $|\eta - \tilde{\eta}| > M_1$ and $\|\Theta(u_*, u^*, u)\|_{L^1(D_u \times D_u \times D_u)} \leq \hat{M}_1$, then, for all $T > 0$:*

$$\|f(t,u) - \tilde{f}(t,u)\|_{C((0,T) \times D_u; L^1(D_u))} \geq \frac{(M_1 - \hat{M}_1)}{1 + (\eta + \tilde{\eta})T} T > 0. \tag{35}$$

Proof. As in Theorem 1, by using the integral formulation of (1) and by straightforward calculations, one has

$$\begin{aligned}
|f(t,u) - \tilde{f}(t,u)| &= \left| \int_0^t \left(\tilde{\eta} \tilde{f}(\tau,u) - \eta f(\tau,u) \right) d\tau \right. \\
&+ \int_0^t \int_{D_u \times D_u} \eta \mathcal{A}(u_*, u^*, u) f(\tau, u_*) f(\tau, u^*) - \tilde{\eta} \tilde{\mathcal{A}}(u_*, u^*, u) \tilde{f}(\tau, u_*) \tilde{f}(\tau, u^*) \, du_* \, du^* \, d\tau \\
&+ F \int_0^t \partial_u \left((\tilde{f}(\tau,u) - f(\tau,u) + f(\tau,u) u \mathbb{E}_1[f](\tau) - \tilde{f}(\tau,u) u \mathbb{E}_1[\tilde{f}](\tau)) \right) d\tau \bigg| \\
&\geq \left| \int_0^t \left(\tilde{\eta} \tilde{f}(\tau,u) - \eta f(\tau,u) \right) d\tau \right| \\
&- \left| \int_0^t \int_{D_u \times D_u} \eta \mathcal{A}(u_*, u^*, u) f(\tau, u_*) f(\tau, u^*) - \tilde{\eta} \tilde{\mathcal{A}}(u_*, u^*, u) \tilde{f}(\tau, u_*) \tilde{f}(\tau, u^*) \, du_* \, du^* \, d\tau \right| \\
&- F \left| \int_0^t \partial_u \left((\tilde{f}(\tau,u) - f(\tau,u) + f(\tau,u) u \mathbb{E}_1[f](\tau) - \tilde{f}(\tau,u) u \mathbb{E}_1[\tilde{f}](\tau)) \right) d\tau \right|
\end{aligned} \tag{36}$$

whence, by integrating the (36) on D_u, we obtain

$$\begin{aligned}
\|f(t,u) - \tilde{f}(t,u)\|_{L^1(D_u)} &\geq \int_{D_u} \left| \int_0^t \left(\tilde{\eta} \tilde{f}(\tau,u) - \eta f(\tau,u) \right) d\tau \right| du \\
&- \int_{D_u} \left| \int_0^t \int_{D_u \times D_u} \eta \mathcal{A}(u_*, u^*, u) f(\tau, u_*) f(\tau, u^*) - \tilde{\eta} \tilde{\mathcal{A}}(u_*, u^*, u) \tilde{f}(\tau, u_*) \tilde{f}(\tau, u^*) \, du_* \, du^* \, d\tau \right| du \\
&- F \int_{D_u} \left| \int_0^t \partial_u \left((\tilde{f}(\tau,u) - f(\tau,u) + f(\tau,u) u \mathbb{E}_1[f](\tau) - \tilde{f}(\tau,u) u \mathbb{E}_1[\tilde{f}](\tau)) \right) d\tau \right|.
\end{aligned} \tag{37}$$

Since $f(t,u) = \tilde{f}(t,u) = 0$ for $u \in \partial D_u$, the third term on the right-hand side of the (37) vanishes.

The first term at the right-hand side of inequality (37) is estimated as follows:

$$\int_{D_u}\left|\int_0^t (\tilde{\eta}\tilde{f}(\tau,u) - \eta f(\tau,u))\, d\tau\right| du \geq \left|\int_0^t \int_{D_u} \tilde{\eta}\tilde{f}(\tau,u) - \eta f(\tau,u)\, du\, d\tau\right| \qquad (38)$$
$$= |\tilde{\eta} - \eta|\, t.$$

Consider now the second term of the right-hand side of inequality (37). First of all,

$$-\int_{D_u}\left|\int_0^t \int_{D_u \times D_u} \eta\, \mathcal{A}(u_*, u^*, u)\, f(\tau, u_*) f(\tau, u^*)\right.$$
$$\left. - \tilde{\eta}\, \tilde{\mathcal{A}}(u_*, u^*, u)\, \tilde{f}(\tau, u_*)\tilde{f}(\tau, u^*)\, du_*\, du^*\, d\tau\right| du \geq \qquad (39)$$
$$-\int_{D_u}\int_0^t \int_{D_u \times D_u}\left| \eta\, \mathcal{A}(u_*, u^*, u)\, f(\tau, u_*) f(\tau, u^*)\right.$$
$$\left. - \tilde{\eta}\, \tilde{\mathcal{A}}(u_*, u^*, u)\, \tilde{f}(\tau, u_*)\tilde{f}(\tau, u^*)\right| du_*\, du^*\, d\tau\, du.$$

By straightforward calculations,

$$\left|\eta\, \mathcal{A}(u_*, u^*, u)\, f(\tau, u_*) f(\tau, u^*) - \tilde{\eta}\, \tilde{\mathcal{A}}(u_*, u^*, u)\, \tilde{f}(\tau, u_*) \tilde{f}(\tau, u^*)\right| =$$
$$= \left|\eta\, \mathcal{A}(u_*, u^*, u)\, f(\tau, u_*) f(\tau, u^*) - \eta\, \mathcal{A}(u_*, u^*, u)\, f(\tau, u_*)\tilde{f}(\tau, u^*)\right.$$
$$+ \eta\, \mathcal{A}(u_*, u^*, u)\, f(\tau, u_*)\tilde{f}(\tau, u^*) + \tilde{\eta}\, \tilde{\mathcal{A}}(u_*, u^*, u)\, f(\tau, u_*)\tilde{f}(\tau, u^*)$$
$$\left.- \tilde{\eta}\, \tilde{\mathcal{A}}(u_*, u^*, u)\, f(\tau, u_*)\tilde{f}(\tau, u^*) - \tilde{\eta}\, \tilde{\mathcal{A}}(u_*, u^*, u)\, \tilde{f}(\tau, u_*)\tilde{f}(\tau, u^*)\right|$$
$$= \left|\eta\, \mathcal{A}(u_*, u^*, u)\, f(\tau, u_*)(f(\tau, u^*) - \tilde{f}(\tau, u^*))\right. \qquad (40)$$
$$- \tilde{\eta}\, \tilde{\mathcal{A}}(u_*, u^*, u)\, \tilde{f}(\tau, u^*)(f(\tau, u_* - f(\tau, u_*)))$$
$$\left. + f(\tau, u_*)\tilde{f}(\tau, u^*)(\eta\, \mathcal{A}(u_*, u^*, u) - \tilde{\eta}\, \tilde{\mathcal{A}}(u_*, u^*, u))\right|$$
$$\leq \eta\, \mathcal{A}(u_*, u^*, u)\, f(\tau, u_*)\left|f(\tau, u^*) - \tilde{f}(\tau, u_*)\right|$$
$$+ \tilde{\eta}\, \tilde{\mathcal{A}}(u_*, u^*, u)\, \tilde{f}(\tau, u^*)\left|f(\tau, u_*) - \tilde{f}(\tau, u_*)\right|$$
$$+ f(\tau, u_*)\tilde{f}(\tau, u^*)\left|\eta\, \mathcal{A}(u_*, u^*, u) - \tilde{\eta}\tilde{\mathcal{A}}(u_*, u^*, u)\right|.$$

In virtue of inequalities (39) and (40),

$$-\int_{D_u}\left|\int_0^t \int_{D_u \times D_u} \eta\, \mathcal{A}(u_*, u^*, u)\, f(\tau, u_*)f(\tau, u^*)\right.$$
$$\left.-\tilde{\eta}\, \tilde{\mathcal{A}}(u_*, u^*, u)\, \tilde{f}(\tau, u_*)\tilde{f}(\tau, u^*)\, du_*\, du^*\, d\tau\right| du \geq$$
$$\geq -\eta \int_0^t \int_{D_u} f(\tau, u_*) \int_{D_u} |f(\tau, u^*) - \tilde{f}(\tau, u^*)| \int_{D_u} \mathcal{A}(u_*, u^*, u)\, du\, du^*\, du_*\, d\tau$$
$$- \tilde{\eta}\int_0^t \int_{D_u} \tilde{f}(\tau, u^*) \int_{D_u} |f(\tau, u_*) - \tilde{f}(\tau, u_*)| \int_{D_u} \mathcal{A}(u_*, u^*, u)\, du\, du_*\, du^*\, d\tau \qquad (41)$$
$$-\int_0^t \int_{D_u}\int_{D_u \times D_u} f(\tau, u_*)\tilde{f}(\tau, u^*)\left|\eta\, \mathcal{A}(u_*, u^*, u) - \tilde{\eta}\, \tilde{\mathcal{A}}(u_*, u^*, u)\right| du_*\, du^*\, du\, d\tau$$
$$= -\eta\, t \|f(t, u) - \tilde{f}(t, u)\|_{L^1(D_u)} - \tilde{\eta}\, t \|f(t, u) - \tilde{f}(t, u)\|_{L^1(D_u)}$$
$$-\int_0^t \int_{D_u}\int_{D_u \times D_u} f(\tau, u_*)\tilde{f}(\tau, u^*)\left|\eta\, \mathcal{A}(u_*, u^*, u) - \tilde{\eta}\, \tilde{\mathcal{A}}(u_*, u^*, u)\right| du_*\, du^*\, du\, d\tau$$

and, using Hölder's inequality,

$$\int_0^t \int_{D_u} \int_{D_u \times D_u} \tilde{f}(\tau, u^*) f(\tau, u_*) \left| \eta \, \mathcal{A}(u_*, u^*, u) - \tilde{\eta} \, \tilde{\mathcal{A}}(u_*, u^*, u) \right| du_* \, du^* \, du \, d\tau \leq$$

$$\leq \int_0^t \int_{D_u} \int_{D_u} \tilde{f}(\tau, u^*) \left(\max_{u_* \in D_u} f(\tau, u_*) \right) \int_{D_u} \left| \eta \, \mathcal{A}(u_*, u^*, u) - \tilde{\eta} \, \tilde{\mathcal{A}}(u_*, u^*, u) \right| du_* \, du^* \, du \, d\tau$$

$$\leq \int_0^t \int_{D_u} \int_{D_u} \left(\max_{u^* \in D_u} \tilde{f}(\tau, u^*) \right) \int_{D_u} \left| \eta \, \mathcal{A}(u_*, u^*, u) - \tilde{\eta} \, \tilde{\mathcal{A}}(u_*, u^*, u) \right| du^* \, du_* \, du \, d\tau \quad (42)$$

$$\leq \int_0^t \int_{D_u \times D_u \times D_u} \left| \eta \, \mathcal{A}(u_*, u^*, u) - \tilde{\eta} \, \tilde{\mathcal{A}}(u_*, u^*, u) \right| du_* \, du^* \, du$$

$$= \| \Theta(u_*, u^*, u) \|_{L^1(D_u \times D_u \times D_u)} \, t.$$

Thanks to relations (41) and (42), inequality (39) becomes

$$- \int_{D_u} \left| \int_0^t \int_{D_u \times D_u} \eta \, \mathcal{A}(u_*, u^*, u) \, f(\tau, u_*) f(\tau, u^*) \right.$$
$$\left. - \tilde{\eta} \, \tilde{\mathcal{A}}(u_*, u^*, u) \, \tilde{f}(\tau, u_*) \tilde{f}(\tau, u^*) \, du_* \, du^* \, d\tau \right| du \geq \quad (43)$$
$$\geq -(\eta + \tilde{\eta}) t \| f(t, u) - \tilde{f}(t, u) \|_{L^1(D_u)} - \| \Theta(u_*, u^*, u) \|_{L^1(D_u \times D_u \times D_u)} t.$$

Finally, by (38) and (43), inequality (37) yields

$$(1 + (\eta + \tilde{\eta}) t) \| f(t, u) - \tilde{f}(t, u) \|_{L^1(D_u)} \geq |\tilde{\eta} - \eta| \, t - \| \Theta(u_*, u^*, u) \|_{L^1(D_u \times D_u \times D_u)} t.$$

Then:

$$\| f(t, u) - \tilde{f}(t, u) \|_{L^1(D_u)} \geq \frac{|\tilde{\eta} - \eta| - \| \Theta(u_*, u^*, u) \|_{L^1(D_u \times D_u \times D_u)}}{1 + (\eta + \tilde{\eta}) t} \, t. \quad (44)$$

Relation (35) is then proved by using the (44), and keeping in mind the fact that $|\eta - \tilde{\eta}| > M_1$ and $\| \Theta(u_*, u^*, u) \|_{L^1(D_u \times D_u \times D_u)} \leq \hat{M}_1$, with $M_1 > \hat{M}_1$:

$$\| f(t, u) - \tilde{f}(t, u) \|_{C((0,T) \times D_u; L^1(D_u))} \geq \frac{|\tilde{\eta} - \eta| - \| \Theta(u_*, u^*, u) \|_{L^1(D_u \times D_u \times D_u)}}{1 + (\eta + \tilde{\eta}) T} T$$

$$\geq \frac{(M_1 - \hat{M}_1)}{1 + (\eta + \tilde{\eta}) T} T.$$

□

Remark 10. *In Theorem 3, the instability is related to the variation of the interaction rate.*

Remark 11. *For instance, if $D_u = [0, \frac{1}{2}]$ is taken into account with $\mathcal{A} = \tilde{\mathcal{A}}$, then the right-hand side of relation (35) is strictly positive, so that the instability of the solutions follows at once.*

4.2. The Discrete Activity Framework

In this section, we want to outline a first step of a study of instability in the discrete framework (4). This is an important issue in view of future numerical analysis.

Theorem 4. *Let $\mathbf{f}(t) = (f_1(t), f_2(t), \ldots, f_n(t))$, $\hat{\mathbf{f}}(t) = (\hat{f}_1(t), \hat{f}_2(t), \ldots, \hat{f}_n(t))$ be the solutions of Equations (22) and (23), respectively. Let $\eta, \hat{\eta}, F \geq 0$ such that $\eta_{hk} \leq \eta$, $\hat{\eta}_{hk} \leq \hat{\eta}$, for $h, k \in \{1, 2, \ldots, n\}$, and $F_i \leq F$, for $i \in \{1, 2, \ldots, n\}$. Furthermore, let*

$$\Gamma(t) := \min_{\mathbf{f}, \hat{\mathbf{f}} \in (C([0,T]))^n} \left\{ \sum_{i=1}^n \left| \int_0^t \sum_{h,k=1}^n \eta_{hk} B_{hk}^i f_h(t) f_k(t) - \hat{\eta}_{hk} \hat{B}_{hk}^i \hat{f}_h(t) \hat{f}_k(t) \right| \right\}.$$

Then,
$$\max_{[0,T]} \|\mathbf{f}(t) - \hat{\mathbf{f}}(t)\|_1 \geq \max_{[0,T]} \frac{\Gamma(t) - (\eta + \hat{\eta})t}{1 + (\sum_{i=1}^n u_i^2 F_i)t}. \quad (45)$$

Proof. Bearing the (27) in mind, and by using the (32), straightforward calculations show, for $i \in \{1, 2, \ldots, n\}$:

$$\left|f_i(t) - \hat{f}_i(t)\right| \geq$$
$$\geq \left|\int_0^t \sum_{h,k=1}^n \eta_{hk} B_{hk}^i f_h(s) f_k(s) - \hat{\eta}_{hk} \hat{B}_{hk}^i \hat{f}_h(s) \hat{f}_k(s) \, ds\right|$$
$$- \left|\int_0^t \left(\hat{f}_i(s) \sum_{k=1}^n \hat{\eta}_{ik} \hat{f}_k(s) - f_i(s) \sum_{k=1}^n \eta_{ik} f_k(s)\right) ds\right| \quad (46)$$
$$- \left(\sum_{i=1}^n u_i^2 F_i\right) \left|\int_0^t \left(\hat{f}_i(s) - f_i(s)\right) ds\right|.$$

By taking the sum on $i \in 1, 2, \ldots, n$ of relations (46), we find

$$\|\mathbf{f}(t) - \hat{\mathbf{f}}(t)\|_1 \geq$$
$$\geq \sum_{i=1}^n \left|\int_0^t \sum_{h,k=1}^n \eta_{hk} B_{hk}^i f_h(s) f_k(s) - \hat{\eta}_{hk} \hat{B}_{hk}^i \hat{f}_h(s) \hat{f}_k(s) \, ds\right|$$
$$- \sum_{i=1}^n \left|\int_0^t \left(\hat{f}_i(s) \sum_{k=1}^n \hat{\eta}_{ik} \hat{f}_k(s) - f_i(s) \sum_{k=1}^n \eta_{ik} f_k(s)\right) ds\right| \quad (47)$$
$$- \sum_{i=1}^n \left(\sum_{i=1}^n u_i^2 F_i\right) \left|\int_0^t \left(\hat{f}_i(s) - f_i(s)\right) ds\right|.$$

Now, observe that

$$\sum_{i=1}^n \left|\int_0^t \left(\hat{f}_i(s) \sum_{k=1}^n \hat{\eta}_{ik} \hat{f}_k(s) - f_i(s) \sum_{k=1}^n \eta_{ik} f_k(s)\right) ds(s)\right| \leq$$
$$\leq \int_0^t \sum_{i=1}^n \hat{f}_i(s) \sum_{k=1}^n \hat{\eta}_{ik} \hat{f}_k(s) \, ds + \int_0^t \sum_{i=1}^n f_i(s) \sum_{k=1}^n \eta_{ik} f_k(s) \, ds \quad (48)$$
$$\leq (\eta + \hat{\eta})t,$$

and

$$\sum_{i=1}^n \left(\sum_{i=1}^n u_i^2 F_i\right) \left|\int_0^t \left(\hat{f}_i(s) - f_i(s)\right) dt\right| \leq \left(\sum_{i=1}^n u_i^2 F_i\right) \int_0^t \|\mathbf{f}(t) - \hat{\mathbf{f}}(t)\|_1. \quad (49)$$

Using these two last relations, inequality (47) may be rewritten in the form

$$\|\mathbf{f}(t) - \hat{\mathbf{f}}(t)\|_1 \geq \Gamma(t) - (\eta + \hat{\eta})t - \left(\sum_{i=1}^n u_i^2 F_i\right) \int_0^t \|\mathbf{f}(t) - \hat{\mathbf{f}}(t)\|_1,$$

so that
$$\|\mathbf{f}(t) - \hat{\mathbf{f}}(t)\|_1 \geq \frac{\Gamma(t) - (\eta + \hat{\eta})t}{1 + (\sum_{i=1}^n u_i^2 F_i)t},$$

and inequality (45) is achieved. □

Remark 12. *By using inequality (45), we see that, if $\Gamma(t) > (\eta + \hat{\eta})t$, then an instability appears in the framework (4). In addition, it is important to note that this is a condition involving the parameters of the system, i.e., interaction rate and transition probability density.*

4.3. Numerical Simulations

This section aims to present some numerical simulations in the framework described by (22). Specifically, the parameters of the system, i.e., interaction rate and transition probability, acquire different values. All the simulations that follow have been performed by using the routine Ode45 of MatLab.

Let $n = 3$, which is three functional subsystems that are taken into account. The initial data are the vector:

$$\mathbf{f}^0 = (3/8, 1/2, 1/8).$$

The interaction rate parameter has the following form:

$$\eta_{hk} = \exp(-\eta\,|h - k|), \qquad \eta > 0.$$

In the first set of simulations, the transition probability is constant, whether the interaction rate varies. Specifically, three cases are considered:

- $\eta = 1$;
- $\eta = 3$;
- $\eta = 6$.

The solution $\mathbf{f}(t) = (f_1(t), f_2(t), f_3(t))$ is of course different from value to value of the interaction rate. Specifically, Figure 1 shows the three plots of the solution $\mathbf{f}(t)$ respectively corresponding to the three different values of η listed above. In addition, Figure 2 offers a comparison between the solutions corresponding to the values $\eta = 3$ and $\eta = 3, 2$, respectively.

In the second set of simulations, the interaction rate is constant, while the transition probability density acquires different real values. Precisely, Figure 3 shows the three plots of the solution $\mathbf{f}(t)$ respectively corresponding to three different values of B_{hk}^i. The considered cases are:

-
$$B_{hk}^i = c_{ihk}\,\frac{1}{s}\,g(|h - i|), \quad i, h, k \in \{1, 2, \ldots, n\},$$

where g is a non-increasing function of $|h - i|$ and s, and the parameters c_{ihk}, for $i, h, k \in \{1, 2, \ldots, n\}$, are positive real numbers, depending on the particular system taken into account;
- $\hat{B}_{hk}^i$ that differs from B_{hk}^i only for $h = 3$ and $k = 2$;
- B_{hk}^i uniform.

Furthermore, Figure 4 shows the solutions corresponding to the values B_{hk}^i and $\hat{B}_{hk}^i$ in the same plot in order to compare their behaviors in time.

It is worth being stressed that the shape of solution strictly depends on the value of the parameters of the system (see Figures 1 and 3) as the results reported in Sections 4.1 and 4.2 show for both the continuous and the discrete framework. Moreover, bearing the Figures 2 and 4 in mind, a small perturbation of a parameter may determine that the related solution has the same shape, but they are not so "close" to each other.

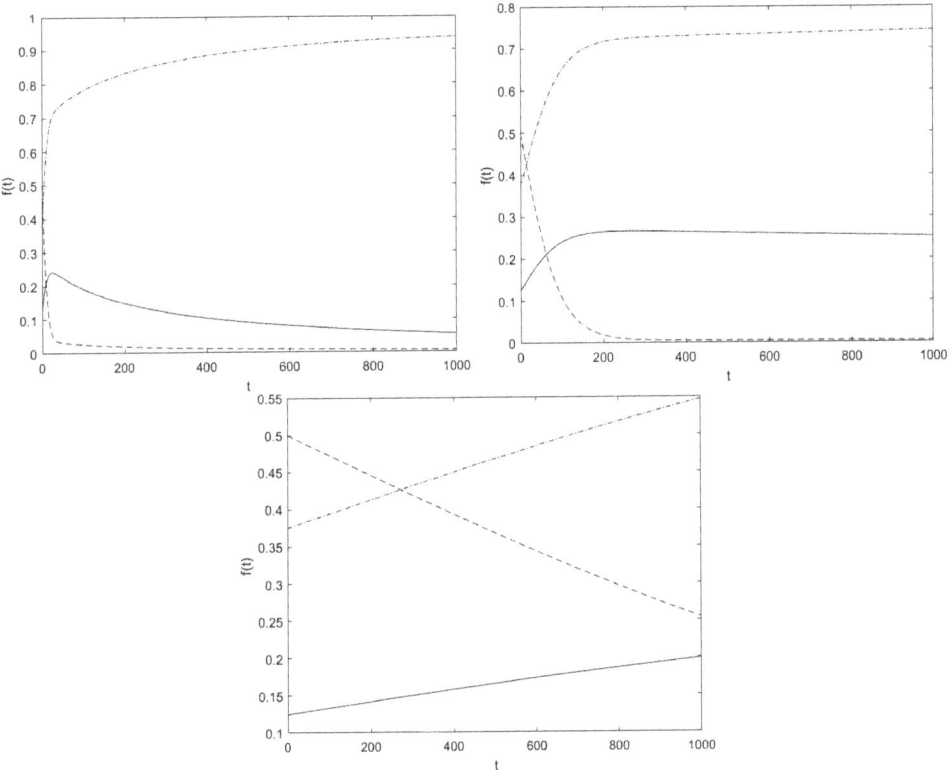

Figure 1. From top left to bottom $\eta = 1, \eta = 3, \gamma = 6$. f_1 dot-dashed, f_2 dashed, f_3 full.

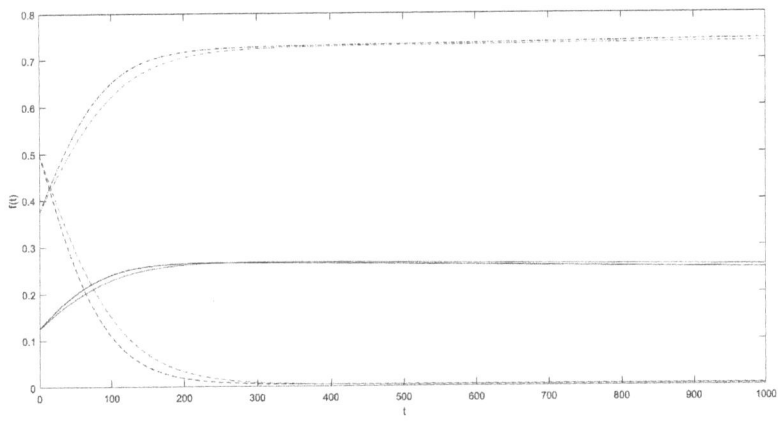

Figure 2. In black the solution for $\eta = 3$, in red the solution for $\eta = 3, 2$.

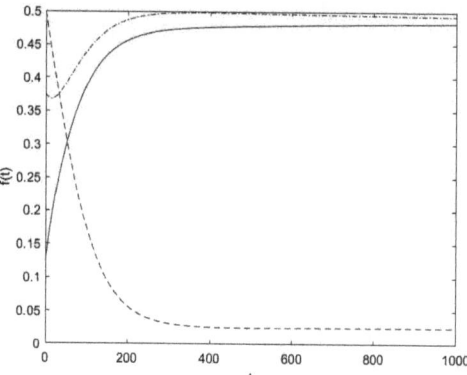

Figure 3. From top left to bottom B_{hk}^i, $\hat{B}_{hk}^i$ perturbed and uniform distribution. f_1 dot-dashed, f_2 dashed, f_3 full.

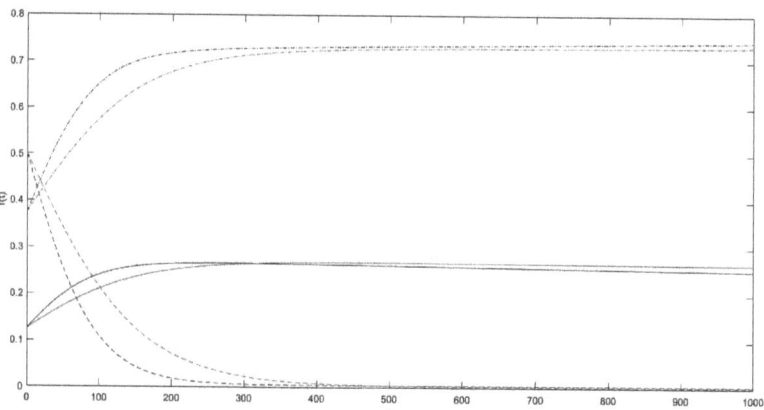

Figure 4. In black the solution for B_{hk}^i, in red the solution for $\hat{B}_{hk}^i$.

5. Conclusions and Research Perspectives

The results proved in Sections 3 and 4 are, in some sense, complementary. The former shows that the difference between two arbitrary solutions of Equation (1), corresponding to two different sets of parameters, i.e., different systems of interaction rates and different probability distributions on the results of interactions, varies with a suitable measure of the difference between the systems of parameters, and the variation is continuous; the latter shows that, if the difference between the interaction rates is sufficiently large in a suitable sense, then the corresponding solutions—though starting from the same initial value—will move at once away from each other and will stay apart at any future time, i.e., their distance has a constant positive lower bound. In addition, the numerical simulations plotted in the figures shown in Section 4.3 seem to give a good visual counterpart of this result.

As far as we are aware, no similar results have been previously reported in the literature about KTAP, perhaps because the study of the dependence of solutions on the perturbations of parameters (interaction rates and transition probabilities) seems to be too difficult in relation to its relevance for applications, so that tackling it is considered as an almost *useless* effort. However, on the contrary, results like the ones found and reported in the present work are probably intended for becoming of the greatest relevance for applications, with special concern with social and economic sciences. In this connection, we can observe that economic interactions in any human society are ruled by the government: in a country in which some commercial transactions are allowed, they will produce exchanges of goods and money, with a subsequent modification of the distribution of wealth; but, in another country, where the same transactions are forbidden, the interaction rate referred to them is zero, and we must expect that the distribution of wealth could not be modified by these transactions, regardless of the values of transition probabilities that are allowed to be the same in both cases. This remark has worked as a suggestion of a search for instability results of the kind of Theorems 3 and 4. Of course, these Theorems cannot be considered as more than a first step on the way towards much more general instability results, for at least the good reason that they only refer to the very special case in which the interaction rates are constant with respect to the couples of states. Accordingly, this research about instability requires to be deepened along at least three lines, which will be the object of future work.

As laid out in the Introduction, KTAP is *not* a theory or simply a model, but a whole scheme of models to describe and—above all—*predict* the behavior of complex systems. In addition, as a matter of fact, our prime scope is its application to human collectivities, in order to suggest some ways to solve the problems raised by many and diffused bad mental habits that control not only human behaviors but also the criteria according to which legislators decide the (inter-)actions that can be allowed and the (inter-)actions that must be forbidden. Laws can modify *both* interaction rates and transition probabilities, so a complete and detailed view of the behaviors they produce could avoid that past mistakes from being repeated in the future.

In this line of thought, first of all, one should find possible conditions of instability in the quite general case in which interaction rates are *arbitrary* functions defined on $D_u \times D_u$: from a purely mathematical viewpoint, this will require in turn a suitable definition of their distance.

Next, one has to find possible instability conditions on the transition probabilities, also in the case in which the interaction rates are left unchanged.

Finally, one has to study the reciprocal influence between the perturbation of interaction rates and the perturbation of transition probabilities. In this connection, it should be noted that Theorem 3 already furnishes a first hint in this direction.

These three lines of search give good and hopefully—on a pragmatic ground—useful perspectives for the development of the study started and reported in the present paper.

Author Contributions: Conceptualization, B.C. and M.M.; methodology, B.C. and M.M.; software, B.C. and M.M.; validation, B.C. and M.M.; formal analysis, B.C. and M.M.; writing—original draft preparation, M.M.; writing—review and editing, M.M. All authors have read and agreed to the published version of the manuscript.

Funding: This research received no external funding.

Institutional Review Board Statement: Not applicable.

Informed Consent Statement: Not applicable.

Data Availability Statement: Not applicable.

Acknowledgments: Marco Menale is supported by the Research Project "ANDROIDS" (AutoNomous DiscoveRy Of depressIve Disorder Signs) for VALERE (VAnviteLli pEr la RicErca), developed by Università degli Studi della Campania "L. Vanvitelli".

Conflicts of Interest: The authors declare no conflict of interest.

References

1. Bar-Yam, Y. *Dynamics of Complex Systems*; CRC Press: Boca Raton, FL, USA, 2019.
2. Bianca, C. Modeling complex systems by functional subsystems representation and thermostatted-KTAP methods. *Appl. Math. Inf. Sci.* **2021**, *6*, 495–499.
3. Cilliers, P. Complexity and Postmodernism: Understanding Complex Systems. *S. Afr. J. Philos.* **1999**, *18*, 275–278. [CrossRef]
4. Morriss, G.P.; Dettmann, C.P. Thermostats: Analysis and application. *Chaos Interdiscip. J. Nonlinear Sci.* **1998**, *8*, 321–336. [CrossRef] [PubMed]
5. Bobylev, A.V.; Cercignani, C. Exact eternal solutions of the Boltzmann equation. *J. Stat. Phys.* **2002**, *106*, 1019–1038. [CrossRef]
6. Cercignani, C. The boltzmann equation. In *The Boltzmann Equation and Its Applications*; Springer: New York, NY, USA, 1988; pp. 40–103.
7. Cercignani, C.; Gabetta, E. *Transport Phenomena and Kinetic Theory: Applications to Gases, Semiconductors, Photons, and Biological Systems*; Springer Science & Business Media: Basel, Switzerland, 2007.
8. Cercignani, C.; Illner, R.; Pulvirenti, M. *The Mathematical Theory of Dilute Gases*; Springer Science & Business Media: Basel, Switzerland, 2013; Volume 106.
9. Bonabeau, E.; Theraulaz, G.; Deneubourg, J.L. Mathematical model of self-organizing hierarchies in animal societies. *Bull. Math. Biol.* **1996**, *58*, 661–717. [CrossRef]
10. Thieme, H.R. *Mathematics in Population Biology*; Princeton University Press: Princeton, NJ, USA, 2018; Volume 12.
11. Giorno, V.; Román-Román, P.; Spina, S.; Torres-Ruiz, F. Estimating a non-homogeneous Gompertz process with jumps as model of tumor dynamics. *Comput. Stat. Data Anal.* **2017**, *107*, 18–31. [CrossRef]
12. Masurel, L.; Bianca, C.; Lemarchand, A. On the learning control effects in the cancer-immune system competition. *Phys. A Stat. Mech. Its Appl.* **2018**, *506*, 462–475. [CrossRef]
13. Pappalardo, F.; Palladini, A.; Pennisi, M.; Castiglione, F.; Motta, S. Mathematical and computational models in tumor immunology. *Math. Model. Nat. Phenom.* **2012**, *7*, 186–203. [CrossRef]
14. Poleszczuk, J.; Macklin, P.; Enderling, H. Agent-based modeling of cancer stem cell driven solid tumor growth. In *Stem Cell Heterogeneity*; Human Press: New York, NY, USA, 2016; pp. 335–346.
15. Spina, S.; Giorno, V.; Román-Román, P.; Torres-Ruiz, F. A stochastic model of cancer growth subject to an intermittent treatment with combined effects: Reduction in tumor size and rise in growth rate. *Bull. Math. Biol.* **2014**, *76*, 2711–2736. [CrossRef]
16. Bianca, C.; Kombargi, A. On the inverse problem for thermostatted kinetic models with application to the financial market. *Appl. Math. Inf. Sci.* **2017**, *11*, 1463–1471. [CrossRef]
17. Bisi, M.; Spiga, G.; Toscani, G. Kinetic models of conservative economies with wealth redistribution. *Commun. Math. Sci.* **2009**, *7*, 901–916. [CrossRef]
18. Carbonaro, B.; Serra, N. Towards mathematical models in psychology: A stochastic description of human feelings. *Math. Model. Methods Appl. Sci.* **2002**, *12*, 1453–1490. [CrossRef]
19. Carbonaro, B.; Giordano, C. A second step towards a stochastic mathematical description of human feelings. *Math. Comput. Model.* **2005**, *41*, 587–614. [CrossRef]
20. Bronson, R.; Jacobson, C. Modeling the dynamics of social systems. *Comput. Math. Appl.* **1990**, *19*, 35–42. [CrossRef]
21. Buonomo B.; Della Marca, R. Modelling information-dependent social behaviors in response to lockdowns: The case of COVID-19 epidemic in Italy. *medRxiv* **2020**. [CrossRef]
22. Fryer, R.G., Jr.; Roland, G. A model of social interactions and endogenous poverty traps. *Ration. Soc.* **2007**, *19*, 335–366. [CrossRef]
23. Kacperski, K. Opinion formation model with strong leader and external impact: A mean field approach. *Phys. A Stat. Mech. Its Appl.* **1999**, *269*, 511–526. [CrossRef]
24. Bianca, C.; Carbonaro, B.; Menale, M. On the Cauchy Problem of Vectorial Thermostatted Kinetic Frameworks. *Symmetry* **2020**, *12*, 517. [CrossRef]

25. Bianca, C. An existence and uniqueness theorem to the Cauchy problem for thermostatted-KTAP models. *Int. J. Math. Anal.* **2012**, *6*, 813–824.
26. Bianca, C.; Mogno, C. Qualitative analysis of a discrete thermostatted kinetic framework modeling complex adaptive systems. *Commun. Nonlinear Sci. Numer. Simul.* **2018**, *54*, 221–232. [CrossRef]
27. Bianca, C.; Menale, M. A Convergence Theorem for the Nonequilibrium States in the Discrete Thermostatted Kinetic Theory. *Mathematics* **2019**, *7*, 673. [CrossRef]
28. Bianca, C.; Menale, M. Existence and uniqueness of nonequilibrium stationary solutions in discrete thermostatted models. *Commun. Nonlinear Sci. Numer. Simul.* **2019**, *73*, 25–34. [CrossRef]
29. Carbonaro, B.; Menale, M. Dependence on the Initial Data for the Continuous Thermostatted Framework. *Mathematics* **2019**, *7*, 612. [CrossRef]
30. Carbonaro, B.; Menale, M. The mathematical analysis towards the dependence on the initial data for a discrete thermostatted kinetic framework for biological systems composed of interacting entities. *AIMS Biophys.* **2020**, *7*, 204.
31. Bianca, C. Existence of stationary solutions in kinetic models with Gaussian thermostats. *Math. Methods Appl. Sci.* **2013**, *2013*, 1768–1775. [CrossRef]
32. Bianca, C.; Menale, M. On the convergence towards nonequilibrium stationary states in thermostatted kinetic models. *Math. Methods Appl. Sci.* **2019**, *42*, 6624–6634. [CrossRef]
33. Bianca, C. Thermostated kinetic equations as models for complex systems in physics and life sciences. *Phys. Life Rev.* **2012**, *9*, 359–399. [CrossRef] [PubMed]
34. Bianca, C.; Mogno, C. Modelling pedestrian dynamics into a metro station by thermostatted kinetic theory methods. *Math. Comput. Model. Dyn. Syst.* **2018**, *24*, 207–235. [CrossRef]
35. Bianca, C.; Menale, M. Mathematical Analysis of a Thermostatted Equation with a Discrete Real Activity Variable. *Mathematics* **2020**, *8*, 57. [CrossRef]
36. Walter, W. *Differential and Integral Inequalities*; Springer-Verlag: Berlin/Heidelberg, Germany, 2012; Volume 55.

Article

Approximations of an Equilibrium Problem without Prior Knowledge of Lipschitz Constants in Hilbert Spaces with Applications

Chainarong Khanpanuk [1], Nuttapol Pakkaranang [2], Nopparat Wairojjana [3],* and Nattawut Pholasa [4],*

[1] Department of Mathematics, Faculty of Science and Technology, Phetchabun Rajabhat University, Phetchabun 67000, Thailand; iprove2000ck@gmail.com
[2] Department of Mathematics, Faculty of Science, King Mongkut's University of Technology Thonburi (KMUTT), Bangkok 10140, Thailand; nuttapol.pak@mail.kmutt.ac.th
[3] Applied Mathematics Program, Faculty of Science and Technology, Valaya Alongkorn Rajabhat University under the Royal Patronage (VRU), 1 Moo 20 Phaholyothin Road, Klong Neung, Klong Luang, Pathumthani 13180, Thailand
[4] School of Science, University of Phayao, Phayao 56000, Thailand
* Correspondence: nopparat@vru.ac.th (N.W.); nattawut_math@hotmail.com (N.P.)

Abstract: The objective of this paper is to introduce an iterative method with the addition of an inertial term to solve equilibrium problems in a real Hilbert space. The proposed iterative scheme is based on the Mann-type iterative scheme and the extragradient method. By imposing certain mild conditions on a bifunction, the corresponding theorem of strong convergence in real Hilbert space is well-established. The proposed method has the advantage of requiring no knowledge of Lipschitz-type constants. The applications of our results to solve particular classes of equilibrium problems is presented. Numerical results are established to validate the proposed method's efficiency and to compare it to other methods in the literature.

Keywords: equilibrium problem; pseudomonotone bifunction; Lipschitz-type conditions; strong convergence theorems; variational inequality problems; fixed-point problem

MSC: 47H05; 47H10; 65Y05; 65K15

1. Introduction

Suppose that $\mathcal{C}$ is a nonempty closed and convex subset of a real Hilbert space $\mathcal{H}$. The inner product and induced norm are denoted by $\langle \cdot, \cdot \rangle$ and $\| \cdot \|$, respectively. Let $f : \mathcal{H} \times \mathcal{H} \to \mathcal{R}$ be a bifunction and $f(y, y) = 0$, for all $y \in \mathcal{C}$. The equilibrium problem (EP) [1,2] for a bifunction f on $\mathcal{C}$ is defined in the following way:

$$\text{Find } u^* \in \mathcal{C} \text{ such that } f(u^*, y) \geq 0, \ \forall y \in \mathcal{C}. \tag{EP}$$

The equilibrium problem is a general mathematical problem in the sense that it unifies various mathematical problems, i.e., fixed-point problems, vector and scalar minimization problems, problems of variational inequality, complementarity problems, Nash equilibrium problems in noncooperative games, saddle point problems, and inverse optimization problems [2–4]. The equilibrium problem is also known as the well-known Ky Fan inequality due to the result [1]. Many authors established and generalized several results on the existence and nature of the solution of the equilibrium problems (see for more detail [1,4,5]). Due to the importance of this problem (EP) in both pure and applied sciences, many researchers studied it in recent years [6–17] and other in [18–22].

Tran et al. in [23] introduced iterative sequence $\{u_n\}$ in the following way:

$$\begin{cases} u_0 \in \mathcal{C}, \\ y_n = \arg\min_{z \in \mathcal{C}}\{\chi f(u_n,z) + \frac{1}{2}\|u_n - z\|^2\}, \\ u_{n+1} = \arg\min_{z \in \mathcal{C}}\{\chi f(y_n,z) + \frac{1}{2}\|u_n - z\|^2\}, \end{cases} \quad (1)$$

where $0 < \chi < \min\{\frac{1}{2c_1}, \frac{1}{2c_2}\}$. This method is also known as the extragradient method in [23] due to the previous contribution of Korpelevich [24] to solve the saddle-point problems. The iterative sequence generated by the above-mentioned method is weakly convergent to the solution with prior knowledge of Lipschitz-type constants. These Lipschitz-like constants are often not known or are difficult to compute. Recently, Hieu et al. [25] introduced an extension of the method (1) for solving the equilibrium problem. Let us consider that $[p]_+ := \max\{p,0\}$ and choose $u_0 \in \mathcal{C}$, $\mu \in (0,1)$ with $\chi_0 > 0$ such that

$$\begin{cases} y_n = \arg\min_{z \in \mathcal{C}}\{\chi_n f(u_n,z) + \frac{1}{2}\|u_n - z\|^2\}, \\ u_{n+1} = \arg\min_{z \in \mathcal{C}}\{\chi_n f(y_n,z) + \frac{1}{2}\|u_n - z\|^2\}, \end{cases} \quad (2)$$

where $\{\chi_n\}$ is updated in the following manner:

$$\chi_{n+1} = \min\left\{\chi_n, \frac{\mu(\|u_n - y_n\|^2 + \|u_{n+1} - y_n\|^2)}{2[f(u_n,u_{n+1}) - f(u_n,y_n) - f(y_n,u_{n+1})]_+}\right\}.$$

Inertial-like methods are well-known two-step iterative methods in which the next iteration is derived from the previous two iterations (see [26,27] for more details). To speed up the iterative sequence convergence rate, an inertial extrapolation term is used. Numerical examples show that inertial effects improve numerical performance in terms of execution time and the expected number of iterations. Recently, many existing methods were established for the case of equilibrium problems (see [28–31] for more details).

In this paper, inspired by the methods in [23,25,26,32], we introduce a general inertial Mann-type subgradient extragradient method to evaluate the approximate solution of the equilibrium problems involving pseudomonotone bifunction. A strong convergence result corresponding to the proposed algorithm is well-established by assuming certain mild conditions. Some of the applications for our main results are considered to solve the fixed-point problems. Lastly, computational results show that the new method is more successful than existing ones [23,33,34].

2. Preliminaries

A *metric projection* $P_\mathcal{C}(u)$ of $u \in \mathcal{H}$ onto a closed and convex subset $\mathcal{C}$ of $\mathcal{H}$ is defined by

$$P_\mathcal{C}(u) = \arg\min_{y \in \mathcal{C}}\{\|y - u\|\}.$$

In this study, the equilibrium problem under the following conditions:

(c1). A bifunction $f : \mathcal{H} \times \mathcal{H} \to \mathcal{R}$ is said to be *pseudomonotone* [3,35] on $\mathcal{C}$ if

$$f(y_1, y_2) \geq 0 \implies f(y_2, y_1) \leq 0, \quad \forall y_1, y_2 \in \mathcal{C}.$$

(c2). A bifunction $f : \mathcal{H} \times \mathcal{H} \to \mathcal{R}$ is said to be Lipschitz-type continuous [36] on $\mathcal{C}$ if there exist constants $c_1, c_2 > 0$ such that

$$f(y_1, y_3) \leq f(y_1, y_2) + f(y_2, y_3) + c_1\|y_1 - y_2\|^2 + c_2\|y_2 - y_3\|^2, \quad \forall y_1, y_2, y_3 \in \mathcal{C}.$$

(c3). $\limsup_{n \to \infty} f(y_n, y) \leq f(q^*, y)$ for all $y \in \mathcal{C}$ and $\{y_n\} \subset \mathcal{C}$ satisfy $y_n \rightharpoonup q^*$.

(c4). $f(u,\cdot)$ is convex and subdifferentiable on $\mathcal{H}$ for each $u \in \mathcal{H}$.

A *cone on* $\mathcal{C}$ at $u \in \mathcal{C}$ is defined by

$$N_{\mathcal{C}}(u) = \{t \in \mathcal{H} : \langle t, y - u \rangle \leq 0, \forall y \in \mathcal{C}\}.$$

Let a convex function $\daleth: \mathcal{C} \to \mathcal{R}$ and *subdifferential of* $\daleth$ at $u \in \mathcal{C}$ is defined by

$$\partial \daleth(u) = \{t \in \mathcal{H} : \daleth(y) - \daleth(u) \geq \langle t, y - u \rangle, \forall y \in \mathcal{C}\}.$$

Lemma 1. [37] *Let* $\daleth: \mathcal{C} \to \mathcal{R}$ *be a subdifferentiable, lower semicontinuous, and convex function on* $\mathcal{C}$. *Then,* $u \in \mathcal{C}$ *is said to be a minimizer of* $\daleth$ *if and only if* $0 \in \partial \daleth(u) + N_{\mathcal{C}}(u)$, *where* $\partial \daleth(u)$ *stands for the subdifferential of* $\daleth$ *at* $u \in \mathcal{C}$ *and* $N_{\mathcal{C}}(u)$ *is a normal cone of* $\mathcal{C}$ *on* u.

Lemma 2. [38] *Assume that* $P_{\mathcal{C}} : \mathcal{H} \to \mathcal{C}$ *be a metric projection such that*
(i) $\|y_1 - P_{\mathcal{C}}(y_2)\|^2 + \|P_{\mathcal{C}}(y_2) - y_2\|^2 \leq \|y_2 - y_1\|^2$, $y_1 \in \mathcal{C}, y_2 \in \mathcal{H}$.
(ii) $y_3 = P_{\mathcal{C}}(y_1)$ *if and only if* $\langle y_1 - y_3, y_2 - y_3 \rangle \leq 0$, $\forall y_2 \in \mathcal{C}$.
(iii) $\|y_1 - P_{\mathcal{C}}(y_1)\| \leq \|y_1 - y_2\|$, $y_2 \in \mathcal{C}, y_1 \in \mathcal{H}$.

Lemma 3. [39] *Assume that* $\{\daleth_n\} \subset (0, +\infty)$ *is a sequence satisfying, i.e.,* $\daleth_{n+1} \leq (1 - \upsilon_n)\daleth_n + \upsilon_n \eth_n$, *for all* $n \in \mathbb{N}$. *Moreover, let* $\{\upsilon_n\} \subset (0,1)$ *and* $\{\eth_n\} \subset \mathcal{R}$ *be two sequences such that* $\lim_{n \to \infty} \upsilon_n = 0$, $\sum_{n=1}^{\infty} \upsilon_n = +\infty$ *and* $\limsup_{n \to \infty} \eth_n \leq 0$. *Then,* $\lim_{n \to \infty} \daleth_n = 0$.

Lemma 4. [40] *Assume that* $\{\daleth_n\}$ *be a sequence of real numbers such that there exists a subsequence* $\{n_i\}$ *of* $\{n\}$ *such that* $\daleth_{n_i} < \daleth_{n_i+1}$ *for all* $i \in \mathbb{N}$. *Then, there is a nondecreasing sequence* $m_k \subset \mathbb{N}$ *such that* $m_k \to \infty$ *as* $k \to \infty$, *and the following conditions are fullfiled by all (sufficiently large) numbers* $k \in \mathbb{N}$:

$$\daleth_{m_k} \leq \daleth_{m_k+1} \text{ and } \daleth_k \leq \daleth_{m_k+1}.$$

In fact, $m_k = \max\{j \leq k : \daleth_j \leq \daleth_{j+1}\}$.

Lemma 5. [41] *For all* $y_1, y_2 \in \mathcal{H}$ *and* $\eth \in \mathcal{R}$, *the following inequalities hold.*
(i) $\|\eth y_1 + (1 - \eth) y_2\|^2 = \eth \|y_1\|^2 + (1 - \eth) \|y_2\|^2 - \eth(1 - \eth) \|y_1 - y_2\|^2$.
(ii) $\|y_1 + y_2\|^2 \leq \|y_1\|^2 + 2\langle y_2, y_1 + y_2 \rangle$.

3. Main Results

We propose an iterative method for solving equilibrium problems involving a pseudomonotone that is based on Tran et al. in [23], and the Mann-type method [32] and the inertial scheme [26]. For clarity in the presentation, we use notation $[t]_+ = \max\{0, t\}$ and follow conventions $\frac{0}{0} = +\infty$ and $\frac{a}{0} = +\infty$ $(a \neq 0)$.

Lemma 6. *A sequence* $\{\chi_n\}$ *generated by* (5) *is monotonically decreasing, converges to* $\chi > 0$, *and has a lower bound* $\min\left\{\frac{\mu}{2\max\{c_1, c_2\}}, \chi_0\right\}$.

Proof. Assume that $f(t_n, z_n) - f(t_n, y_n) - f(y_n, z_n) > 0$ such that

$$\frac{\mu(\|t_n - y_n\|^2 + \|z_n - y_n\|^2)}{2[f(t_n, z_n) - f(t_n, y_n) - f(y_n, z_n)]} \geq \frac{\mu(\|t_n - y_n\|^2 + \|z_n - y_n\|^2)}{2[c_1 \|t_n - y_n\|^2 + c_2 \|z_n - y_n\|^2]}$$

$$\geq \frac{\mu}{2\max\{c_1, c_2\}}. \qquad (3)$$

This implies that $\{\chi_n\}$ has a lower bound $\min\left\{\frac{\mu}{2\max\{c_1, c_2\}}, \chi_0\right\}$. Moreover, there exists a fixed real number $\chi > 0$, such that $\lim_{n \to \infty} \chi_n = \chi$. □

Lemma 7. *Suppose that Conditions (c1)–(c4) are satisfied. Then, sequence $\{u_n\}$ generated by the Algorithm 1 is a bounded sequence.*

Algorithm 1 (Explicit Accelerated Strong Convergence Iterative Scheme)

STEP 0: Choose $u_{-1}, u_0 \in \mathcal{C}, \phi > 0, \chi_0 > 0, \{\rho_n\} \subset (a,b) \subset (0, 1-\varrho_n)$ and $\{\varrho_n\} \subset (0,1)$ satisfies the following conditions:

$$\lim_{n \to \infty} \varrho_n = 0 \text{ and } \sum_{n=1}^{+\infty} \varrho_n = +\infty.$$

STEP 1: Compute $t_n = u_n + \phi_n(u_n - u_{n-1})$ and choose ϕ_n such that

$$0 \leq \phi_n \leq \hat{\phi}_n \text{ and } \hat{\phi}_n = \begin{cases} \min\left\{\frac{\phi}{2}, \frac{\varsigma_n}{\|u_n - u_{n-1}\|}\right\} & \text{if } u_n \neq u_{n-1}, \\ \frac{\phi}{2} & \text{otherwise,} \end{cases} \quad (4)$$

where $\varsigma_n = \circ(\varrho_n)$, i.e., $\lim_{n \to \infty} \frac{\varsigma_n}{\varrho_n} = 0$.

STEP 2: Compute

$$y_n = \arg\min_{y \in \mathcal{C}}\{\chi_n f(t_n, y) + \frac{1}{2}\|t_n - y\|^2\}.$$

If $t_n = y_n$, then STOP the sequence. Else, go to STEP 3.

STEP 3: Construct a half-space $\mathcal{H}_n = \{z \in \mathcal{H} : \langle t_n - \chi_n \omega_n - y_n, z - y_n \rangle \leq 0\}$ where $\omega_n \in \partial_2 f(t_n, y_n)$ and compute

$$z_n = \arg\min_{y \in \mathcal{H}_n}\{\chi_n f(y_n, y) + \frac{1}{2}\|t_n - y\|^2\}.$$

STEP 4: Compute $u_{n+1} = (1 - \rho_n - \varrho_n)u_n + \rho_n z_n$.

STEP 5: Compute

$$\chi_{n+1} = \min\left\{\chi_n, \frac{\mu\|t_n - y_n\|^2 + \mu\|z_n - y_n\|^2}{2[f(t_n, z_n) - f(t_n, y_n) - f(y_n, z_n)]_+}\right\}. \quad (5)$$

Set $n := n + 1$ and go back to **Step 1**.

Proof. From the value of z_n, we have

$$0 \in \partial_2\left\{\chi_n f(y_n, y) + \frac{1}{2}\|t_n - y\|^2\right\}(z_n) + N_{\mathcal{H}_n}(z_n).$$

For $\omega \in \partial f(y_n, z_n)$ there exists $\overline{\omega} \in N_{\mathcal{H}_n}(z_n)$ such that

$$\chi_n \omega + z_n - t_n + \overline{\omega} = 0.$$

This implies that

$$\langle t_n - z_n, y - z_n \rangle = \chi_n \langle \omega, y - z_n \rangle + \langle \overline{\omega}, y - z_n \rangle, \ \forall y \in \mathcal{H}_n.$$

Due to $\overline{\omega} \in N_{\mathcal{H}_n}(z_n)$, it implies that $\langle \overline{\omega}, y - z_n \rangle \leq 0$ for each $y \in \mathcal{H}_n$. Thus, we have

$$\langle t_n - z_n, y - z_n \rangle \leq \chi_n \langle \omega, y - z_n \rangle, \ \forall y \in \mathcal{H}_n. \quad (6)$$

Moreover, $\omega \in \partial f(y_n, z_n)$ and owing to the subdifferential, we have

$$f(y_n, y) - f(y_n, z_n) \geq \langle \omega, y - z_n \rangle, \ \forall y \in \mathcal{H}. \quad (7)$$

From Expressions (6) and (7), we obtain

$$\chi_n f(y_n, y) - \chi_n f(y_n, z_n) \geq \langle t_n - z_n, y - z_n \rangle, \ \forall y \in \mathcal{H}_n. \tag{8}$$

Due to the definition of $\mathcal{H}_n$, we have

$$\chi_n \langle \omega_n, z_n - y_n \rangle \geq \langle t_n - y_n, z_n - y_n \rangle. \tag{9}$$

Now, using $\omega_n \in \partial f(t_n, y_n)$, we obtain

$$f(t_n, y) - f(t_n, y_n) \geq \langle \omega_n, y - y_n \rangle, \ \forall y \in \mathcal{H}.$$

By letting $y = z_n$, we obtain

$$f(t_n, z_n) - f(t_n, y_n) \geq \langle \omega_n, z_n - y_n \rangle, \ \forall y \in \mathcal{H}. \tag{10}$$

Combining Expressions (9) and (10), we obtain

$$\chi_n \{f(t_n, z_n) - f(t_n, y_n)\} \geq \langle t_n - y_n, z_n - y_n \rangle. \tag{11}$$

By substituting $y = u^*$ in Expression (8), we obtain

$$\chi_n f(y_n, u^*) - \chi_n f(y_n, z_n) \geq \langle t_n - z_n, u^* - z_n \rangle. \tag{12}$$

Since $u^* \in Ep(f, \mathcal{C})$, we have $f(u^*, y_n) \geq 0$. From the pseudomonotonicity of bifunction f, we achieve $f(y_n, u^*) \leq 0$. It follows from Expression (12) that

$$\langle t_n - z_n, z_n - u^* \rangle \geq \chi_n f(y_n, z_n). \tag{13}$$

From the description of χ_{n+1}, we obtain

$$f(t_n, z_n) - f(t_n, y_n) - f(y_n, z_n) \leq \frac{\mu \|t_n - y_n\|^2 + \mu \|z_n - y_n\|^2}{2\chi_{n+1}} \tag{14}$$

From (13) and (14), we obtain

$$\langle t_n - z_n, z_n - u^* \rangle \geq \chi_n \{f(t_n, z_n) - f(t_n, y_n)\}$$
$$- \frac{\mu \chi_n}{2\chi_{n+1}} \|t_n - y_n\|^2 - \frac{\mu \chi_n}{2\chi_{n+1}} \|z_n - y_n\|^2. \tag{15}$$

Combining Expressions (11) and (15), we have

$$\langle t_n - z_n, z_n - u^* \rangle \geq \langle t_n - y_n, z_n - y_n \rangle$$
$$- \frac{\mu \chi_n}{2\chi_{n+1}} \|t_n - y_n\|^2 - \frac{\mu \chi_n}{2\chi_{n+1}} \|z_n - y_n\|^2. \tag{16}$$

We have the given formula in place:

$$-2\langle t_n - z_n, z_n - u^* \rangle = -\|t_n - u^*\|^2 + \|z_n - t_n\|^2 + \|z_n - u^*\|^2. \tag{17}$$

$$2\langle y_n - t_n, y_n - z_n \rangle = \|t_n - y_n\|^2 + \|z_n - y_n\|^2 - \|t_n - z_n\|^2. \tag{18}$$

Combining (16)–(18), we obtain

$$\|z_n - u^*\|^2 \leq \|t_n - u^*\|^2 - \left(1 - \frac{\mu \chi_n}{\chi_{n+1}}\right) \|t_n - y_n\|^2 - \left(1 - \frac{\mu \chi_n}{\chi_{n+1}}\right) \|z_n - y_n\|^2. \tag{19}$$

Since $\chi_n \to \chi$, then there is number $\Im \in (0, 1-\mu)$ that

$$\lim_{n\to\infty}\left(1 - \frac{\mu\chi_n}{\chi_{n+1}}\right) = 1 - \mu > \Im > 0.$$

Thus, there exists a finite number $n_1 \in \mathbb{N}$, such that

$$\left(1 - \frac{\mu\chi_n}{\chi_{n+1}}\right) > \Im > 0, \; \forall n \geq n_1. \tag{20}$$

From Expression (19), we obtain

$$\|u_{n+1} - u^*\|^2 \leq \|t_n - u^*\|^2, \; \forall n \geq n_1. \tag{21}$$

From Expression (4), we have $\phi_n \|u_n - u_{n-1}\| \leq \varsigma_n$, for all $n \in \mathbb{N}$ and $\lim_{n\to\infty}\left(\frac{\varsigma_n}{\varrho_n}\right) = 0$ implies that

$$\lim_{n\to\infty} \frac{\phi_n}{\varrho_n}\|u_n - u_{n-1}\| \leq \lim_{n\to\infty} \frac{\varsigma_n}{\varrho_n} = 0. \tag{22}$$

From Expression (21) and $\{t_n\}$, we have

$$\|z_n - u^*\| \leq \|t_n - u^*\| = \|u_n + \phi_n(u_n - u_{n-1}) - u^*\|$$
$$\leq \|u_n - u^*\| + \phi_n\|u_n - u_{n-1}\|$$
$$\leq \|u_n - u^*\| + \varrho_n \frac{\phi_n}{\varrho_n}\|u_n - u_{n-1}\|$$
$$\leq \|u_n - u^*\| + \varrho_n \beth_1, \tag{23}$$

where for some fixed $\beth_1 > 0$ and

$$\frac{\phi_n}{\varrho_n}\|u_n - u_{n-1}\| \leq \beth_1, \; \forall n \geq 1. \tag{24}$$

It is given that $u^* \in Ep(f, \mathcal{C})$ and by definition of $\{u_{n+1}\}$, we have

$$\|u_{n+1} - u^*\| = \|(1 - \rho_n - \varrho_n)u_n + \rho_n z_n - u^*\|$$
$$= \|(1 - \rho_n - \varrho_n)(u_n - u^*) + \rho_n(z_n - u^*) - \varrho_n u^*\|$$
$$\leq \|(1 - \rho_n - \varrho_n)(u_n - u^*) + \rho_n(z_n - u^*)\| + \varrho_n\|u^*\|. \tag{25}$$

Next, we compute

$$\|(1 - \rho_n - \varrho_n)(u_n - u^*) + \rho_n(z_n - u^*)\|^2$$
$$= (1 - \rho_n - \varrho_n)^2\|u_n - u^*\|^2 + \rho_n^2\|z_n - u^*\|^2 + 2\langle(1 - \rho_n - \varrho_n)(u_n - u^*), \rho_n(z_n - u^*)\rangle$$
$$\leq (1 - \rho_n - \varrho_n)^2\|u_n - u^*\|^2 + \rho_n^2\|z_n - u^*\|^2 + 2\rho_n(1 - \rho_n - \varrho_n)\|u_n - u^*\|\|z_n - u^*\|$$
$$\leq (1 - \rho_n - \varrho_n)^2\|u_n - u^*\|^2 + \rho_n^2\|z_n - u^*\|^2$$
$$+ \rho_n(1 - \rho_n - \varrho_n)\|u_n - u^*\|^2 + \rho_n(1 - \rho_n - \varrho_n)\|z_n - u^*\|^2$$
$$\leq (1 - \rho_n - \varrho_n)(1 - \varrho_n)\|u_n - u^*\|^2 + \rho_n(1 - \varrho_n)\|z_n - u^*\|^2 \tag{26}$$
$$\leq (1 - \rho_n - \varrho_n)(1 - \varrho_n)\|u_n - u^*\|^2 + \rho_n(1 - \varrho_n)(\|u_n - u^*\| + \varrho_n\beth_1)^2$$
$$\leq (1 - \varrho_n)^2\|u_n - u^*\|^2 + \varrho_n^2\beth_1^2 + 2\varrho_n\beth_1(1 - \varrho_n)\|u_n - u^*\|^2. \tag{27}$$

The above expression implies that

$$\|(1 - \rho_n - \varrho_n)(u_n - u^*) + \rho_n(z_n - u^*)\| \leq (1 - \varrho_n)\|u_n - u^*\| + \varrho_n\beth_1. \tag{28}$$

Combining Expressions (25) and (28), we obtain

$$\|u_{n+1} - u^*\| \leq (1 - \varrho_n)\|u_n - u^*\| + \varrho_n \beth_1 + \varrho_n\|u^*\|$$
$$\leq \max\left\{\|u_n - u^*\|, \beth_1 + \|u^*\|\right\}$$
$$\leq \vdots$$
$$\leq \max\left\{\|u_0 - u^*\|, \beth_1 + \|u^*\|\right\}. \quad (29)$$

Therefore, we conclude that $\{u_n\}$ is bounded sequence. □

Theorem 1. *Let $\{u_n\}$ be a sequence generated by Algorithm 1, and Conditions (c1)–(c4) are satisfied. Then, $\{u_n\}$ strongly converges to $u^* = P_{Ep(f,\mathcal{C})}(0)$.*

Proof. By using definition of $\{u_{n+1}\}$, we have

$$\|u_{n+1} - u^*\|^2 = \|(1 - \rho_n - \varrho_n)u_n + \rho_n z_n - u^*\|^2$$
$$= \|(1 - \rho_n - \varrho_n)(u_n - u^*) + \rho_n(z_n - u^*) - \varrho_n u^*\|^2$$
$$= \|(1 - \rho_n - \varrho_n)(u_n - u^*) + \rho_n(z_n - u^*)\|^2 + \varrho_n^2\|u^*\|^2$$
$$- 2\langle (1 - \rho_n - \varrho_n)(u_n - u^*) + \rho_n(z_n - u^*), \varrho_n u^*\rangle. \quad (30)$$

From Expression (26), we have

$$\|(1 - \rho_n - \varrho_n)(u_n - u^*) + \rho_n(z_n - u^*)\|^2$$
$$\leq (1 - \rho_n - \varrho_n)(1 - \varrho_n)\|u_n - u^*\|^2 + \rho_n(1 - \varrho_n)\|z_n - u^*\|^2. \quad (31)$$

Combining Expressions (30) and (31) (for some $\beth_2 > 0$), we obtain

$$\|u_{n+1} - u^*\|^2$$
$$\leq (1 - \rho_n - \varrho_n)(1 - \varrho_n)\|u_n - u^*\|^2 + \rho_n(1 - \varrho_n)\|z_n - u^*\|^2 + \varrho_n \beth_2$$
$$\leq (1 - \rho_n - \varrho_n)(1 - \varrho_n)\|u_n - u^*\|^2 + \varrho_n \beth_2$$
$$+ \rho_n(1 - \varrho_n)\left[\|t_n - u^*\|^2 - \left(1 - \frac{\mu\chi_n}{\chi_{n+1}}\right)\|t_n - y_n\|^2 - \left(1 - \frac{\mu\chi_n}{\chi_{n+1}}\right)\|z_n - y_n\|^2\right]. \quad (32)$$

From Expression (23), we have

$$\|t_n - u^*\|^2 \leq \|u_n - u^*\|^2 + \varrho_n \beth_3, \quad (33)$$

for some $\beth_3 > 0$. Substituting (33) into (32), we obtain

$$\|u_{n+1} - u^*\|^2$$
$$\leq (1 - \rho_n - \varrho_n)(1 - \varrho_n)\|u_n - u^*\|^2 + \varrho_n \beth_2$$
$$+ \rho_n(1 - \varrho_n)\left[\|u_n - u^*\|^2 + \varrho_n \beth_3 - \left(1 - \frac{\mu\chi_n}{\chi_{n+1}}\right)\|t_n - y_n\|^2 - \left(1 - \frac{\mu\chi_n}{\chi_{n+1}}\right)\|z_n - y_n\|^2\right]$$
$$= (1 - \varrho_n)^2\|u_n - u^*\|^2 + \varrho_n \beth_2 + \rho_n(1 - \varrho_n)\varrho_n \beth_3$$
$$- \rho_n(1 - \varrho_n)\left[\left(1 - \frac{\mu\chi_n}{\chi_{n+1}}\right)\|t_n - y_n\|^2 + \left(1 - \frac{\mu\chi_n}{\chi_{n+1}}\right)\|z_n - y_n\|^2\right]$$
$$\leq \|u_n - u^*\|^2 + \varrho_n \beth_4 - \rho_n(1 - \varrho_n)\left[\left(1 - \frac{\mu\chi_n}{\chi_{n+1}}\right)\|t_n - y_n\|^2 + \left(1 - \frac{\mu\chi_n}{\chi_{n+1}}\right)\|z_n - y_n\|^2\right], \quad (34)$$

for some $\beth_4 > 0$. It is given that $u^* = P_{Ep(f,\mathcal{C})}(0)$ and by using Lemma 2 (ii) ($Ep(f,\mathcal{C})$ is a convex and closed set ([23,34])), we obtain

$$\langle u^*, u^* - y\rangle \leq 0, \ \forall y \in Ep(f,\mathcal{C}). \quad (35)$$

The remainder of the proof shall be taken into account in the following two parts:

Case 1: Assume that there is a fixed number $n_2 \in \mathbb{N}$ ($n_2 \geq n_1$) such as

$$\|u_{n+1} - u^*\| \leq \|u_n - u^*\|, \quad \forall n \geq n_2. \tag{36}$$

It implies that $\lim_{n \to \infty} \|u_n - u^*\|$ exists, and due to (34), we obtain

$$\rho_n(1-\varrho_n)\left[\left(1 - \frac{\mu\chi_n}{\chi_{n+1}}\right)\|t_n - y_n\|^2 + \left(1 - \frac{\mu\chi_n}{\chi_{n+1}}\right)\|z_n - y_n\|^2\right]$$
$$\leq \|u_n - u^*\|^2 + \varrho_n J_4 - \|u_{n+1} - u^*\|^2. \tag{37}$$

Due to the existence of $\lim_{n \to \infty} \|u_n - u^*\|$, $\varrho_n \to 0$ and $\chi_n \to \chi$, we infer that

$$\lim_{n \to \infty} \|y_n - t_n\| = \lim_{n \to \infty} \|y_n - z_n\| = 0. \tag{38}$$

We can calculate that

$$\lim_{n \to \infty} \|z_n - t_n\| \leq \lim_{n \to \infty} \|t_n - y_n\| + \lim_{n \to \infty} \|y_n - z_n\| = 0. \tag{39}$$

It follows that

$$\|u_{n+1} - u_n\| = \|(1 - \rho_n - \varrho_n)u_n + \rho_n z_n - u_n\|$$
$$= \|u_n - \varrho_n u_n + \rho_n z_n - \rho_n u_n - u_n\|$$
$$\leq \rho_n \|z_n - u_n\| + \varrho_n \|u_n\|. \tag{40}$$

The term is referred to above that

$$\lim_{n \to \infty} \|u_{n+1} - u_n\| = 0. \tag{41}$$

Thus, this implies that $\{y_n\}$ and $\{z_n\}$ are bounded. The reflexivity of $\mathcal{H}$ and the boundedness of $\{u_n\}$ guarantee that there is a subsequence $\{u_{n_k}\}$, such that $\{u_{n_k}\} \rightharpoonup \hat{x} \in \mathcal{H}$ as $k \to \infty$. Next, our aim to prove that $\hat{x} \in Ep(f, \mathcal{C})$. Using (8), due to χ_{n+1} and (11), we write

$$\begin{aligned}\chi_{n_k} f(y_{n_k}, y) &\geq \chi_{n_k} f(y_{n_k}, z_{n_k}) + \langle t_{n_k} - z_{n_k}, y - z_{n_k}\rangle \\ &\geq \chi_{n_k} f(t_{n_k}, z_{n_k}) - \chi_{n_k} f(t_{n_k}, y_{n_k}) - \frac{\mu\chi_{n_k}}{2\chi_{n_k+1}}\|t_{n_k} - y_{n_k}\|^2 \\ &\quad - \frac{\mu\chi_{n_k}}{2\chi_{n_k+1}}\|y_{n_k} - z_{n_k}\|^2 + \langle t_{n_k} - z_{n_k}, y - z_{n_k}\rangle \\ &\geq \langle t_{n_k} - y_{n_k}, z_{n_k} - y_{n_k}\rangle - \frac{\mu\chi_{n_k}}{2\chi_{n_k+1}}\|t_{n_k} - y_{n_k}\|^2 \\ &\quad - \frac{\mu\chi_{n_k}}{2\chi_{n_k+1}}\|y_{n_k} - z_{n_k}\|^2 + \langle t_{n_k} - z_{n_k}, y - z_{n_k}\rangle,\end{aligned} \tag{42}$$

while y is an any arbitrary member in $\mathcal{H}_n$. It continues from (38) and (39) that the right-hand side approaches to zero. From $\chi > 0$, Condition (c3) and $y_{n_k} \rightharpoonup \hat{x}$, we have

$$0 \leq \limsup_{k \to \infty} f(y_{n_k}, y) \leq f(\hat{x}, y), \quad \forall y \in \mathcal{H}_n. \tag{43}$$

The following is that $f(\hat{x}, y) \geq 0$, $\forall y \in \mathcal{C}$; thus $\hat{x} \in Ep(f, \mathcal{C})$. It continues from that

$$\limsup_{n \to \infty} \langle u^*, u^* - u_n\rangle = \limsup_{k \to \infty} \langle u^*, u^* - u_{n_k}\rangle = \langle u^*, u^* - \hat{x}\rangle \leq 0. \tag{44}$$

Due to $\lim_{n \to \infty} \|u_{n+1} - u_n\| = 0$., we can deduce that

$$\limsup_{n \to \infty} \langle u^*, u^* - u_{n+1}\rangle \leq \limsup_{k \to \infty} \langle u^*, u^* - u_n\rangle + \limsup_{k \to \infty} \langle u^*, u_n - u_{n+1}\rangle \leq 0. \tag{45}$$

Next, consider the following value

$$\begin{aligned}
\|t_n - u^*\|^2 &= \|u_n + \phi_n(u_n - u_{n-1}) - u^*\|^2 \\
&= \|u_n - u^* + \phi_n(u_n - u_{n-1})\|^2 \\
&= \|u_n - u^*\|^2 + \phi_n^2\|u_n - u_{n-1}\|^2 + 2\langle u_n - u^*, \phi_n(u_n - u_{n-1})\rangle \\
&\leq \|u_n - u^*\|^2 + \phi_n^2\|u_n - u_{n-1}\|^2 + 2\phi_n\|u_n - u^*\|\|u_n - u_{n-1}\| \\
&= \|u_n - u^*\|^2 + \phi_n\|u_n - u_{n-1}\|[2\|u_n - u^*\| + \phi_n\|u_n - u_{n-1}\|] \\
&\leq \|u_n - u^*\|^2 + \phi_n\|u_n - u_{n-1}\|\beth_5,
\end{aligned}$$ (46)

Substituting $q_n = (1-\rho_n)u_n + \rho_n z_n$, we have

$$u_{n+1} = q_n - \varrho_n u_n = (1-\varrho_n)q_n - \varrho_n(u_n - q_n) = (1-\varrho_n)q_n - \varrho_n \rho_n(u_n - z_n).$$ (47)

where $u_n - q_n = u_n - (1-\rho_n)u_n - \rho_n z_n = \rho_n(u_n - z_n)$. Consider that

$$\begin{aligned}
\|q_n - u^*\|^2 &= \|(1-\rho_n)u_n + \rho_n z_n - u^*\|^2 \\
&= \|(1-\rho_n)(u_n - u^*) + \rho_n(z_n - u^*)\|^2 \\
&= (1-\rho_n)^2\|u_n - u^*\|^2 + \rho_n^2\|z_n - u^*\|^2 + 2\langle (1-\rho_n)(u_n - u^*), \rho_n(z_n - u^*)\rangle \\
&\leq (1-\rho_n)^2\|u_n - u^*\|^2 + \rho_n^2\|z_n - u^*\|^2 + 2\rho_n(1-\rho_n)\|u_n - u^*\|\|z_n - u^*\| \\
&\leq (1-\rho_n)^2\|u_n - u^*\|^2 + \rho_n^2\|z_n - u^*\|^2 + \rho_n(1-\rho_n)\|u_n - u^*\|^2 + \rho_n(1-\rho_n)\|z_n - u^*\|^2 \\
&= (1-\rho_n)\|u_n - u^*\|^2 + \rho_n\|z_n - u^*\|^2 \\
&\leq (1-\rho_n)\|u_n - u^*\|^2 + \rho_n\|t_n - u^*\|^2 \\
&\leq (1-\rho_n)\|u_n - u^*\|^2 + \rho_n[\|u_n - u^*\|^2 + \phi_n\|u_n - u_{n-1}\|\beth_5] \\
&\leq \|u_n - u^*\|^2 + \phi_n\|u_n - u_{n-1}\|\beth_5.
\end{aligned}$$ (48)

Next, consider that

$$\begin{aligned}
\|u_{n+1} - u^*\|^2 &= \|(1-\varrho_n)q_n + \rho_n \varrho_n(z_n - u_n) - u^*\|^2 \\
&= \|(1-\varrho_n)(q_n - u^*) + [\rho_n \varrho_n(z_n - u_n) - \varrho_n u^*]\|^2 \\
&\leq (1-\varrho_n)^2\|q_n - u^*\|^2 + 2\langle \rho_n \varrho_n(z_n - u_n) - \varrho_n u^*, (1-\varrho_n)(q_n - u^*) + \rho_n \varrho_n(z_n - u_n) - \varrho_n u^*\rangle \\
&= (1-\varrho_n)^2\|q_n - u^*\|^2 + 2\langle \rho_n \varrho_n(z_n - u_n) - \varrho_n u^*, q_n - \varrho_n q_n - \varrho_n(u_n - q_n) - u^*\rangle \\
&= (1-\varrho_n)\|q_n - u^*\|^2 + 2\rho_n \varrho_n\langle z_n - u_n, u_{n+1} - u^*\rangle + 2\varrho_n\langle u^*, u^* - u_{n+1}\rangle \\
&\leq (1-\varrho_n)\|q_n - u^*\|^2 + 2\rho_n \varrho_n\|z_n - u_n\|\|u_{n+1} - u^*\| + 2\varrho_n\langle u^*, u^* - u_{n+1}\rangle
\end{aligned}$$ (49)

for some $\beth_5 > 0$. Combining Expressions (46), (48), and (49), we obtain

$$\begin{aligned}
\|u_{n+1} - u^*\|^2 &\leq (1-\varrho_n)\|u_n - u^*\|^2 + (1-\varrho_n)\phi_n\|u_n - u_{n-1}\|\beth_5 \\
&\quad + 2\rho_n \varrho_n\|z_n - u_n\|\|u_{n+1} - u^*\| + 2\varrho_n\langle u^*, u^* - u_{n+1}\rangle \\
&\leq (1-\varrho_n)\|u_n - u^*\|^2 + \varrho_n\left[\frac{\phi_n}{\varrho_n}(1-\varrho_n)\|u_n - u_{n-1}\|\beth_5 \right. \\
&\quad \left. + 2\rho_n\|z_n - u_n\|\|u_{n+1} - u^*\| + 2\langle u^*, u^* - u_{n+1}\rangle\right].
\end{aligned}$$ (50)

Due to (45), (50), and the implemented Lemma 3, we conclude that $\|u_n - u^*\| \to 0$ as $n \to \infty$.

Case 2: Assume there is a subsequence $\{n_i\}$ of $\{n\}$ that

$$\|u_{n_i} - u^*\| \leq \|u_{n_{i+1}} - u^*\|, \ \forall i \in \mathbb{N}.$$

Using Lemma 4, there is a $\{m_k\} \subset \mathbb{N}$ sequence, such as $\{m_k\} \to \infty$,

$$\|u_{m_k} - u^*\| \leq \|u_{m_{k+1}} - u^*\| \quad \text{and} \quad \|u_k - u^*\| \leq \|u_{m_{k+1}} - u^*\|, \text{ for all } k \in \mathbb{N}.$$ (51)

Similar to Case 1, Relation (37) gives that

$$\rho_{m_k}(1-\varrho_{m_k})\left[\left(1-\frac{\mu\chi_{m_k}}{\chi_{m_k+1}}\right)\|t_{m_k}-y_{m_k}\|^2+\left(1-\frac{\mu\chi_{m_k}}{\chi_{m_k+1}}\right)\|z_{m_k}-y_{m_k}\|^2\right] \\ \leq \|u_{m_k}-u^*\|^2+\varrho_{m_k}\mathbb{J}_4-\|u_{m_k+1}-u^*\|^2. \tag{52}$$

Due to $\varrho_{m_k} \to 0$ and $\chi_{m_k} \to \chi$, we deduce the following:

$$\lim_{n\to\infty}\|t_{m_k}-y_{m_k}\| = \lim_{n\to\infty}\|z_{m_k}-y_{m_k}\| = 0. \tag{53}$$

It continues on from that

$$\begin{aligned}\|u_{m_k+1}-u_{m_k}\| &= \|(1-\rho_{m_k}-\varrho_{m_k})u_{m_k}+\rho_{m_k}z_{m_k}-u_{m_k}\| \\ &= \|u_{m_k}-\varrho_{m_k}u_{m_k}+\rho_{m_k}z_{m_k}-\rho_{m_k}u_{m_k}-u_{m_k}\| \\ &\leq \rho_{m_k}\|z_{m_k}-u_{m_k}\|+\varrho_{m_k}\|u_{m_k}\| \longrightarrow 0.\end{aligned} \tag{54}$$

We use the same reasoning as that in Case 1:

$$\limsup_{k\to\infty}\langle u^*, u^*-u_{m_k+1}\rangle \leq 0. \tag{55}$$

Now, using Expressions (50) and (51), we have

$$\begin{aligned}\|u_{m_k+1}-u^*\|^2 &\leq (1-\varrho_{m_k})\|u_{m_k}-u^*\|^2+\varrho_{m_k}\left[\frac{\phi_{m_k}}{\varrho_{m_k}}(1-\varrho_{m_k})\|u_{m_k}-u_{m_k-1}\|\mathbb{J}_5\right. \\ &\quad \left.+2\rho_{m_k}\|z_{m_k}-u_{m_k}\|\|u_{m_k+1}-u^*\|+2\langle u^*, u^*-u_{m_k+1}\rangle\right]. \\ &\leq (1-\varrho_{m_k})\|u_{m_k+1}-u^*\|^2+\varrho_{m_k}\left[\frac{\phi_{m_k}}{\varrho_{m_k}}(1-\varrho_{m_k})\|u_{m_k}-u_{m_k-1}\|\mathbb{J}_5\right. \\ &\quad \left.+2\rho_{m_k}\|z_{m_k}-u_{m_k}\|\|u_{m_k+1}-u^*\|+2\langle u^*, u^*-u_{m_k+1}\rangle\right].\end{aligned} \tag{56}$$

It implies that

$$\begin{aligned}\|u_{m_k+1}-u^*\|^2 &\leq \left[\frac{\phi_{m_k}}{\varrho_{m_k}}(1-\varrho_{m_k})\|u_{m_k}-u_{m_k-1}\|\mathbb{J}_5\right. \\ &\quad \left.+2\rho_{m_k}\|z_{m_k}-u_{m_k}\|\|u_{m_k+1}-u^*\|+2\langle u^*, u^*-u_{m_k+1}\rangle\right].\end{aligned} \tag{57}$$

Since $\varrho_{m_k}\to 0$, and $\|u_{m_k}-u^*\|$ is bounded. Thus, with Expressions (55) and (57), we have

$$\|u_{m_k+1}-u^*\|^2 \to 0, \text{ as } k\to\infty. \tag{58}$$

The above implies that

$$\lim_{n\to\infty}\|u_k-u^*\|^2 \leq \lim_{n\to\infty}\|u_{m_k+1}-u^*\|^2 \leq 0. \tag{59}$$

As a result, $u_n \to u^*$. This completes the proof of the theorem. □

By letting $\phi_n = 0$, we obtain a strong convergence of the result in [25].

Corollary 1. *Let $f : \mathcal{C}\times\mathcal{C} \to \mathcal{R}$ be a bifunction satisfying Conditions (c1)–(c4). Choosing $u_0 \in \mathcal{C}, \chi_0 > 0, \{\rho_n\} \subset (a,b) \subset (0, 1-\varrho_n)$ and $\{\varrho_n\} \subset (0,1)$ satisfies the following conditions:*

$$\lim_{n\to\infty}\varrho_n = 0 \text{ and } \sum_n^\infty \varrho_n = +\infty.$$

Let $\{u_n\}$ be a sequence that is generated in the following manner:

$$\begin{cases} y_n = \arg\min_{y \in C}\{\chi_n f(u_n, y) + \frac{1}{2}\|u_n - y\|^2\}, \\ z_n = \arg\min_{y \in \mathcal{H}_n}\{\chi_n f(y_n, y) + \frac{1}{2}\|u_n - y\|^2\}, \\ u_{n+1} = (1 - \rho_n - \varrho_n)u_n + \rho_n z_n, \end{cases} \quad (60)$$

where $\mathcal{H}_n = \{z \in \mathcal{H} : \langle u_n - \chi_n \omega_n - y_n, z - y_n \rangle \leq 0\}$ and $\omega_n \in \partial_2 f(u_n, y_n)$. The step size is updated in the following way:

$$\chi_{n+1} = \min\left\{\chi_n, \frac{\mu\|u_n - y_n\|^2 + \mu\|z_n - y_n\|^2}{2[f(u_n, z_n) - f(u_n, y_n) - f(y_n, z_n)]_+}\right\}.$$

Then, sequence $\{u_n\}$ converges strongly to $u^* \in Ep(f, C)$.

4. Applications to Solve Fixed-Point Problems

We propose our results to focus on fixed-point problems regarding κ-strict pseudo-contraction mapping. The fixed-point problem (FPP) for $\mathcal{S} : \mathcal{H} \to \mathcal{H}$ is defined in the following manner:

$$\text{Find } u^* \in C \text{ such that } \mathcal{S}(u^*) = u^*. \quad \text{(FPP)}$$

We assume that the following conditions were met:

(c1*) A mapping $\mathcal{S} : C \to C$ is said to be κ-strict pseudocontraction [42] on C if

$$\|Ty_1 - Ty_2\|^2 \leq \|y_1 - y_2\|^2 + \kappa\|(y_1 - Ty_1) - (y_2 - Ty_2)\|^2, \quad \forall y_1, y_2 \in C;$$

(c2*) A mapping that is weakly sequentially continuous on C if

$$\mathcal{S}(y_n) \rightharpoonup \mathcal{S}(q^*) \text{ for any sequence in } C \text{ satisfying } y_n \rightharpoonup q^*.$$

If we consider that mapping $\mathcal{S}$ is weakly continuous and a κ-strict pseudocontraction, then $f(u, y) = \langle u - \mathcal{S}u, y - u \rangle$ satisfies the conditions (c1)–(c4) (see [43]) and $2c_1 = 2c_2 = \frac{3-2\kappa}{1-\kappa}$. The values of y_n and z_n in Algorithm 1 can be written as follows:

$$\begin{cases} y_n = \arg\min_{y \in C}\{\chi_n f(t_n, y) + \frac{1}{2}\|t_n - y\|^2\} = P_C[t_n - \chi_n(t_n - \mathcal{S}(t_n))], \\ z_n = \arg\min_{y \in \mathcal{H}_n}\{\chi_n f(y_n, y) + \frac{1}{2}\|t_n - y\|^2\} = P_{\mathcal{H}_n}[t_n - \chi_n(y_n - \mathcal{S}(y_n))]. \end{cases} \quad (61)$$

Corollary 2. *Suppose C is a nonempty, convex, and closed subset of a Hilbert space $\mathcal{H}$ and $\mathcal{S} : C \to C$ is weakly continuous and κ-strict pseudocontraction with solution set $\text{Fix}(\mathcal{S}) \neq \emptyset$. Let $u_{-1}, u_0 \in C$, $\phi > 0$, $\chi_0 > 0$, $\{\rho_n\} \subset (a, b) \subset (0, 1 - \varrho_n)$ and $\{\varrho_n\} \subset (0, 1)$ fulfill the items, i.e., $\lim_{n \to \infty} \varrho_n = 0$ and $\sum_{n=1}^{\infty} \varrho_n = +\infty$. Moreover, choose ϕ_n satisfying $0 \leq \phi_n \leq \hat{\phi}_n$ such that*

$$\hat{\phi}_n = \begin{cases} \min\left\{\frac{\phi}{2}, \frac{\varsigma_n}{\|u_n - u_{n-1}\|}\right\} & \text{if } u_n \neq u_{n-1}, \\ \frac{\phi}{2} & \text{else,} \end{cases} \quad (62)$$

where $\varsigma_n = o(\varrho_n)$, i.e., $\lim_{n \to \infty} \frac{\varsigma_n}{\varrho_n} = 0$. Assume that $\{u_n\}$ is the sequence generated in the following manner:

$$\begin{cases} t_n = u_n + \phi_n(u_n - u_{n-1}), \\ y_n = P_C[t_n - \chi_n(t_n - \mathcal{S}(t_n))], \\ z_n = P_{\mathcal{H}_n}[t_n - \chi_n(y_n - \mathcal{S}(y_n))], \\ u_{n+1} = (1 - \rho_n - \varrho_n)u_n + \rho_n z_n, \end{cases}$$

where $\mathcal{H}_n = \{z \in \mathcal{H} : \langle (1-\chi_n)t_n + \chi_n \mathcal{S}(t_n) - y_n, z - y_n \rangle \leq 0\}$. Compute

$$\chi_{n+1} = \min\left\{\chi_n, \frac{\mu \|t_n - y_n\|^2 + \mu \|z_n - y_n\|^2}{2[\langle (t_n - y_n) - [T(t_n) - T(y_n)], z_n - y_n \rangle]_+}\right\}$$

Then, $\{u_n\}$ strongly converges to $u^* \in Fix(\mathcal{S}, \mathcal{C})$.

Corollary 3. *Suppose $\mathcal{C}$ to be a convex and closed subset of a Hilbert space $\mathcal{H}$ and $\mathcal{S} : \mathcal{C} \to \mathcal{C}$ is weakly continuous and κ-strict pseudocontraction with solution set $Fix(\mathcal{S}) \neq \varnothing$. Let $u_0 \in \mathcal{C}$, $\chi_0 > 0$, $\{\rho_n\} \subset (a,b) \subset (0, 1-\varrho_n)$ and $\{\varrho_n\} \subset (0,1)$ fulfills the requirement, i.e., $\lim_{n\to\infty} \varrho_n = 0$ and $\sum_{n=1}^{\infty} \varrho_n = +\infty$. Assume that $\{u_n\}$ is the sequence formed as follows:*

$$\begin{cases} y_n = P_\mathcal{C}[u_n - \chi_n(u_n - \mathcal{S}(u_n))], \\ z_n = P_{\mathcal{H}_n}[u_n - \chi_n(y_n - \mathcal{S}(y_n))], \\ u_{n+1} = (1 - \rho_n - \varrho_n)u_n + \rho_n z_n, \end{cases}$$

where $\mathcal{H}_n = \{z \in \mathcal{H} : \langle (1-\chi_n)u_n + \chi_n \mathcal{S}(u_n) - y_n, z - y_n \rangle \leq 0\}$. Compute

$$\chi_{n+1} = \min\left\{\chi_n, \frac{\mu \|u_n - y_n\|^2 + \mu \|z_n - y_n\|^2}{2[\langle (u_n - y_n) - [T(u_n) - T(y_n)], z_n - y_n \rangle]_+}\right\}$$

Then, sequence $\{u_n\}$ converges strongly to $u^* \in Fix(\mathcal{S}, \mathcal{C})$.

5. Applications to Solve Variational-Inequality Problems

Next, we consider the application of our results in the problem of classical variational inequalities [44,45]. The variational-inequality problem (VIP) for an operator $\mathcal{L} : \mathcal{H} \to \mathcal{H}$ is stated in the following manner:

$$\text{Find } u^* \in \mathcal{C} \text{ such that } \langle \mathcal{L}(u^*), y - u^* \rangle \geq 0, \ \forall y \in \mathcal{C}. \tag{VIP}$$

We assume that the following conditions were met:

($\mathcal{L}1$) The solution set of problem (VIP) denoted by $VI(\mathcal{L}, \mathcal{C})$ is nonempty.
($\mathcal{L}2$) An operator $\mathcal{L} : \mathcal{H} \to \mathcal{H}$ is said to be pseudomonotone if

$$\langle \mathcal{L}(y_1), y_2 - y_1 \rangle \geq 0 \implies \langle \mathcal{L}(y_2), y_1 - y_2 \rangle \leq 0, \ \forall y_1, y_2 \in \mathcal{C}.$$

($\mathcal{L}3$) An operator $\mathcal{L} : \mathcal{H} \to \mathcal{H}$ is said to be Lipschitz continuous through $L > 0$, such that

$$\|\mathcal{L}(y_1) - \mathcal{L}(y_2)\| \leq L\|y_1 - y_2\|, \ \forall y_1, y_2 \in \mathcal{C};$$

($\mathcal{L}4$) $\limsup_{n \to \infty} \langle \mathcal{L}(y_n), y - y_n \rangle \leq \langle \mathcal{L}(q^*), y - q^* \rangle$ for all $y \in \mathcal{C}$ and $\{y_n\} \subset \mathcal{C}$ satisfy $y_n \rightharpoonup q^*$.

If we define $f(u,y) := \langle \mathcal{L}(u), y - u \rangle$ for all $u, y \in \mathcal{C}$. Then, problem (EP) becomes the problem of variational inequalities described above where $L = 2c_1 = 2c_2$. From the above value of the bifunction f, we have

$$\begin{cases} y_n = \arg\min_{y \in \mathcal{C}} \{\chi_n f(t_n, y) + \frac{1}{2}\|t_n - y\|^2\} = P_\mathcal{C}(t_n - \chi_n \mathcal{L}(t_n)), \\ z_n = \arg\min_{y \in \mathcal{H}_n} \{\chi_n f(y_n, y) + \frac{1}{2}\|t_n - y\|^2\} = P_{\mathcal{H}_n}(t_n - \chi_n \mathcal{L}(y_n)). \end{cases} \tag{63}$$

Corollary 4. *Suppose that $\mathcal{L} : \mathcal{C} \to \mathcal{H}$ is a function satisfying the assumptions ($\mathcal{L}1$)–($\mathcal{L}4$). Let $u_{-1}, u_0 \in \mathcal{C}$, $\phi > 0$, $\chi_0 > 0$, $\{\rho_n\} \subset (a,b) \subset (0, 1-\varrho_n)$ and $\{\varrho_n\} \subset (0,1)$ satisfies the* items, i.e., $\lim_{n\to\infty} \varrho_n = 0$ and $\sum_{n=1}^{\infty} \varrho_n = +\infty$. Moreover, choose ϕ_n satisfying $0 \leq \phi_n \leq \hat{\phi}_n$, such that

$$\hat{\phi}_n = \begin{cases} \min\left\{\frac{\phi}{2}, \frac{\varsigma_n}{\|u_n - u_{n-1}\|}\right\} & \text{if } u_n \neq u_{n-1}, \\ \frac{\phi}{2} & \text{else,} \end{cases} \quad (64)$$

where $\varsigma_n = \circ(\varrho_n)$, i.e., $\lim_{n\to\infty} \frac{\varsigma_n}{\varrho_n} = 0$. Assume that $\{u_n\}$ is the sequence generated in the following manner:

$$\begin{cases} t_n = u_n + \phi_n(u_n - u_{n-1}), \\ y_n = P_\mathcal{C}(t_n - \chi_n \mathcal{L}(t_n)), \\ z_n = P_{\mathcal{H}_n}(t_n - \chi_n \mathcal{L}(y_n)), \\ u_{n+1} = (1 - \rho_n - \varrho_n)u_n + \rho_n z_n, \end{cases}$$

where $\mathcal{H}_n = \{z \in \mathcal{H} : \langle t_n - \chi_n \mathcal{L}(t_n) - y_n, z - y_n \rangle \leq 0\}$. Compute

$$\chi_{n+1} = \min\left\{\chi_n, \frac{\mu\|t_n - y_n\|^2 + \mu\|z_n - y_n\|^2}{2[\langle \mathcal{L}(t_n) - \mathcal{L}(y_n), z_n - y_n \rangle]_+}\right\}.$$

Then, sequences $\{u_n\}$ converge strongly to $u^* \in VI(\mathcal{L}, \mathcal{C})$.

Corollary 5. *Suppose that $\mathcal{L} : \mathcal{C} \to \mathcal{H}$ is a function meeting conditions $(\mathcal{L}1)$–$(\mathcal{L}4)$. Let $u_0 \in \mathcal{C}$, $\chi_0 > 0$, $\{\rho_n\} \subset (a,b) \subset (0, 1 - \varrho_n)$ and $\{\varrho_n\} \subset (0,1)$ satisfies the conditions, i.e., $\lim_{n\to\infty} \varrho_n = 0$ and $\sum_{n=1}^{\infty} \varrho_n = +\infty$. Assume that $\{u_n\}$ is the sequence generated in the following manner:*

$$\begin{cases} y_n = P_\mathcal{C}(u_n - \chi_n \mathcal{L}(u_n)), \\ z_n = P_{\mathcal{H}_n}(u_n - \chi_n \mathcal{L}(y_n)), \\ u_{n+1} = (1 - \rho_n - \varrho_n)u_n + \rho_n z_n, \end{cases}$$

where $\mathcal{H}_n = \{z \in \mathcal{H} : \langle u_n - \chi_n \mathcal{L}(u_n) - y_n, z - y_n \rangle \leq 0\}$.
Compute

$$\chi_{n+1} = \min\left\{\chi_n, \frac{\mu\|u_n - y_n\|^2 + \mu\|z_n - y_n\|^2}{2[\langle \mathcal{L}(u_n) - \mathcal{L}(y_n), z_n - y_n \rangle]_+}\right\}.$$

Then, sequences $\{u_n\}$ converge strongly to $u^* \in VI(\mathcal{L}, \mathcal{C})$.

Remark 1. *Condition $(\mathcal{L}4)$ could be exempted when $\mathcal{L}$ is monotone. Indeed, this condition, which is a particular case of Condition (c3), is only used to prove (43). Without Condition $(\mathcal{L}4)$, inequality (42) can be obtained by imposing monotonocity on $\mathcal{L}$. In that case,*

$$\langle \mathcal{L}(y), y - y_n \rangle \geq \langle \mathcal{L}(y_n), y - y_n \rangle, \quad \forall y \in \mathcal{C}. \quad (65)$$

By allowing $f(u,y) = \langle \mathcal{L}(u), y - u \rangle$ in (42), we have

$$\limsup_{k\to\infty} \langle \mathcal{L}(y_{n_k}), y - y_{n_k} \rangle \geq 0, \quad \forall y \in \mathcal{H}_n. \quad (66)$$

Combining (65) with (66), we conclude that

$$\limsup_{k\to\infty} \langle \mathcal{L}(y), y - y_{n_k} \rangle \geq 0, \quad \forall y \in \mathcal{C}. \quad (67)$$

Let $y_t = (1-t)z + ty$, for every $t \in [0,1]$. By using the convexity of set $\mathcal{C}$, $y_t \in \mathcal{C}$ for every $t \in (0,1)$. Since $y_{n_k} \rightharpoonup z \in \mathcal{C}$ and $\langle \mathcal{L}(y), y - z \rangle \geq 0$ for every $y \in \mathcal{C}$, we have

$$0 \leq \langle \mathcal{L}(y_t), y_t - z \rangle = t\langle \mathcal{L}(y_t), y - z \rangle. \quad (68)$$

Therefore, $\langle \mathcal{L}(y_t), y - z \rangle \geq 0, t \in (0,1)$. Since $y_t \to z$ as $t \to 0$ and due to $\mathcal{L}$ continuity, we have $\langle \mathcal{L}(z), y - z \rangle \geq 0$, for each $y \in \mathcal{C}$, which provides $z \in VI(\mathcal{L}, \mathcal{C})$.

Remark 2. *From Remark 1, it can be concluded that Corollaries 4 and 5 still hold, even if we remove Condition ($\mathcal{L}$4) in the case of monotone operators.*

6. Numerical Illustrations

Numerical results are presented in this section to demonstrate the efficiency of our proposed method. The MATLAB codes were run in MATLAB version 9.5 (R2018b) on an Intel(R) Core(TM)i5-6200 CPU PC @ 2.30 GHz 2.40 GHz, RAM 4.00 GB.

Example 1. *Let there be m companies that manufacture the same product. Assume vector u of each item u_i represents the quantity of the material produced by a company i. We consider that cost function P to be a declining affine function that relies on $\mu = \sum_{i=1}^{m} u_i$, i.e., $P_i(\mu) = \phi_i - \psi_i S$, where $\phi_i > 0, \psi_i > 0$. The formula for profit of every company i is taken as $F_i(u) = P_i(S)u_i - q_i(u_i)$, where $q_i(u_i)$ is the tax value and cost for developing item u_i. Moreover, consider that $C_i = [u_i^{\min}, u_i^{\max}]$ is the set of actions related to each company i, and the plan to figure out the model as $\mathcal{C} := \mathcal{C}_1 \times \mathcal{C}_2 \times \cdots \times \mathcal{C}_m$. In addition, each member wants to achieve its peak turnover by a good level of production on the basis that the performance of other firms is an input parameter. The commonly used modelling methodology is based on the famous Nash equilibrium principle. A point $u^* \in \mathcal{C} = \mathcal{C}_1 \times \mathcal{C}_2 \times \cdots \times \mathcal{C}_m$ is the level of equilibrium of the model if*

$$F_i(u^*) \geq F_i(u^*[u_i]), \forall u_i \in \mathcal{C}_i, \forall i = 1, 2, \cdots, m,$$

wile $u^*[u_i]$ is obtain from u^* by letting ζ_i^* with u_i. Furthermore, we consider $f(u,y) := \Delta(u,y) - \Delta(u,u)$ while $\Delta(u,y) := -\sum_{i=1}^{m} F_i(u[y_i])$. An equilibrium level of the model is defined by

$$\text{Find } u^* \in \mathcal{C} : f(u^*, z) \geq 0, \forall z \in \mathcal{C}.$$

Bifunction f converts into the following form (see [23]):

$$f(u,y) = \langle Pu + Qy + c, y - u \rangle$$

where $c \in \mathcal{R}^m$ and P, Q matrices of order m. Matrix P is positive semidefinite, and matrix $Q - P$ is negative semidefinite with Lipschitz-type constants $c_1 = c_2 = \frac{1}{2}\|P - Q\|$ (see [23]) for details. P, Q are taken randomly. (Two diagonal matrices randomly A_1 and A_2 take elements from $[0, 2]$ and $[-2, 0]$ respectively. Randomly $O_1 = \text{RandOrthMat}(m)$ and $O_2 = \text{RandOrthMat}(m)$ orthogonal matrices are generated. Then, a positive semidefinite matrix $B_1 = O_1 A_1 O_1^T$ and a negative semidefinite matrix $B_2 = O_2 A_2 O_2^T$ are achieved. Lastly, set $Q = B_1 + B_1^T, S = B_2 + B_2^T$ and $P = Q - S$.). The constraint set $C \subset \mathcal{R}^m$ be defined by

$$\mathcal{C} := \{u \in \mathcal{R}^m : -10 \leq u_i \leq 10\}.$$

*Numerical explanations for the first 200 iterations of three methods are considered in Figures 1–6 and Table 1 by letting initial points $u_0 = u_{-1} = (1, 1, \cdots, 1, 1)^T$. For Algorithm 3.2 (**mAlg2**) in [34]: $\chi = \frac{1}{4c_1}$ and $\rho_n = \frac{1}{100(n+2)}$; For Algorithm (**mAlg3**) in (60): $\chi_0 = 0.20, \mu = 0.70$, $\varrho_n = \frac{1}{100(n+2)}, \rho_n = 0.5(1 - \varrho_n)$; For Algorithm 1 (**mAlg1**): $\chi_0 = 0.20, \mu = 0.70, \phi = 0.60$, $\varsigma_n = \frac{1}{(n+1)^2}, \varrho_n = \frac{1}{100(n+2)}$ and $\rho_n = 0.5(1 - \varrho_n)$.*

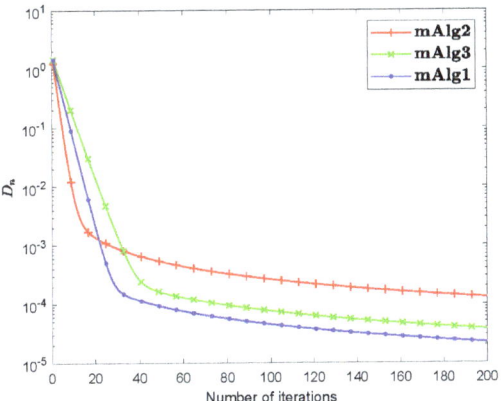

Figure 1. Algorithm 1 compared to Algorithm (60) and Algorithm 3.2 in [34] for $\mathcal{R}^5$.

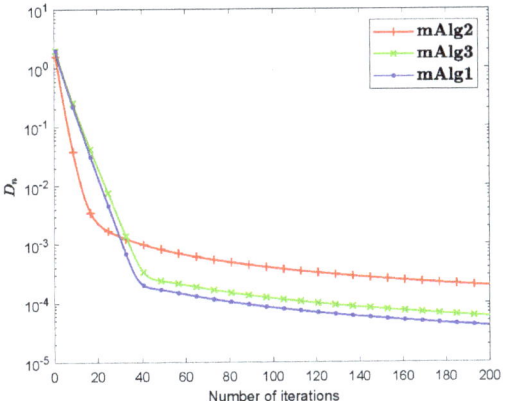

Figure 2. Algorithm 1 compared to Algorithm (60) and Algorithm 3.2 in [34] for $\mathcal{R}^{10}$.

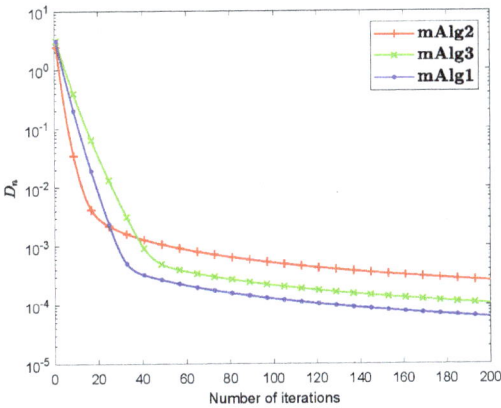

Figure 3. Algorithm 1 compared to Algorithm (60) and Algorithm 3.2 in [34] for $\mathcal{R}^{20}$.

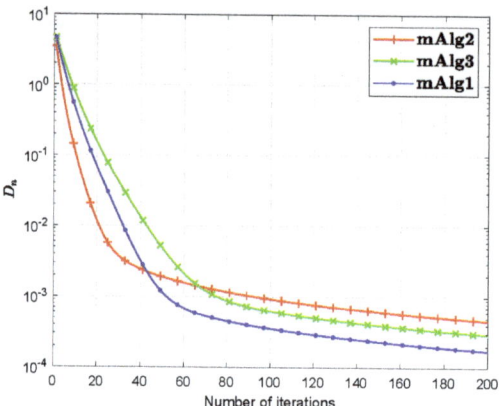

Figure 4. Algorithm 1 compared to Algorithm (60) and Algorithm 3.2 in [34] for $\mathcal{R}^{50}$.

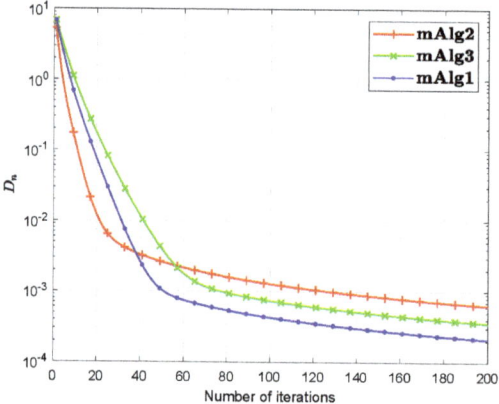

Figure 5. Algorithm 1 compared to Algorithm (60) and Algorithm 3.2 in [34] for $\mathcal{R}^{100}$.

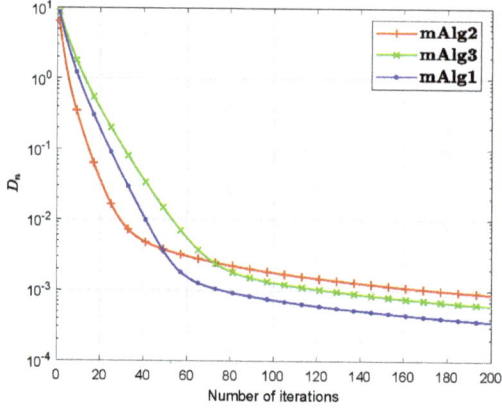

Figure 6. Algorithm 1 compared to Algorithm (60) and Algorithm 3.2 in [34] for $\mathcal{R}^{200}$.

Table 1. Figures 1–6 execution time required for first 200 iterations.

	Execution Time in Seconds		
m	mAlg2	mAlg3	mAlg1
5	2.55846812	2.73622248	2.923849848
10	2.89823133	2.99853685	3.341848537
20	3.23847254	3.51835212	3.332562246
50	3.93645046	4.05462157	4.084188882
100	4.57837436	5.32873548	5.723835682
200	5.86241836	6.28194713	6.825465869

Example 2. *Assume that set $C \subset L^2([0,1])$ is defined by*

$$C := \{u \in L^2([0,1]) : \|u\| \leq 1\}.$$

Let us define an operator $\mathcal{L} : C \to \mathcal{H}$, such that

$$\mathcal{L}(u)(t) = \int_0^1 [u(t) - H(t,s)f(u(s))]ds + g(t),$$

where $H(t,s) = \frac{2tse^{(t+s)}}{e\sqrt{e^2-1}}$, $f(u) = \cos(u)$ and $g(t) = \frac{2te^t}{e\sqrt{e^2-1}}$. In the above $\mathcal{H} = L^2([0,1])$ is a Hilbert space with inner product $\langle u, y \rangle = \int_0^1 u(t)y(t)dt$, $\forall u, y \in \mathcal{H}$ and induced norm is $\|u\| = \sqrt{\int_0^1 |u(t)|^2 dt}$. Numerical explanations for the first 200 iterations of three methods are considered in Figures 7–10 by letting initial points $u_0 = u_{-1} = (1, 1, \cdots, 1, 1)^T$. For Algorithm 3.2 (mAlg2) in [34]: $\chi = \frac{1}{3c_1}$ and $\rho_n = \frac{1}{100(n+2)}$; For Algorithm (mAlg3) in (60): $\chi_0 = 0.50$, $\mu = 0.50$, $\varrho_n = \frac{1}{100(n+2)}$, $\rho_n = 0.7(1 - \varrho_n)$; For Algorithm 1 (mAlg1): $\chi_0 = 0.50$, $\mu = 0.50$, $\phi = 0.70$, $\varsigma_n = \frac{1}{(n+1)^2}$, $\varrho_n = \frac{1}{100(n+2)}$ and $\rho_n = 0.7(1 - \varrho_n)$.

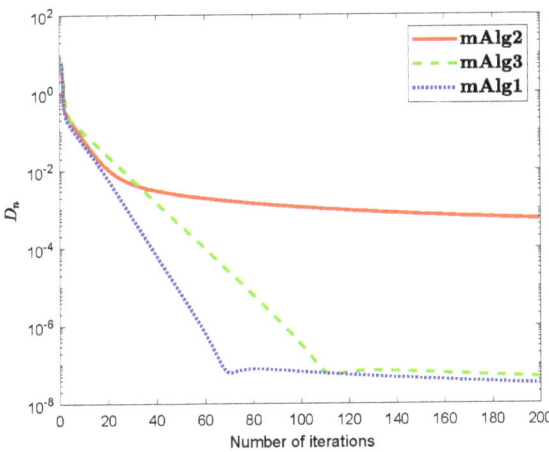

Figure 7. Algorithm 1 compared to Algorithm (60) and Algorithm 3.2 in [34] for $u_0 = 1 + t + 2t^2$.

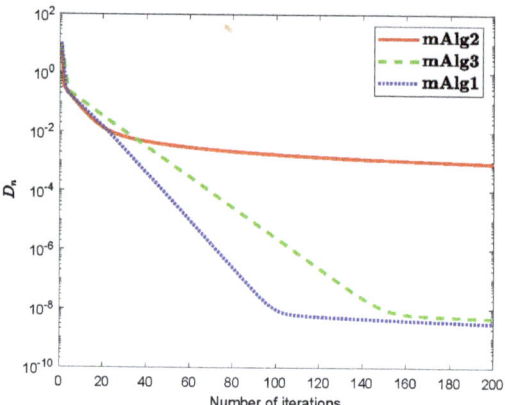

Figure 8. Algorithm 1 compared to Algorithm (60) and Algorithm 3.2 in [34] for $u_0 = 1 + 2t + 3e^t$.

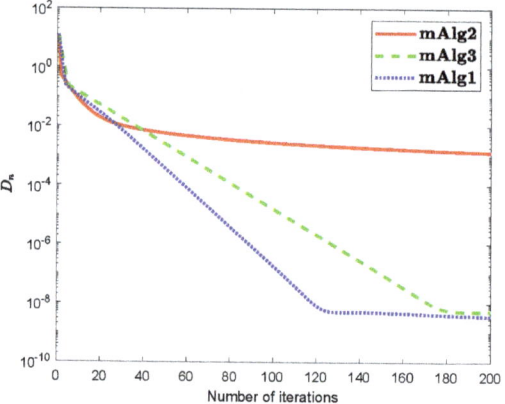

Figure 9. Algorithm 1 compared to Algorithm (60) and Algorithm 3.2 in [34] for $u_0 = 1 + 2t + \sin(t)$.

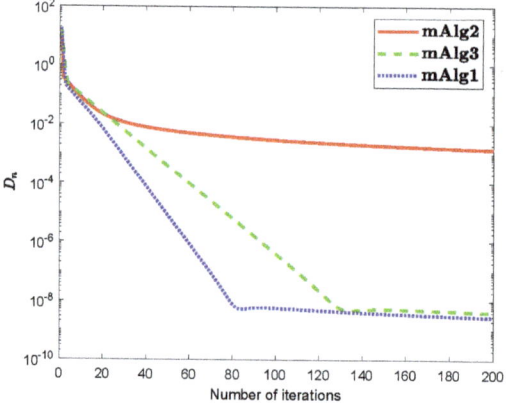

Figure 10. Algorithm 1 compared to Algorithm (60) and Algorithm 3.2 in [34] for $u_0 = 1 + 3t^2 + \cos(t)$.

7. Conclusions

We studied a Mann-type extragradient-like scheme for determining the numerical solution of equilibrium problem involving pseudomonotone function and also prove a strong convergent theorem. Computational conclusions were established to illustrate the computational performance of our algorithms relative to other approaches. Such computational experiments showed that the inertial effect increases the efficacy of the iterative method in this sense.

Author Contributions: Formal analysis, C.K.; funding acquisition, N.P. (Nuttapol Pakkaranang), N.P. (Nattawut Pholasa) and C.K.; investigation, N.W., N.P. (Nuttapol Pakkaranang) and C.K.; methodology, C.K.; project administration, C.K., N.P. (Nattawut Pholasa) and C.K.; resources, N.P. (Nattawut Pholasa) and C.K.; software, N.P. (Nuttapol Pakkaranang); supervision, N.P. (Nattawut Pholasa); Writing—original draft, N.W. and N.P. (Nuttapol Pakkaranang); Writing—review and editing, N.P. (Nuttapol Pakkaranang). All authors have read and agreed to the published version of the manuscript.

Funding: The APC was funded by University of Phayao.

Institutional Review Board Statement: Not applicable.

Informed Consent Statement: Not applicable.

Acknowledgments: Chainarong Khanpanuk would like to thank Phetchabun Rajabhat University. Nattawut Pholasa was financial supported by University of Phayao.

Conflicts of Interest: The authors declare no conflict of interest.

References

1. Fan, K. *A Minimax Inequality and Applications, Inequalities III*; Shisha, O., Ed.; Academic Press: New York, NY, USA, 1972.
2. Muu, L.; Oettli, W. Convergence of an adaptive penalty scheme for finding constrained equilibria. *Nonlinear Anal. Theory Methods Appl.* **1992**, *18*, 1159–1166. [CrossRef]
3. Blum, E. From optimization and variational inequalities to equilibrium problems. *Math. Stud.* **1994**, *63*, 123–145.
4. Bigi, G.; Castellani, M.; Pappalardo, M.; Passacantando, M. Existence and solution methods for equilibria. *Eur. J. Oper. Res.* **2013**, *227*, 1–11. [CrossRef]
5. Antipin, A. Equilibrium programming: Proximal methods. *Comput. Math. Math. Phys.* **1997**, *37*, 1285–1296.
6. Giannessi, F.; Maugeri, A.; Pardalos, P.M. *Equilibrium Problems: Nonsmooth Optimization and Variational Inequality Models*; Kluwer Academic Publisher: Dordrecht, The Netherlands, 2006; Volume 58.
7. Dafermos, S. Traffic Equilibrium and Variational Inequalities. *Transp. Sci.* **1980**, *14*, 42–54. [CrossRef]
8. ur Rehman, H.; Kumam, P.; Cho, Y.J.; Yordsorn, P. Weak convergence of explicit extragradient algorithms for solving equilibirum problems. *J. Inequalities Appl.* **2019**, *2019*. [CrossRef]
9. Ur Rehman, H.; Pakkaranang, N.; Kumam, P.; Cho, Y.J. Modified subgradient extragradient method for a family of pseudomonotone equilibrium problems in real a Hilbert space. *J. Nonlinear Convex Anal.* **2020**, *21*, 2011–2025.
10. Ur Rehman, H.; Kumam, P.; Dong, Q.L.; Cho, Y.J. A modified self-adaptive extragradient method for pseudomonotone equilibrium problem in a real Hilbert space with applications. *Math. Methods Appl. Sci.* **2020**, *44*, 3527–3547. [CrossRef]
11. Ur Rehman, H.; Kumam, P.; Sitthithakerngkiet, K. Viscosity-type method for solving pseudomonotone equilibrium problems in a real Hilbert space with applications. *AIMS Math.* **2021**, *6*, 1538–1560. [CrossRef]
12. Ur Rehman, H.; Kumam, P.; Gibali, A.; Kumam, W. Convergence analysis of a general inertial projection-type method for solving pseudomonotone equilibrium problems with applications. *J. Inequalities Appl.* **2021**, *2021*. [CrossRef]
13. Rehman, H.U.; Kumam, P.; Dong, Q.L.; Peng, Y.; Deebani, W. A new Popov's subgradient extragradient method for two classes of equilibrium programming in a real Hilbert space. *Optimization* **2020**, 1–36. [CrossRef]
14. Ur Rehman, H.; Kumam, P.; Shutaywi, M.; Alreshidi, N.A.; Kumam, W. Inertial Optimization Based Two-Step Methods for Solving Equilibrium Problems with Applications in Variational Inequality Problems and Growth Control Equilibrium Models. *Energies* **2020**, *13*, 3292. [CrossRef]
15. Ferris, M.C.; Pang, J.S. Engineering and Economic Applications of Complementarity Problems. *Siam Rev.* **1997**, *39*, 669–713. [CrossRef]
16. Ur Rehman, H.; Kumam, P.; Cho, Y.J.; Suleiman, Y.I.; Kumam, W. Modified Popov's explicit iterative algorithms for solving pseudomonotone equilibrium problems. *Optim. Methods Softw.* **2020**, *36*, 82–113. [CrossRef]
17. Ur Rehman, H.; Kumam, P.; Abubakar, A.B.; Cho, Y.J. The extragradient algorithm with inertial effects extended to equilibrium problems. *Comput. Appl. Math.* **2020**, *39*. [CrossRef]

18. Ur Rehman, H.; Pakkaranang, N.; Hussain, A.; Wairojjana, N. A modified extra-gradient method for a family of strongly pseudomonotone equilibrium problems in real Hilbert spaces. *J. Math. Comput. Sci.* **2020**, *22*, 38–48. [CrossRef]
19. Wairojjana, N.; ur Rehman, H.; Argyros, I.K.; Pakkaranang, N. An Accelerated Extragradient Method for Solving Pseudomonotone Equilibrium Problems with Applications. *Axioms* **2020**, *9*, 99. [CrossRef]
20. Wairojjana, N.; ur Rehman, H.; la Sen, M.D.; Pakkaranang, N. A General Inertial Projection-Type Algorithm for Solving Equilibrium Problem in Hilbert Spaces with Applications in Fixed-Point Problems. *Axioms* **2020**, *9*, 101. [CrossRef]
21. Wairojjana, N.; Pakkaranang, N.; Ur Rehman, H.; Pholasa, N.; Khanpanuk, T. Strong Convergence of Extragradient-Type Method to Solve Pseudomonotone Variational Inequalities Problems. *Axioms* **2020**, *9*, 115. [CrossRef]
22. Wairojjana, N.; Younis, M.; Ur Rehman, H.; Pakkaranang, N.; Pholasa, N. Modified Viscosity Subgradient Extragradient-Like Algorithms for Solving Monotone Variational Inequalities Problems. *Axioms* **2020**, *9*, 118. [CrossRef]
23. Tran, D.Q.; Dung, M.L.; Nguyen, V.H. Extragradient algorithms extended to equilibrium problems. *Optimization* **2008**, *57*, 749–776. [CrossRef]
24. Korpelevich, G. The extragradient method for finding saddle points and other problems. *Matecon* **1976**, *12*, 747–756.
25. Hieu, D.V.; Quy, P.K.; Vy, L.V. Explicit iterative algorithms for solving equilibrium problems. *Calcolo* **2019**, *56*. [CrossRef]
26. Polyak, B. Some methods of speeding up the convergence of iteration methods. *USSR Comput. Math. Math. Phys.* **1964**, *4*, 1–17. [CrossRef]
27. Beck, A.; Teboulle, M. A Fast Iterative Shrinkage-Thresholding Algorithm for Linear Inverse Problems. *SIAM J. Imaging Sci.* **2009**, *2*, 183–202. [CrossRef]
28. Hung, P.G.; Muu, L.D. The Tikhonov regularization extended to equilibrium problems involving pseudomonotone bifunctions. *Nonlinear Anal. Theory Methods Appl.* **2011**, *74*, 6121–6129. [CrossRef]
29. Konnov, I. Application of the Proximal Point Method to Nonmonotone Equilibrium Problems. *J. Optim. Theory Appl.* **2003**, *119*, 317–333. [CrossRef]
30. Moudafi, A. Proximal point algorithm extended to equilibrium problems. *J. Nat. Geom.* **1999**, *15*, 91–100.
31. Oliveira, P.; Santos, P.; Silva, A. A Tikhonov-type regularization for equilibrium problems in Hilbert spaces. *J. Math. Anal. Appl.* **2013**, *401*, 336–342. [CrossRef]
32. Mann, W.R. Mean value methods in iteration. *Proc. Am. Math. Soc.* **1953**, *4*, 506–506. [CrossRef]
33. Censor, Y.; Gibali, A.; Reich, S. The Subgradient Extragradient Method for Solving Variational Inequalities in Hilbert Space. *J. Optim. Theory Appl.* **2010**, *148*, 318–335. [CrossRef]
34. Hieu, D.V. Halpern subgradient extragradient method extended to equilibrium problems. *Rev. Real Acad. Cienc. Exactas Físicas Nat. Ser. Matemáticas* **2016**, *111*, 823–840. [CrossRef]
35. Bianchi, M.; Schaible, S. Generalized monotone bifunctions and equilibrium problems. *J. Optim. Theory Appl.* **1996**, *90*, 31–43. [CrossRef]
36. Mastroeni, G. On Auxiliary Principle for Equilibrium Problems. In *Nonconvex Optimization and Its Applications*; Springer: New York, NY, USA, 2003; pp. 289–298. [CrossRef]
37. Tiel, J.V. *Convex Analysis: An Introductory Text*, 1st ed.; Wiley: New York, NY, USA, 1984.
38. Kreyszig, E. *Introductory Functional Analysis with Applications*, 1st ed.; Wiley: New York, NY, USA, 1989.
39. Xu, H.K. Another control condition in an iterative method for nonexpansive mappings. *Bull. Aust. Math. Soc.* **2002**, *65*, 109–113. [CrossRef]
40. Maingé, P.E. Strong Convergence of Projected Subgradient Methods for Nonsmooth and Nonstrictly Convex Minimization. *Set Valued Anal.* **2008**, *16*, 899–912. [CrossRef]
41. Bauschke, H.H.; Combettes, P.L. *Convex Analysis and Monotone Operator Theory in Hilbert Spaces*, 2nd ed.; CMS Books in Mathematics; Springer International Publishing: New York, NY, USA, 2017.
42. Browder, F.; Petryshyn, W. Construction of fixed points of nonlinear mappings in Hilbert space. *J. Math. Anal. Appl.* **1967**, *20*, 197–228. [CrossRef]
43. Wang, S.; Zhang, Y.; Ping, P.; Cho, Y.; Guo, H. New extragradient methods with non-convex combination for pseudomonotone equilibrium problems with applications in Hilbert spaces. *Filomat* **2019**, *33*, 1677–1693. [CrossRef]
44. Stampacchia, G. Formes bilinéaires coercitives sur les ensembles convexes. *Comptes Rendus Hebd. Seances Acad. Sci.* **1964**, *258*, 4413.
45. Konnov, I.V. On systems of variational inequalities. *Russ. Math. C/C Izv. Vyss. Uchebnye Zaved. Mat.* **1997**, *41*, 77–86.

Article
Mathematical Approach for System Repair Rate Analysis Used in Maintenance Decision Making

Nataša Kontrec [1,*], Stefan Panić [1], Biljana Panić [2], Aleksandar Marković [1] and Dejan Stošović [3]

[1] Faculty of Sciences and Mathematics, University of Priština in Kosovska Mitrovica, 38220 Kosovska Mitrovica, Serbia; stefan.panic@pr.ac.rs (S.P.); aleksandar.markovic@pr.ac.rs (A.M.)
[2] Faculty of Organizational Sciences, University of Belgrade, 11000 Belgrade, Serbia; biljana.panic@fon.bg.ac.rs
[3] Faculty of Technical Sciences, University of Priština in Kosovska Mitrovica, 38220 Kosovska Mitrovica, Serbia; dejan.stosovic@pr.ac.rs
* Correspondence: natasa.kontrec@pr.ac.rs

Abstract: Reliability, the number of spare parts and repair time have a great impact on system availability. In this paper, we observed a repairable system comprised of several components. The aim was to determine the repair rate by emphasizing its stochastic nature. A model for the statistical analysis of the component repair rate in function of the desired level of availability is presented. Furthermore, based on the presented model, the approach for the calculation of probability density functions of maximal and minimal repair times for a system comprised of observed components was developed as an important measure that unambiguously defines the total annual repair time. The obtained generalized analytical expressions that can be used to predict the total repair time for an observed entity are the main contributions of the manuscript. The outputs of the model can be useful for making decisions in which time interval repair or replacement should be done to maintain the system and component availability. In addition to planning maintenance activities, the presented models could be used for service capacity planning and the dynamic forecasting of system characteristics.

Keywords: repair rate; mathematical modeling; availability; probability; maintenance

1. Introduction

Maintenance comprises a set of procedures and methods for keeping a system in an operational state or returning the system to a functional state after failure [1]. Depending on the activity and time for their implementation, maintenance can be corrective or preventive. Corrective maintenance implies a set of activities to be undertaken after a system stopped working, i.e., stopped performing its main function. In other words, the corrective maintenance activities are to be implemented only in cases of failure occurring as a result of an error (human, procedural or an error made by testing equipment), due to deterioration, environmental effect or damage caused by improper handling. In this category of maintenance, repairing or replacing a part, not before the exact moment of failure, is considered more efficient. The preventive maintenance implies periodical checking of the system's conditions and parameters to prevent the occurrence of failure. This concept of preventive maintenance is based on the supervision and control of the system's conditions while it is still in function and on the undertaking of those activities which delay the occurrence of failure and keep the system in its operational state. Since the occurrences of unplanned failures and damage to the system are almost unavoidable, even in cases when a system is regularly maintained, corrective maintenance should not be disregarded. All activities, whether in the form of preventive or corrective maintenance, require a certain period for their implementation. This time frame is usually called downtime and refers to a period when the observed component or system is not available. Because a large number of factors influence the duration of delay, these can be divided into waiting and active downtimes.

The waiting downtimes are delays that occur due to waiting for spare parts, administrative procedures, deliveries, staff, etc. The active downtime refers to the time used for the repair or replacement of a component or system. As numerous factors can affect the time for repair, we can conclude that it is a random variable. Systems or components are divided into repairable and non-repairable. In non-repairable systems, the distribution of time to failure is most commonly observed. Many authors have dealt with these issues in their papers [2–4]. On the other hand, there are repairable systems, i.e., systems that can be returned to their functional state with certain activities, after the occurrence of failure. The key performance measures of both repairable and non-repairable systems are availability and reliability. The availability is defined as a probability that a system will perform its function in a time [5]. When it comes to military aircraft and weapon industry, availability can be defined as "a measure of the degree to which an item is in an operable state and can be committed at the start of a mission when the mission is called for at an unknown (random) point in time" [6].

Maintenance contracts are most likely utilized when the system's availability is vital. Their characteristic is that no specific maintenance activities such as servicing, repairs and required materials are paid for, but only the performances of the system result from the undertaking maintenance activities. This concept originates from the military industry, i.e., it is related to the maintenance of military aircraft and weapon systems. These types of contracts are called performance-based logistic (PBL) contracts. In other words, it is a strategy utilized in complex systems to lower maintenance expenses and increase their reliability and availability [7]. Maintenance contracts have also found their use in civilian companies, under the name performance-based contracts (PBC) [8]. In practice, when the airplane's engine is serviced under the PBL contract, maintenance is not charged by the number of working hours used for engine repair or by the number of used spare parts, but by the time during which the airplane is available after repairs i.e., number of hours the engine is in the operational state [9]. Kang et al. [10] have observed systems whose maintenance was regulated with PBL contracts. They concluded that the mean time between failures (MTBF), mean time To repair (MTTR) and the number of spare parts have the greatest impact on availability. Evaluating the availability of a certain component or system is a common topic in the related literature. Inherited availability and methods for its evaluation in repairable systems have been researched in papers [11–13]. Papers [9,14–16] provided major contributions concerning the issue of calculating the availability of repairable systems and operating under the maintenance contracts. Moreover, some control problems with interval analysis are presented in [17,18] and some statistical analyses that can be used for this purpose are presented in [19,20]. A similar issue was researched in paper [21], in which it was concluded that the repair time and reliability have a significantly greater effect on the system's availability than the number of spare parts in the inventory. Thus, according to reviewed literature, it can be concluded that the reliability and repair rate have the greatest impact on availability. In this paper, we observed a system modeled using an alternating renewal process and we analyzed the system repair rate in order to provide support in decision-making process when it comes to system maintenance planning. The main contribution of our paper is the new method that relies on the observation of the annual repair rate. This method was based on the determination of the maximal repair rate of units that compose a corresponding entity, including its magnitude and performance measures that unambiguously define the total repair rate of an observed entity and which have not been discussed in renewal theory literature to date. We calculated two new parameters that are interesting for the observation, maximal and minimal repair rate of each unit that constitute the corresponding entity by observing them as stochastic variables. In this way, we obtained analytical expressions that can be used to predict the total repair time of the corresponding entity. These generalized PDF expressions are our main contribution.

2. Mathematical Method for System Repair Rate Analysis

The stochastic modeling of a component or system repair time is not new and has already been justified in the paper [22]. There, the author emphasizes the importance of the stochastic modeling of the system maintenance by observing a one-unit reparable system with a non-negligible repair time. The homogeneous, nonhomogeneous and compound Poisson process for system maintenance modeling was investigated. In this paper, we used a model presented in [23], where the authors studied similar systems such as that in [22] with the assumption that the MTBF is Rayleigh distributed. Based on these assumptions, they investigated the repair rate in dependence of the desired level of availability. Only repairable components and systems were taken into consideration, i.e., the systems that alternate between successive up and down intervals. Thus, the alternating renewal process [24] was used to model such a system. This process can be observed as a series of independent and non-negative random variables such as the time to failure and time to repair. It was assumed that each time the failure occurs, the component will be restored and start to behave the same as the new one. Notably, we only observed perfect repair, although in the literature and in practice, there are two types of repair: perfect and imperfect. While perfect repair means that the unit can be reused in the state "as good as new", imperfect repair is defined as an action after which the unit is not "as good as new" but it is in usable/operational condition.

The purpose of maintenance contracts is to reduce the costs and increase system availability that can be further calculated as expected operative time $E(t)$ and a renewal cycle ($E[T] + E[R]$ i.e., sum of the expected operative time and time to repair) [25]:

$$A = \lim_{t \to \infty} A(t) = \frac{E[T]}{E[T] + E[R]}, \tag{1}$$

The expected operative time is a random variable which, if probability density function exists, can be calculated as

$$E[t] = \int_0^\infty t p(t) dt. \tag{2}$$

In [4], the authors assumed that this variable is Rayleigh distributed with the following probability density function (PDF):

$$p(t) = \frac{2t}{\sigma} \exp\left(\frac{-t^2}{\sigma}\right). \tag{3}$$

Assuming that the expected time to failure of the component is a Rayleigh-distributed random variable, the authors provided the expression for the PDF of the repair rate in dependence of unit's availability as

$$p(\mu) = \frac{8A^2}{(1-A)^2 \mu^3 \pi \sigma_0} \exp\left(\frac{-4A^2}{(1-A)^2 \mu^2 \pi \sigma_0}\right), \tag{4}$$

where A is availability, μ is the repair rate, and $\sigma_0 = E(\sigma)$. Based on Equation (4), the cumulative probability density function CDF can be expressed as

$$F(\mu) = \int_0^\mu p(\mu) d\mu = 1 - \exp\left(\frac{-4A^2}{(1-A^2)\mu^2 \pi \sigma_0}\right) \tag{5}$$

Based on these calculations for a single component or subsystem, in this paper, we proposed a new model for calculating the maximal and minimal repair rate of the system comprised of two or more components. The PDF function of the first component is:

$$p_1(\mu) = \frac{8A_1^2}{\left(1-A_1^2\right)\mu^3 \pi \sigma_{0_1}} \exp\left(\frac{-4A_1^2}{\left(1-A_1^2\right)\mu^2 \pi \sigma_{0_1}}\right), \tag{6}$$

where A_1 is the set level of availability of the first unit and, i.e., the mathematical expectation of the Rayleigh-distributed parameter for that component. The cumulative density function (CDF) is then:

$$F_1(\mu) = 1 - \exp\left(\frac{-4A_1^2}{\left(1-A_1^2\right)\mu^2 \pi \sigma_{0_1}}\right) \tag{7}$$

Using the same equation, we can determine the PDF of the second unit $p_2(\mu)$ with availability A_2 and $\sigma_{0_2} = E(\sigma)$ as

$$p_2(\mu) = \frac{8A_2^2}{\left(1-A_2^2\right)\mu^3 \pi \sigma_{0_2}} \exp\left(\frac{-4A_2^2}{\left(1-A_2^2\right)\mu^2 \pi \sigma_{0_2}}\right) \tag{8}$$

and the CDF:

$$F_2(\mu) = 1 - \exp\left(\frac{-4A_2^2}{\left(1-A_2^2\right)\mu^2 \pi \sigma_{0_2}}\right). \tag{9}$$

For a system composed of two parts, we can calculate the maximal repair rate as $\mu_{max} = \max(\mu_1, \mu_2)$. In that case, the PDF is:

$$p_{\mu_{max}}(\mu) = p(\mu_1 > \mu_2) \vee p(\mu_2 > \mu_1) = p_1(\mu)F_2(\mu) + p_2(\mu)F_1(\mu) \tag{10}$$

while the CDF is:

$$F_{\mu_{max}}(\mu) = F_1(\mu)F_2(\mu). \tag{11}$$

Furthermore, when a system is comprised of n parts, then the repair rate can be calculated as $\mu_{max} = \max(\mu_1, \mu_2, \ldots, \mu_n)$. The general form of the repair rate's PDF is then:

$$p_{\mu_{max}}(\mu) = \sum_{\substack{i=1 \\ i \neq j}}^{n} p_i(\mu) \prod_{j=1}^{n} F_j(\mu) \tag{12}$$

and the general form of the CDF is:

$$F_{\mu_{max}}(\mu) = \prod_{i=1}^{n} F_i(\mu) \tag{13}$$

Similarly, we can calculate the minimal repair rate as $\mu_{min} = \min(\mu_1, \mu_2)$, so the PDF is:

$$p_{\mu_{min}}(\mu) = p_1(\mu)(1 - F_2(\mu)) + p_2(\mu)(1 - F_1(\mu)) \tag{14}$$

while the CDF is:

$$F_{\mu_{min}}(\mu) = 1 - (1 - F_1(\mu))(1 - F_2(\mu)) \tag{15}$$

General forms of the PDF and CDF equations when the system is comprised of n parts and repair rate is $\mu = \min(\mu_1, \mu_2, \ldots, \mu_n)$ are:

$$p_{\mu_{\min}}(\mu) = \sum_{\substack{i=1 \\ i \neq j}}^{n} p_i(\mu) \prod_{j=1}^{n} (1 - F_j(\mu)) \qquad (16)$$

and the general form of the CDF for this system is:

$$F_{\mu_{\min}}(\mu) = 1 - \prod_{i=1}^{n} (1 - F_i(\mu)). \qquad (17)$$

3. Numerical Results and Discussions

To verify the model presented in the previous section, we used the data calculated in [10,21]. In these papers, the authors observed the unmanned aerial vehicle (UAV), i.e., its three major components: engine, propeller and avionics. The available data of the major interest for our paper are as follows:

- Each UAV is supposed to have 120 flight hour per month, which further means 1440 flight hours per year;
- MTBF for UAV's engine is 750 flight hours, while for avionics this is 1000 h and 500 h for propeller per year.
- Thus, based on that in [21], the failure rate was calculated as 1.92 (failures per year) for the engine, 2.88 for the propeller and 1.44 for avionics.

As can be seen, all three figures present the PDF of the UAV's engine, propeller and avionics, respectively, for different values of availability (any other value could also be selected).

As the availability A increases, the maximum PDF values moves to the right, which further means that maximum PDF values are obtained for higher annual repair rate values, i.e., as A grows a higher value of μ is needed to achieve maximum PDF. For example, as it can be seen on Figure 1 that present maximum annual repair rate in dependence of availability for UAV's engine, to achieve the availability of 85% the number of repairs per year should be around 10, for availability of 90% the repair rate should be around 15 and for 95% it should be around 30. The same interpretation could be given for Figures 2 and 3 that presents the dependence of the annual repair rate of availability for UAV's propeller and avionics.

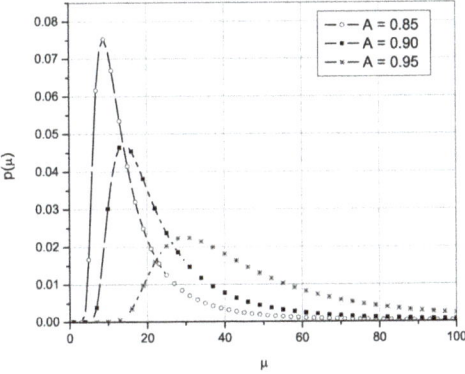

Figure 1. PDF of UAV's engine repair rate.

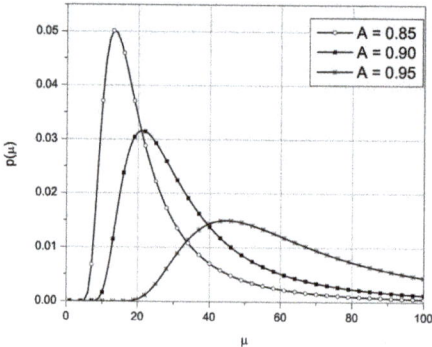

Figure 2. PDF of UAV's propeller repair rate.

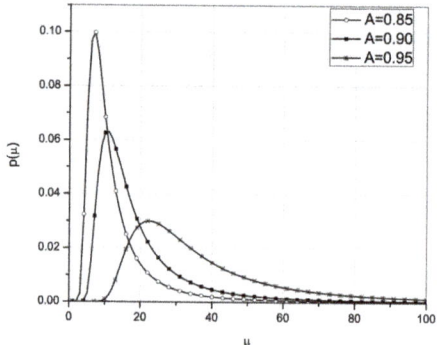

Figure 3. PDF of UAV's avionics repair rate.

Similarly, we can calculate the annual expected time for the maximum and minimum repair rate of the UAV system. We have to take into consideration all three critical components of the UAV system: engine, propeller and avionics. Here, we set the availability at $A = 0.80, A = 0.85, A = 0.9, A = 0.95$. Since in this case we are observing a system comprised of three critical UAV components, the PDF maximal repair rate can be calculated as: $\mu_{max} = \max(\mu_1, \mu_2, \mu_3)$ According to the Equation (12), the PDF of the repair rate is:

$$p_{\mu_{max}}(\mu) = p_1(\mu)F_2(\mu)F_3(\mu) + p_2(\mu)F_1(\mu)F_3(\mu) + p_3(\mu)F_1(\mu)F_2(\mu) \qquad (18)$$

while the CDF is:

$$F_{\mu_{max}}(\mu) = F_1(\mu)F_2(\mu)F_3(\mu). \qquad (19)$$

Figures 4 and 5 represent the PDF and CDF, respectively, of the UAV's repair rate depending on time; the repair rate was calculated as the maximum of its components' repair rate and based on the presented equations. The desired level of availability is set to 80%, 85%, 90% and 95%. Actually, in Figure 4, we observed the magnitude of the maximum repair rate of the system comprised of the engine, propeller and avionics. It can be seen that the maximum PDF of this parameter shifts to the right again as the value of availability increases, which means that with the higher values of A, the maximum value of the repair rate of the whole system is more likely to take on a higher value. Figure 5 shows that for smaller values of A, the range of values that the maximum repair rate could take is smaller, but when the availability increases, the range that maximum repair rate values can take also increases.

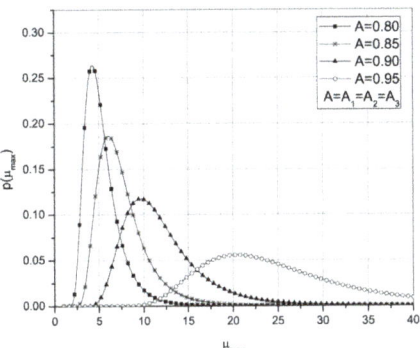

Figure 4. PDF of repair rate for $\mu = \max(\mu_1, \mu_2, \mu_3)$.

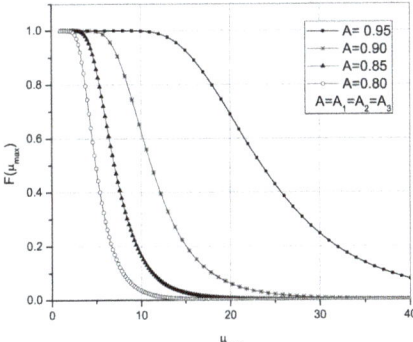

Figure 5. CDF of repair rate for $\mu = \max(\mu_1, \mu_2, \mu_3)$.

Similarly, we can determine the minimal repair rate of the observed UAV, comprised of three critical components as $\mu_{\min} = \min(\mu_1, \mu_2, \mu_3)$, so the PDF is:

$$p_{\mu_{\min}}(\mu) = p_1(\mu)(1-F_2(\mu))(1-F_3(\mu)) + p_2(\mu)(1-F_1(\mu))(1-F_3(\mu)) + \\ + p_3(\mu)(1-F_1(\mu))(1-F_2(\mu)), \quad (20)$$

while the CDF is:

$$F_{\mu_{\min}}(\mu) = 1 - (1-F_1(\mu))(1-F_2(\mu))(1-F_3(\mu)) \quad (21)$$

Figures 6 and 7 represent the PDF and CDF, respectively, of the UAV's repair rate depending on time and the repair rate is calculated as the minimum of its components' repair rate. The desired level of availability is set to 80%, 85%, 90% and 95% as in the previous example. From Figures 6 and 7, we can also see that as the A parameter increases, the maximum PDF values migrate to the right as in previous case. However when compared to Figures 4 and 5 we can see that for the same values of parameter A, maximum PDF values are obtained for lower annual repair rate values and this repair rate value represents lower system performance bound for observed entity. The presented figures show the probability that the repairs conducted in a certain time frame will provide the desired level of system availability.

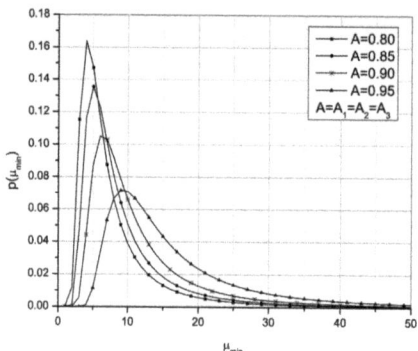

Figure 6. PDF of repair rate for $\mu_{\min} = \min(\mu_1, \mu_2, \mu_3)$.

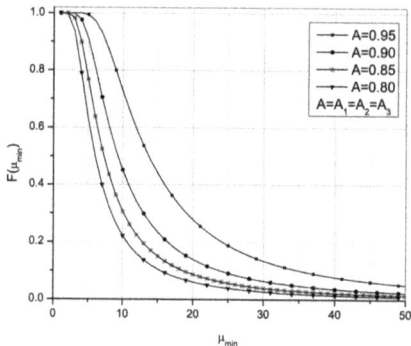

Figure 7. CDF of repair rate for $\mu_{\min} = \min(\mu_1, \mu_2, \mu_3)$.

4. Conclusions

The analysis presented in this paper can be applied to other repairable systems, not only to ones where the component time to failure is modeled with Rayleigh distribution. After determining the characteristics of the repair rate of an individual unit, the statistical analysis of the repair rate of a system consisting of several components is presented, which was the main contribution of this paper. Actually, a novel method for the determination of the maximal and minimal repair rate of the entity comprised of the observed units is presented. The obtained generalized PDF expressions can be used to predict total repair time. The presented method provides two new measures that comprehensively define the total repair time and have not been studied in this way before. In the numerical section, the proposed model was applied to a UVA system consisting of three key components: the motor, propeller, and avionics. PDFs of repair rate for each component, as well as the PDF and CDF maximum and minimum repair rates for the entire UAV system. The obtained information is graphically presented and it can be concluded that as that maximum PDF values are obtained for higher annual repair rate values as the availability increases, i.e., as A grows, a higher value of repair rate is needed to achieve a maximum PDF. The similar behavior is noticed when we observed the minimal annual repair rate but in this case the maximum PDF values are obtained for lower annual repair rate values and this repair rate value represent a lower system performance bound for an observed entity. Based on that, we can predict the time interval by which the maintenance action will have to be successfully completed in order to achieve the desired level of availability. Although we set availability to certain levels, numerical analysis can be repeated with different availability values.

Author Contributions: Conceptualization, N.K. and B.P.; methodology, S.P. and N.K.; validation, N.K., B.P. and S.P.; data curation, A.M. and D.S.; writing—original draft preparation, N.K.; writing—review and editing, S.P., B.P. and A.M.; visualization, D.S.; supervision, S.P.; All authors have read and agreed to the published version of the manuscript.

Funding: This research received no external funding.

Institutional Review Board Statement: Not applicable.

Informed Consent Statement: Not applicable.

Acknowledgments: This work is supported by Faculty of Sciences and Mathematics, University of Prišitna in Kosovska Mitrovica, project no. IJ-0203.

Conflicts of Interest: The authors declare no conflict of interest.

References

1. Huang, X.; Han, X. Method for fuzzy maintainability index demonstration in lognormal distribution. *J. Syst. Eng. Electron.* **2008**, *30*, 375–378.
2. Chung, W.K. Reliability analysis of repairable and non-repairable systems with common-cause failures. *Microelectron. Reliab.* **1989**, *29*, 545–547. [CrossRef]
3. Liu, Y.; Kapur, K.C. Reliability measures for dynamic multistate nonrepairable systems and their applications to system performance evaluation. *IIE Trans.* **2007**, *38*, 511–520. [CrossRef]
4. Kontrec, N.; Milovanović, G.; Panić, S.; Milošević, H. A Reliability-Based Approach to Nonrepairable Spare Part Forecasting in Aircraft Maintenance System. *Math. Probl. Eng.* **2015**, *2015*, 1–7. [CrossRef]
5. Barlow, R.E.; Proschan, F. *Statistical Theory of Reliability and Life Testing. Probability Models*; Holt, Rinehart and Winston: New York, NY, USA, 1975.
6. United States Department of Defense Guide for Achieving Reliability, Availability, and Maintainability. 2005. Available online: https://www.acqnotes.com/Attachments/DoD%20Reliability%20Availability%20and%20Maintainability%20(RAM)%20Guide.pdf (accessed on 30 January 2021).
7. Randall, W.; Pohlen, T.; Hanna, J. Evolving a theory of performance-based logistics using insights from service dominant logic. *J. Bus. Logist.* **2010**, *31*, 35–61. [CrossRef]
8. Phillips, E.H. Performance based logistics: A whole new approach. *Aviat. Week. Space Technol.* **2005**, *163*, 52–55.
9. Nowicki, D.; Kumar, U.D.; Steudel, H.J.; Verma, D. Spares provisioning under performance-based logistics contract: Profit-centric approach. *Oper. Res. Soc.* **2008**, *59*, 342–352. [CrossRef]
10. Kang, K.; Doerr, K.H.; Boudreau, M.; Apte, U. A decision support model for valuing proposed improvements in component reliability. *Mil. Oper. Res.* **2010**, *15*, 55–68. [CrossRef]
11. Claasen, S.J.; Joubert, J.W.; Yadavalli, V.S.S. Interval estimation of the availability of a two unit standby system with non instantaneous switch over and "dead time". *Pakistan J. Stat. Oper. Res.* **2004**, *20*, 115–122.
12. Hwan Cha, J.; Sangyeol, L.; Jongwoo, J. Sequential confidence interval estimation for system availability. *Qual. Reliab. Eng. Int.* **2005**, *22*, 165–176.
13. Ke, J.C.; Chu, Y.K. Nonparametric analysis on system availability: Confidence bound and power function. *J. Math. Stat.* **2007**, *3*, 181–187. [CrossRef]
14. Kang, K.; Doerr, K.H.; Sanchez, S.M. A Design of Experiments Approach to Readiness Risk Analysis. In Proceedings of the Simulation Conference WSC 06, Proceedings of the Winter 2006, Monterey, CA, USA, 3–6 December 2006; pp. 1332–1339.
15. Kim, S.H.; Cohen, M.A.; Netessine, S. Performance contract in after-sales service supply chains. *Manag. Sci.* **2007**, *53*, 1843–1858. [CrossRef]
16. Oner, K.B.; Kiesmuller, G.P.; van Houtum, G.J. Optimization of component reliability in the design phase of capital goods. *Eur. J. Oper. Res.* **2010**, *205*, 615–624. [CrossRef]
17. Treanţă, S. On a global efficiency criterion in multiobjective variational control problems with path-independent curvilinear integral cost functionals. *Ann. Oper. Res.* **2020**. [CrossRef]
18. Treanţă, S. On Modified Interval-Valued Variational Control Problems with First-Order PDE Constraints. *Symmetry* **2020**, *12*, 472. [CrossRef]
19. Stojanovic, V.; Kevkic, T.; Ljajko, E.; Jelic, G. Noise-Indicator ARMA Model with Application in Fitting Physically-Based Time Series. *UPB Sci. Bull.-Ser. A Appl. Math. Phys.* **2018**, *81*, 257–264.
20. Randjelovic, M.; Stojanovic, V.; Kevkic, T. Noise-indicator autoregressive conditional heteroskedastic process with application in modeling actual time series. *UPB Sci. Bull.-Ser. A Appl. Math. Phys.* **2019**, *81*, 77–84.
21. Mirzahosseinian, H.; Piplani, R. A study of repairable parts inventory system operating under peformance-based contract. *Eur. J. Oper. Res.* **2011**, *214*, 256–261. [CrossRef]
22. Andrzejczak, K. Stochastic modelling of the repairable system. *J. KONBiN* **2015**, *35*, 5–14. [CrossRef]

23. Kontrec, N.; Panić, S.; Petrović, M.; Milošević, H. A stochastic model for estimation of repair rate for system operating under performance based logistics. *Eksploat. Niezawodn.* **2018**, *20*, 68–72. [CrossRef]
24. Wolstenholme, L.C. *Reliability Modeling. A Stochastic Approach*; CRC Press: Boca Raton, FL, USA, 1999; ISBN 9781584880141.
25. Ross, S.M. *Applied Probability Models with Optimization Applications*; Dover Publication, Inc.: New York, NY, USA, 2013; ISBN 9780486318646.

Article

Numerical Solution of an Interval-Based Uncertain SIR (Susceptible–Infected–Recovered) Epidemic Model by Homotopy Analysis Method

Emmanuel A. Bakare [1,†], Snehashish Chakraverty [2,†] and Radovan Potucek [3,*,†]

1. Department of Mathematics, Federal University, Oye Ekiti, Ado Ekiti 371 104, Ekiti State, Nigeria; emmanuel.bakare@fuoye.edu.ng
2. Department of Mathematics, National Institute of Technology Rourkela, Odisha 769001, India; chakravertys@nitrkl.ac.in
3. Department of Mathematics and Physics, University of Defence, Kounicova 65, 662 10 Brno, Czech Republic
* Correspondence: radovan.potucek@unob.cz; Tel.: +420-973-44-3056
† These authors contributed equally to this work.

Abstract: This work proposes an interval-based uncertain Susceptible–Infected–Recovered (SIR) epidemic model. The interval model has been numerically solved by the homotopy analysis method (HAM). The SIR epidemic model is proposed and solved under different uncertain intervals by the HAM to obtain the numerical solution of the model. Furthermore, the SIR ODE model was transformed into a stochastic differential equation (SDE) model and the results of the stochastic and deterministic models were compared using numerical simulations. The results obtained were compared with the numerical solution and found to be in good agreement. Finally, various simulations were done to discuss the solution.

Keywords: homotopy analysis method; uncertainty; interval analysis; simulation; stochastic; susceptible; infected; recovered

1. Introduction

Interval analysis is a method developed by mathematicians in the 1950s as a way of handling bounds or rounding errors and measurement errors in mathematical computation. It is useful in formulating numerical methods that yield desirable results. In short, it defines each value as a range of possibilities. This work aims to formulate interval arithmetic that solves upper and lower endpoints for the range of values of a particular function in one or more variables. These limitations are not necessarily the supremum or infimum since the exact solution of those values can be very intractable or even impossible. The treatments of interval arithmetic for real intervals of quantities with the form $[u,v] = \{x \in \mathbb{R} : u \leq x \leq v\}$, where $u = -\infty$ and $v = \infty$, are permitted. The permission is based on the fact that if one of the real intervals is infinite, we would have an unbounded interval, and if both are infinite, we would have the extended real number system. Considering the classical calculation with real numbers, simple arithmetic operations and functions on elementary intervals must initially be defined. It is after this that complicated functions can be evaluated from the basic elements. In interval arithmetic, we state the range of possible outcomes explicitly.

Thus, the results are no longer stated as numbers but as intervals, which denotes imprecise values. With the size of the intervals, we express the extent of uncertainties, which are similar to error bars to a metric. The evaluations of the outer bounds of intervals are enabled by simple arithmetic operations, for example, basic arithmetic and trigonometric. Interval arithmetic was introduced by [1] as an approach to bound rounding errors in mathematical computation. The theory of interval analysis emerged considering the computation of both the exact solution and the error term as a single entity, that is, the interval. Though a simple idea, it is a very powerful technique with numerous applications in mathematics, computer science, and engineering.

In their survey, they discussed the basic concepts of interval arithmetic and some of its extensions. They also reviewed the successful application of this theory in computer science, in particular. The authors of [2] investigated the solution of linear and nonlinear ordinary differential equations with the fuzzy initial condition. They proposed two Euler-type methods to obtain a numerical solution to the problem. They also compared their solution with existing results. They observed that the results obtained were tighter than the results from the existing method. The authors of [3] also investigated the numerical solution of n-th order fuzzy differential equations in the fuzzy environment using a homotopy perturbation method (HPM). They used triangular fuzzy convex normalized sets for the fuzzy parameter and variables.

They also compared their results obtained with the existing solution in terms of plots to show the efficiency of their method. The authors of [4] gave an overview of applications of interval arithmetic and discussed the verification methods for linear and nonlinear systems of equations. They also then discussed item software in the field and gave some historical remarks. The authors of [5] provided algorithms for computing the operations of interval arithmetic. They generated data that are sufficiently detailed to convert directly to a program to efficiently implement the interval operations. Finally, they extended these results to the case of general intervals, which are defined as connected sets of rules that are not necessarily closed. For this present work, we considered an interval-based uncertain epidemic model.

A related mathematical model was proposed first for SIR transmission dynamics and then the HAM was applied to find the solution. This method employs the concept of the homotopy from topology to generate a convergent series solution of nonlinear systems. The convergent series solution of nonlinear systems was enabled by utilizing a homotopy–MacLaurin series to deal with the nonlinearity in the system. The HAM is much better than most of the existing analytic approximation method because most of the existing methods are valid only for weakly nonlinear problems [6]. It overcomes the restrictions of all other analytic approximation methods and is valid for highly nonlinear problems [6]. The HAM is always valid even if small physical parameters exist or not, it provides an easy way to guarantee the convergence of approximation series, and lastly, it provides sufficient freedom to choose the equation type of sub-problems and the base function of solutions [6]. The strength of the HAM to naturally exhibit convergence of the series solution is strange in most analytic and semi-analytic approaches to nonlinear PDEs [7]. Recently, [8] used the HAM approach to solve the SIS and SIR models of [9]. The authors of [10], extended the work of [8] to solve the SIR epidemic model in the presence of a constant vaccination strategy. The authors of [7] also applied the HAM to solve the SIR epidemic model. They obtained an explicit analytic solution of the coupled nonlinear differential equations describing the epidemic model proposed. They also compared the numerical results, which showed that the two results are in good agreement. The authors of [11] studied a new approach for solving the SIR epidemic model using the HAM that was based on dividing the entire domain into subintervals.

Other works on the homotopy analysis method with the SIR model can be found in [12–20]. The aim of this work was to obtain the numerical solution of an interval-based uncertain SIR epidemic model using the HAM and comparing their stochastic version. The homotopy analysis method (HAM) has been applied here to study the solution of the epidemic model under uncertain intervals. The results obtained by the HAM were compared with the approximate solution and were found to be in strong agreement. We have also developed the stochastic version of the SIR epidemic model presented in this paper in order to measure the effect of randomness of the variables in the model. To the best of our knowledge, no work has been done in the area of an interval-based uncertain SIR epidemic model and very few works have been done on the stochastic model of SIR epidemic models so far. The paper is organized as follows: in Section 2, preliminaries and basic definitions are presented. In Section 3, the presentation of the proposed model is made. In Section 4, we describe the interval-based uncertain model. In Section 5, we present the homotopy analysis approach to a non-linear system, while in Section 6, we present the solution of the SIR epidemic model by the HAM. In Section 7, we present the solution, numerical results, and a discussion on the interval-based uncertain SIR epidemic model. Section 8 showcases the stochastic version of the model. In Section 9, the graphical illustrations of our results are discussed. In Section 10, the numerical solution of the SDE model are discussed. In Section 11, we present the discussion, conclusion, and possible extensions, and finally, the references are presented.

2. Preliminaries

In this section, we present some notations, definitions, and preliminaries that are used further in this paper.

A. Interval Arithmetic [1]

Interval arithmetic is defined on the sets of intervals, instead of sets of real numbers. Interval arithmetic defines a set of operations on intervals, as follows:

$$Y * W = \{x : \exists u \in Y \land \exists v \in W : x = u * v\},$$

where u and v are intervals.

B. Closed Interval [1]

A closed interval, denoted by $[m, n]$, is the set of real numbers given by

$$[m, n] = \{x \in \mathbb{R} : m \leq x \leq n\}.$$

C. Endpoint notation, interval equality [1]

Two intervals, A and B, are said to be equal if they are the same sets. Hence, operationally this occurs if their corresponding endpoints are equal; $A = B$ if $\underline{A} = \underline{B}$ and $\overline{A} = \overline{B}$. Here $\underline{A}$, represents the left endpoint of an interval A while $\overline{A}$ represents the right endpoint of an interval A, such that $A = [\underline{A}, \overline{A}]$.

D. Midpoint of A [1]

The midpoint of A is given by

$$m(A) = \frac{1}{2}(\underline{A} + \overline{A}).$$

E. Interval Arithmetic and Operations [1]

The key point in the definition of arithmetic operations is that computing intervals are computing with sets. Let

$$A = [a : a \in A] \text{ and } B = [b : b \in B]$$

Then, the following properties hold:

(i) The sum of two intervals, A and B, is the set

$$A + B = \{a + b : a \in A, b \in B\}.$$

(ii) The difference of two intervals, A and B, is the set

$$A - B = \{a - b : a \in A, b \in B\}.$$

(iii) The product of A and B is given by

$$A * B = \{a * b : a \in A, b \in B\}.$$

(iv) The quotient of A/B is defined as

$$\frac{A}{B} = \left\{\frac{a}{b} : a \in A,\ b \in B,\ b \neq 0\right\}.$$

3. Model Formulation

A population comprising three kinds of individuals, denoted by S (susceptible human), I (infected human), and R (recovered human), are considered. The susceptible human $((S(t))$ is the number of susceptible humans at time t, that is, humans who are vulnerable and are yet to contract the disease but have a probability of contracting it. The infected human $((I(t))$ is the population of the infected and infectious persons who have the disease and can transmit it to others, while the recovered human $((R(t))$ is the population of recovered humans who cannot get the disease or transmit it, because they have natural immunity, they have recovered from the disease and are immune to re-infection, they have been placed in isolation, or they have died.

The population of susceptible humans is generated through the reduction of the rate of transmission β with the infected, such that the rate of change of the population of susceptible humans is given by the following:

$$\frac{dS}{dt} = -\beta S I,\ \beta > 0. \tag{1}$$

The rate of change of the population of infected humans is increased by the rate of transmission β with the susceptible, and reduced by the rate at which the infected population becomes isolate or recovered γ. Hence it is given by

$$\frac{dI}{dt} = -\beta S I - \gamma I,\ \beta > 0,\ \gamma > 0. \tag{2}$$

The population of recovered humans is generated by the rate at which the infected population becomes isolated or recovered. Hence it is given by

$$\frac{dR}{dt} = \gamma I,\ \gamma > 0. \tag{3}$$

Hence, the governing equation by [9] related to the present model is given by

$$\frac{dS}{dt} = -\beta S I,$$

$$\frac{dI}{dt} = -\beta S I - \gamma I, \tag{4}$$

$$\frac{dR}{dt} = \gamma I.$$

Subject to the initial conditions,

$$S(0) = S_0,\ I(0) = I_0,\ R(0) = R_0.$$

4. Interval-Based Uncertain Model

As mentioned in the introduction, if we assumed that the parameters involved in a model are given in terms of an interval then it will become an interval-based model and the solution has to be handled carefully. As such, let us suppose that we have the rate of transmission β and the rate at which the infected population become isolated or recovered γ in terms of intervals $\tilde{\beta} = \left[\underline{\beta}, \overline{\beta}\right]$ and $\tilde{\gamma} = \left[\underline{\gamma}, \overline{\gamma}\right]$ then the corresponding interval model may be written as

$$\frac{dS}{dt} = -\tilde{\beta} S I,$$

$$\frac{dI}{dt} = -\tilde{\beta}SI - \tilde{\gamma}I, \qquad (5)$$

$$\frac{dR}{dt} = \tilde{\gamma}I,$$

with the initial conditions

$$S_0 = S(0), \ I_0 = I(0), \ R_0 = R(0)$$

where S, I, and R are all now in interval form.

It may be noted from the open literature that the involved parameters, such as β and γ, are usually given in term of some ranges, so we have investigated the problem considering those ranges in terms of intervals. Hence, the intervals of β and γ are taken as the following:

(i) $\tilde{\beta} = [0.01, 0.03]$,
(ii) $\tilde{\gamma} = [0.005, 0.015]$.

Next, the above interval model has been solved by the homotopy analysis method (HAM). We provide, in the next section, some mathematical results.

5. Mathematical Results

We assume here that all parameters in Equation (5) are positive intervals. For the SIR model (5) to be meaningful biologically, we need to prove that all its stated variables are non-negative (except S) for all time, that is, the solutions of the Equation (5) with non-negative initial data will remain non-negative for all time $t > 0$.

Proposition 1. If the initial values $S \geq 0$, $I \geq 0$, $R \geq 0$, then the solutions $(S(t), I(t), R(t))$ of the model (5) are non-negative for all $t \geq 0$.

Proof. Let $\Omega_p = \{t > 0; S(t) > 0, I(t) > 0, R(t) > 0\}$, We say that, from the equations of Equation (5) that

$$\frac{dS}{dt} = -\tilde{\beta}SI,$$

where $\tilde{\beta} = [\underline{\beta}, \overline{\beta}]$ and $\tilde{\gamma} = [\underline{\gamma}, \overline{\gamma}]$. Therefore,

$$\frac{d}{dt} S(t) \exp\left[\left(\tilde{\beta}I\right)t\right] = 0.$$

Hence, $S(t_1) \exp\left[\left(\tilde{\beta}I\right)t\right] - S(0) = 0$, so $S(t_1) = -S_0 \exp\left[\left(\tilde{\beta}I\right)t_1\right] < 0$.
Then,

$$\frac{d}{dt} I(t) \exp\left[\left(-\tilde{\beta}S + \tilde{\gamma}\right)t\right] = 0,$$

$$I(t) \exp\left[\left(-\tilde{\beta}S + \tilde{\gamma}\right)t\right] - I(0) = 0,$$

$$I(t_1) = I(0) \exp\left[-\left(-\tilde{\beta}S + \tilde{\gamma}\right)t_1\right].$$

Such that

$$I(t_1) = I(0) \exp\left[\left(\tilde{\beta}S - \tilde{\gamma}\right)t_1\right] > 0.$$

Similarly,

$$R(t_1) = R(0)(\tilde{\gamma}I)t_1 > 0.$$

Hence, the solutions $(S(t), I(t), R(t))$ of the Equation (5) are non-negative (except S) for all $t > 0$. □

Proposition 2. Suppose Equation (5) has a unique interval-based positive solution (S, I, R) defined on a horizon of infinite time.

Proof. We established that Equation (5) can be rewritten in the following form:

$$\frac{dS}{dt} = f_1(S, I, R)S, \frac{dI}{dt} = f_2(S, I, R)I,$$

where $S + I + R = 1$, so $R = 1 - S - I$.

The functions f_1 and f_2 are C^∞. Thus, according to the Cauchy–Lipschitz theorem [21], Equation (5) has a unique positive solution (S, I) on the infinite time horizon, whenever $S_0 > 0$, $I_0 > 0$. □

Corollary 1. *The compact domain* $\Omega_1 = \{(S, I, R) \in \Omega : 0 \leq S + I + R \leq 1\}$ *is positively invariant and attracts all trajectories from* Ω.

Proposition 3. *The domain* Ω *is positively invariant through the positive semi-wave produced by Equation (5).*

Proof. Equation (5) can be rewritten as follows:

$$\frac{d}{dt}\begin{pmatrix} S \\ I \\ R \end{pmatrix} = \begin{pmatrix} f_1(S, I, R) \\ f_2(S, I, R) \\ f_3(S, I, R) \end{pmatrix} = F(S, I, R).$$

Applying the assumption that

$$f_1(S = 0, I, R) = 0 \text{ for } (I, R) \geq 0,$$
$$f_1(S = c, I, R) = -\tilde{\beta}cI \leq 0 \text{ for } (I, R) \geq 0,$$
$$f_2(S, I = 0, R) = 0 \text{ for } (S, R) \geq 0,$$
$$f_3(S, I, R = 0) = \tilde{\gamma}I \geq 0 \text{ for } (S, I) \geq 0.$$

Thus, the field remains on the domain Ω.

In contrast, we show that by setting $S(t)$ and $I(t)$ as continuous intervals, such that $S(0) = S_0 > 0$ and $I(0) = I_0 > 0$, if $I(\tilde{t}) < 0$, then by the intermediate value theorem, there exists $\tau_1 \in [0, \tilde{t}]$, such that $I(\tau_1) = 0$. By applying the second equation of Equation (5), we obtain $I(t) = I(\tau_1)e^g = 0$ for $t \geq t_0$, where g is the base of $-\tilde{\beta}I(t) - \tilde{\gamma}I$. Therefore, $I(t) = 0$ for $t \geq \tau_1$ which is a contradiction. We apply the same arguments to $S(t)$. We show this according to Proposition 4. □

Proposition 4 ([22]). *Suppose* $S(t), I(t), R(t)$ *is a solution of Equation (5), then* $S(t) \geq 0$, $I(t) \geq 0$ *and* $R(t) \geq 0$ *for all* $t > 0$.

If we add the first two equations of Equation (5) together, we obtain

$$\frac{d}{dt}[S(t) + I(t)] = -\tilde{\gamma}I \leq 0.$$

Now, by applying Proposition 4 and $S(0) + I(0) = N$, we have $S(t) + I(t) \leq N$. From Proposition 4, we also have $N - S(t) - I(t) = R(t)$. Hence, we conclude that $R(t) \geq 0$. From Proposition 4, we have $S(t) + I(t) \leq 0$, which implies that $N - S(t) - I(t) \leq N$. Therefore, $R(t) = N - S(t) - I(t) \leq N$ because

$$\frac{dI}{dt} \div \frac{dS}{dt} = \frac{dI}{dS} = \left(\frac{\tilde{\gamma}I}{\tilde{\beta}SI} - 1\right)dS,$$

then
$$\int_0^t dI = \int_0^t \left(\frac{\tilde{\gamma}}{\tilde{\beta}}S - 1\right) dS, I(t) - I_0 = \frac{\tilde{\gamma}}{\tilde{\beta}} \log_e S_0 - S(t) - S_0.$$

Therefore,
$$I(S(t)) = I_0 + S_0 + \frac{\tilde{\gamma}}{\tilde{\beta}} \log_e \left(\frac{S(t)}{S_0}\right).$$

The quantity $\frac{\tilde{\gamma}}{\tilde{\beta}}S - 1$ is positive if $S < \frac{\tilde{\gamma}}{\tilde{\beta}}$. From Equation (5), it is clear that $I(0) = -\infty$ and $I(S_0) = I(0) > 0$.

Hence, there exists a point S_∞, uniquely, $0 < S_\infty < S_0$ such that $I(S_\infty) = 0$ and $I(S) > 0$ for $S_\infty < S \leq S_0$. The point $(S_\infty, 0)$ is called the equilibrium point of the first two Equations of (5) since both dS/dt and dI/dt vanish at $t = 0$. We show this according to Proposition 5.

Proposition 5 ([22]). *If $(S(t), I(t), R(t))$ is a solution of the interval base uncertain model Equation (5) then $S(t) + I(t) \geq N$, and $0 \leq R(t) \leq N$ for all $t > 0$.*

By dividing $\frac{dS}{dt}$ by $\frac{dR}{dt}$, which yields
$$\frac{dS}{dR} = \frac{-\tilde{\beta}SI}{\tilde{\gamma}I} = \frac{-\tilde{\beta}S}{\tilde{\gamma}}.$$

therefore, $\int_0^t \frac{dS}{S} = \int_0^t \left(-\frac{\tilde{\beta}}{\tilde{\gamma}}\right) dR$. By the initial condition, we obtain
$$\log_e \left(\frac{S(t)}{S_0}\right) = \log_e S(t) - \log_e S(0) = -\frac{\tilde{\beta}}{\tilde{\gamma}} \int_0^t dR,$$
$$\log_e \left(\frac{S(t)}{S_0}\right) = -\frac{\tilde{\beta}}{\tilde{\gamma}}[R(t) - R(0)],$$
$$\log_e \left(\frac{S(t)}{S_0}\right) = -\frac{\tilde{\beta}}{\tilde{\gamma}}R(t) + \frac{\tilde{\beta}}{\tilde{\gamma}}R_0.$$

So that $\frac{S(t)}{S_0} = e^{-\frac{\tilde{\beta}}{\tilde{\gamma}}R(t)} \cdot e^{\frac{\tilde{\beta}}{\tilde{\gamma}}R_0}$, and
$$S(t) = S_0 e^{-\frac{\tilde{\beta}}{\tilde{\gamma}}[R(t) - R(0)]}.$$

From Proposition 5, $0 < R(t) \leq N$ and we have that $S_0 e^{-\frac{\tilde{\beta}N}{\tilde{\gamma}}} \leq S_0 e^{-\frac{\tilde{\beta}}{\tilde{\gamma}}[R(t) - R(0)]} \leq S_0$. Because $S_0 > 0$, we conclude that $0 < S(t) \leq S_0$ for all $t \geq 0$. We show this according to Lemma 1.

Lemma 1 ([22]). *Suppose $(S(t), I(t), R(t))$ be a solution of Equation (5) in the domain $\Gamma_2 = \{(S, I) : S \geq 0, I \geq 0, S + I \leq N\}$ then $0 < S(t) \leq S_0$ and $S(t) = S_0 e^{\frac{\tilde{\beta}(R(t) - R(0))}{\tilde{\gamma}}} \geq S_0 e^{-\frac{\tilde{\beta}N}{\tilde{\gamma}}}$ for all $t \geq 0$.*

Recall that from the first equation of Equation (5) and Proposition 5, we have $\frac{dS}{dt} = -\tilde{\beta}SI \leq 0$ and we say $S(t)$ is a decreasing function, then $\lim_{t \to \infty} S(t) = S_\infty$, such that S_∞ is a finite number. Recall also from Equation (5), the third equation $\frac{dR}{dt} = \tilde{\gamma}I \geq 0$ and we say $R(t)$ is an increasing function. Hence, by Proposition 5, $\lim_{t \to \infty} R(t) = R_\infty$, then R_∞ is a finite number. We show this according to Lemma 2.

Lemma 2 ([22]). *If $(S(t), I(t), R(t))$ is a solution of Equation (5), then $S(t) \to S_\infty$ as $R(t) \to R_\infty$ as $t \to \infty$, such that S_∞ and R_∞ are finite numbers.*

Recall from Proposition 5 and from Lemma 2, $\lim_{t\to\infty} R(t) = R_\infty$. So that $\lim_{t\to\infty} \int_0^t I(m)dm = \frac{R_\infty}{\gamma}$. Therefore, $\int \frac{dR}{dt} = \int \tilde{\gamma} I dt = R(t) = \tilde{\gamma} \int_0^t I(m)dm$ implies that $\lim_{t\to\infty} \frac{R(t)}{\tilde{\gamma}} = \lim_{t\to\infty} \int_0^t I(m)dm$. Then, $\lim_{t\to\infty} \int_0^t I(m)dm$ converges. Therefore, $\sum_{v=0}^\infty I(v)$ is convergent and $\lim_{t\to\infty} I(t) = 0$.

Alternately, we integrate the first equation of Equation (5):

$$\int_0^\infty \frac{dS}{dt} dt = -\tilde{\beta} \int_0^\infty S(t)I(t)dt.$$

Because $S_\infty - S_0 = -\tilde{\beta} \int_0^\infty S(t)I(t)dt$ and $S_0 - S_\infty = \tilde{\beta} \int_0^\infty S(t)I(t)dt$, then

$$S_0 - S_\infty \geq \tilde{\beta} \int_0^\infty S(t)I(t)dt,$$

which implies that $I(t)$ is integrable in the interval $[0, \infty)$, and hence, $\lim_{t\to\infty} I(t) = 0$. We show this according to Lemma 3.

Lemma 3 ([20]). *If $(S(t), I(t), R(t))$ is a solution of Equation (5) then $I(t) \to 0$ as $t \to \infty$.*

We hereby present below the procedure for the HAM for the benefit of finding the numerical solution of our interval-based uncertain model. Consider a nonlinear equation of the form

$$A[v(t)] = 0, \qquad (6)$$

where A is a linear operator, t denotes the time, and $v(t)$ is an unknown function. Let $v_0(t)$ denote an initial approximation of $v(t)$ and Z denote an auxiliary linear operator [21]. We construct the zero-order deformation equation

$$(1-q)Z[\varphi(t;q) - \vartheta_0(t)] = qh_1 H(t) A(t;p), \qquad (7)$$

where $q \in [0,1]$ is the embedding parameter and $h \neq 0$ is a non-zero auxiliary function. When $q = 0$ and $q = 1$, the zero-order deformation equation becomes, respectively,

$$\varphi(t;0) = \vartheta_0(t) \qquad (8)$$

and

$$\varphi(t;1) = \vartheta_0(t). \qquad (9)$$

Thus, as q increases from 0 to 1, the solution $\varphi(t;q)$ varies continuously from the initial approximation $\vartheta_0(t)$ of the exact solution $\vartheta(t)$. Such a kind of continuous variation is called deformation in topology. Expanding $\varphi(t;p)$ by the Taylor series in the power series of q, we have

$$\varphi(t;q) = \vartheta_0(t) + \sum_{m=1}^\infty \vartheta_m q^m, \qquad (10)$$

where

$$\vartheta_m(t) = \frac{1}{m!} \frac{\partial^m \varphi(t;q)}{\partial q^m} \qquad (11)$$

is the deformation derivative. If the auxiliary linear operator A, the initial approximation $v_0(t)$, the auxiliary parameter h_I and the auxiliary function $H(t)$ are properly chosen so that

(i) the solution $\varphi(t;q)$ of the zero-order deformation Equation (6) exists for all $q \in [0,1]$,
(ii) the deformation derivative (11) exists for all $m = 1, 2, \ldots$,
(iii) the series (10) converge at $q = 1$,

then, we have the series solution:

$$\varphi(t;1) = \vartheta_0(t) + \sum_{m=1}^\infty \vartheta_m(t). \qquad (12)$$

Define the vector as

$$\vec{\vartheta}_m(t) = \{\vartheta_0(t), \vartheta_1(t), \ldots, \vartheta_m(t)\}. \tag{13}$$

According to the definition (10), the governing equation can be derived from the zero-order deformation Equation (6). Differentiating (6) m-times with respect to the embedding parameter q, then by setting $q = 0$, and finally, dividing by m, we have the m-th order deformation equation

$$Z[b_m(t) - \lambda_m \vartheta_{m-1}(t)] = hH(t)P_m\left(\vec{\vartheta}_{m-1}(t)\right), \tag{14}$$

where

$$P_m\left(\vec{\vartheta}_{m-1}(t)\right) = \frac{1}{(m-1)!} \frac{\partial^{m-1} A[\varphi(t;q)]}{\partial q^{m-1}}, \tag{15}$$

$$\lambda_m = \begin{cases} 0 \text{ if } m \leq 1, \\ 1 \text{ if } m > 1. \end{cases} \tag{16}$$

Note that according to definition (16), the right-hand side of (15) depends only on $\vec{\vartheta}_{m-1}(t)$. Thus, we easily gain the series $\vartheta_1(t), \vartheta_2(t), \ldots$ by solving the linear higher-order deformation Equation (15) using the well-known symbolic computation software such as Maple, Matlab, or Mathematica. Prediction and controlling the infection was studied in detail not only in [22] but also in other papers, for example [4,23–36]. We discuss in the next section the homotopy analysis method.

6. Homotopy Analysis Method

For this section, we solved the interval-based uncertain model (5) by considering intervals of the transmission as $\tilde{\beta} = [0.01, 0.03]$ and the interval of recovery as $\tilde{\gamma} = [0.005, 0.015]$, respectively. To solve the interval-based uncertain model Equation (5) by the HAM, we consider the first equation in the interval-based uncertain model Equation (5) and choose the linear operator

$$A[S(t;q)] = \frac{dS(t;q)}{dt} \tag{17}$$

with the property that

$$A[\alpha_1] = 0, \tag{18}$$

where α_1 is a constant of integration. The inverse operator A^{-1} is given by

$$A^{-1}(\cdot) = \int_0^t (\cdot) dt. \tag{19}$$

Let the nonlinear operator be defined as

$$A[S(t;q)] = \frac{dS(t;q)}{dt} - \beta S(t;q) I(t;q). \tag{20}$$

The proper selection of the auxiliary parameter and function during the implementation of the HAM method can yield uniformly valid and accurate solutions [19].

By constructing the zero-order deformation equation we have the following:

$$(1-q)A[S(t;q) - S_0(t;q)] = qh_1 H(t) A[S(t;p)], \tag{21}$$

where

(i) if $q = 0$ then $S(t;0) = S_0(t)$,
(ii) if $q = 1$ then $S(t;1) = S(t)$.

Therefore, we have the m-th order deformation equation

$$A[S_{h,m}(t) - \lambda S_{m-1}(t)] = h_1 H(t) P\left(\vec{S}_{m-1}(t)\right), \ m \geq 1, \tag{22}$$

where

$$P_m\left(\vec{S}_{m-1}(t)\right) = \frac{d^{m-1} S_{m-1}(t)}{dt} - \tilde{\beta} SI. \tag{23}$$

The solution of the m-th order deformation Equation (22) for $m > 1$ and using $h_1 = -1$ and $H(t) = 1$ is given by

$$S_m(t) = \lambda_m S_{m-1}(t) - \int_\infty^t \left[\frac{d^{m-1} S_{m-1}(t)}{dt} + \beta \sum_{k=0}^{m-1} S_k(t) I_{m-1-k}(t)\right] dt, \ m \geq 1. \tag{24}$$

Following earlier steps, we obtain

$$I_m(t) = \lambda_m I_{m-1}(t) - \int_\infty^t \left[\frac{d^{m-1} I_{m-1}(t)}{dt} - \beta \sum_{k=0}^{m-1} S_k(t) I_{m-1-k}(t) + \gamma I_{m-1}(t)\right] dt \tag{25}$$

and

$$R_m(t) = \lambda_m R_{m-1}(t) - \int_\infty^t \left[\frac{d^{m-1} R_{m-1}(t)}{dt} - \gamma I_{m-1}(t)\right] dt, \tag{26}$$

where $m \geq 1$ in both last equations.

7. Numerical Results and Discussion

In this section, we present the results of the homotopy analysis method for solving an interval-based uncertain model. The solutions of the interval-based uncertain model with interval $\tilde{\beta} = [0.01, 0.03]$ and constant value $\gamma = 0.01$ in Table A1, and with interval $\tilde{\gamma} = [0.01, 0.015]$ and constant value $\beta = 0.01$ in Table A2. Tables A3 and A4 present the minimum, maximum, and midpoints of the susceptible, infected, and recovered human population with intervals of β and γ. The results of the HAM show strong agreement with the approximation technique. In Table A3, we present the result obtained by the Runge–Kutta of fourth order method for the susceptible, infected, and recovered humans. Then, we observed that the results are in good agreement with the homotopy analysis method (HAM) in Table A4.

In Table A1, we present the result of the susceptible, infected, and recovered humans, where β is considered an interval and γ is given as a constant. In Table A2, β is considered a constant and γ is given as an interval. It is observed from Table A1 that as time increases, the lower bound (minimum) and the upper bound (maximum) are decreasing for the susceptible human population. It is also detected that the lower bound (minimum) and the upper bound (maximum) of both the infected and recovered human populations increase with time.

In Table A2, it is observed that the same situation seems to be occurring in both the lower bound (minimum) and the upper bound (maximum) for the susceptible humans. It is also noticed that the lower bound (minimum) and the upper bound (maximum) of both the infected and recovered human populations increase with time.

It is seen from Tables A3 and A4 that the lower bound (minimum) and the upper bound of the susceptible population is decreasing with time, as seen from Tables A1 and A2. At the same time, the lower bound (minimum) and the upper bound (maximum) of both the infected and recovered human populations increase with time. In Tables A3 and A4, the interval [$\beta = 0.02, \gamma = 0.01$] denotes the center for the intervals $\tilde{\beta} = [0.01, 0.03]$ and $\tilde{\gamma} = [0.01, 0.015]$, while in Tables A1 and A2, the interval [$\beta = 0.02, \gamma = 0.01$] represents the center for β and constant value γ. In the next section we discuss the stochastic version of the model.

8. Stochastic Version of the Model

In this part, we denote the complete probability space with a filtration $\{\mathfrak{F}_t\}_{t\geq 0}$ with $(\Omega, \mathfrak{F}, Q)$ and it satisfies the condition that it is increasing and continuous while $\mathfrak{F}_0$ have every Q-empty sets. We introduce randomness into Equation (5) and assume that the white noise depends on the size of the corresponding populations where we applied the corresponding pattern $f_i(S(t).I(t), R(t))dW(t)$, such that f_i represents the intensity of the random perturbation $i \in [1,3]$ and $W(t)_{t\geq 0}$ is a single dimensional Brownian motion that is defined on a complete probability space $(\Omega, \mathfrak{F}, \{\mathfrak{F}_t\}_{t\geq 0}, Q)$. Then, the stochastic model of the SIR system (5) is described by the stochastic differential equations (SDEs):

$$dS = \left(-\tilde{\beta}SI\right)dt + f_1 S(t)dW(t),$$
$$dI = \left(\tilde{\beta}SI - \tilde{\gamma}I\right)dt + f_2 I(t)dW(t), \quad (27)$$
$$dR = \tilde{\gamma}Idt + f_3 R(t)dW(t).$$

Let $X(t) = (S(t), I(t), R(t))$. Then, we can rewrite Equation (5) in the pattern of a single dimensional SDE of the form

$$dX(t) = F(X(t), t)dt + G(X(t), t)dW(t)$$

such that $F: \mathbb{R}_+^2 \times \mathbb{R}_+^2 \to \mathbb{R}_+^2$, which is given by

$$F = \begin{pmatrix} -\tilde{\beta}SI \\ \tilde{\beta}SI - \tilde{\gamma}I \\ \tilde{\gamma}I \end{pmatrix} \quad (28)$$

and the function $G: \mathbb{R}_+^2 \times \mathbb{R}_+^2 \to \mathbb{R}_+^2$ is given by

$$G = \begin{pmatrix} f_1 S(t) \\ f_2 I(t) \\ f_3 R(t) \end{pmatrix}. \quad (29)$$

In the next section, we discuss the graphical illustration of our results.

9. Graphical Illustration of Our Results

Figure 1 shows the plot of the maximum, midpoint, and the minimum of the susceptible human intervals $\tilde{\beta} = [0.01, 0.03]$ and $\tilde{\gamma} = [0.01, 0.015]$. It reveals that as the maximum value is decreasing, the midpoint is also decreasing, as is the minimum point. It is clearly seen from the plot that the uncertainty lies between the upper and lower bounds. Figure 2 shows the plot of the maximum, midpoint, and the minimum of the infected human intervals $\tilde{\beta} = [0.01, 0.03]$ and $\tilde{\gamma} = [0.01, 0.015]$. It reveals that as the maximum value is increasing, the midpoint is also increasing, as is the minimum point. It is clearly seen from the plot that the uncertainty lies between the upper and lower bounds. Figure 3 shows the plot of the maximum, midpoint, and the minimum of the recovered human intervals $\tilde{\beta} = [0.01, 0.03]$ and $\tilde{\gamma} = [0.01, 0.015]$. It reveals that as the maximum value is increasing, the midpoint is also increasing, as is the minimum point. It is clearly seen from the plot that the uncertainty lies between the upper and lower bounds. We discuss in Section 10 the numerical solutions of the stochastic differential equation model.

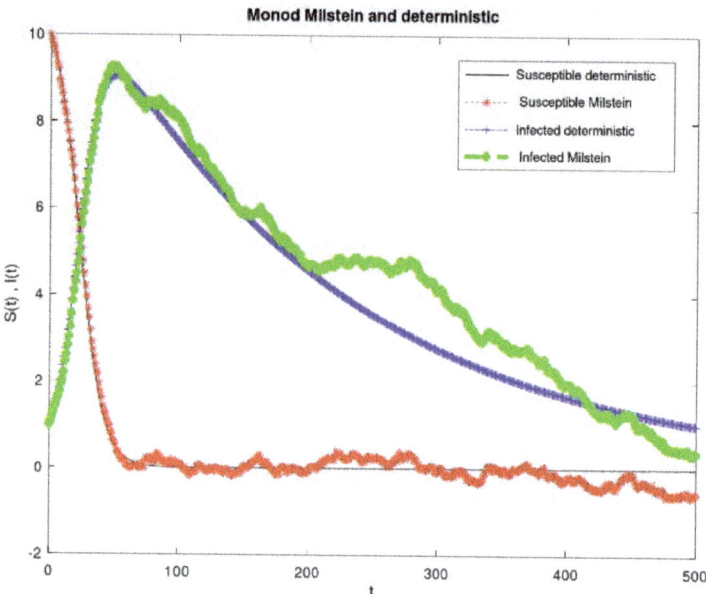

Figure 1. Plot of the dynamic behaviors of the susceptible and infected populations under the intervals $\widetilde{\beta} = 0.01$ and $\widetilde{\gamma} = 0.01$.

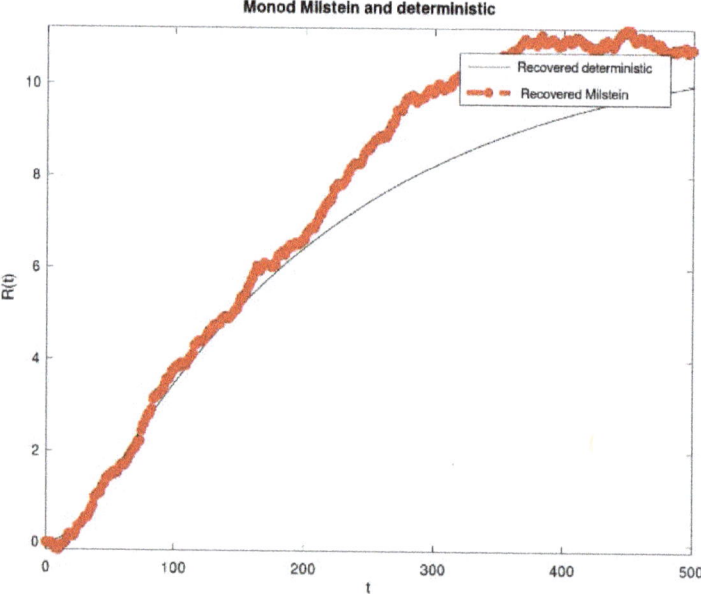

Figure 2. Plot of the dynamic behaviors of the susceptible and infected populations under the intervals $\widetilde{\beta} = 0.01$ and $\widetilde{\gamma} = 0.01$.

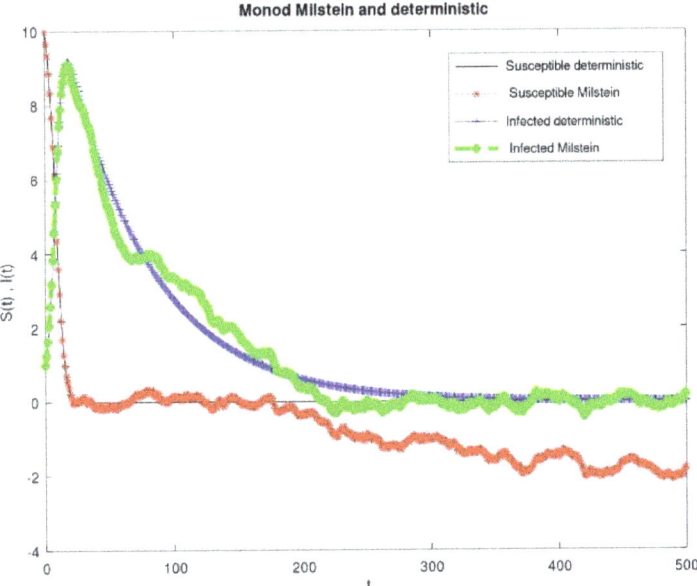

Figure 3. Plot of the dynamic behaviors of the susceptible and infected populations under the intervals $\tilde{\beta} = 0.03$ and $\tilde{\gamma} = 0.015$.

10. Numerical Solution of the SDE Model

In this section, we present the simulation of the SDE model (27) with the use of the Milstein method given the parameter value intervals $\tilde{\beta} = [0.01, 0.03]$ and $\tilde{\gamma} = [0.01, 0.015]$. We obtained our numerical results of the SDE model for 500 runs of the stochastic model simulation and the results of the corresponding deterministic model are presented in Figures 1–4, in which we display the time series solution of all the variables in the SDE model. It was obtained that in the Figures 1–4, the SDE model simulations are lower than their deterministic model simulation.

Figure 1 shows the simulations of the dynamic behaviors of the susceptible and the infected populations under the intervals $\tilde{\beta} = 0.01$ and $\tilde{\gamma} = 0.005$. It was observed that the stochastic simulations of the susceptible and the infected populations were lower than their deterministic simulations. Figure 2 shows the simulations of the dynamic behaviors of the recovered population under the intervals $\tilde{\beta} = 0.01$ and $\tilde{\gamma} = 0.005$. It was observed that the stochastic simulations of the recovered population were higher than their deterministic simulations. Figure 3 shows the simulations of the dynamic behaviors of the susceptible and the infected populations under the intervals $\tilde{\beta} = 0.03$ and $\tilde{\gamma} = 0.015$. It was observed that the stochastic simulations of the susceptible and the infected populations were lower than their deterministic simulations. Figure 4 shows the simulations of the dynamic behaviors of the recovered populations under the intervals $\tilde{\beta} = 0.03$ and $\tilde{\gamma} = 0.015$. It was observed that the stochastic simulations of the recovered population were lower than their deterministic simulations.

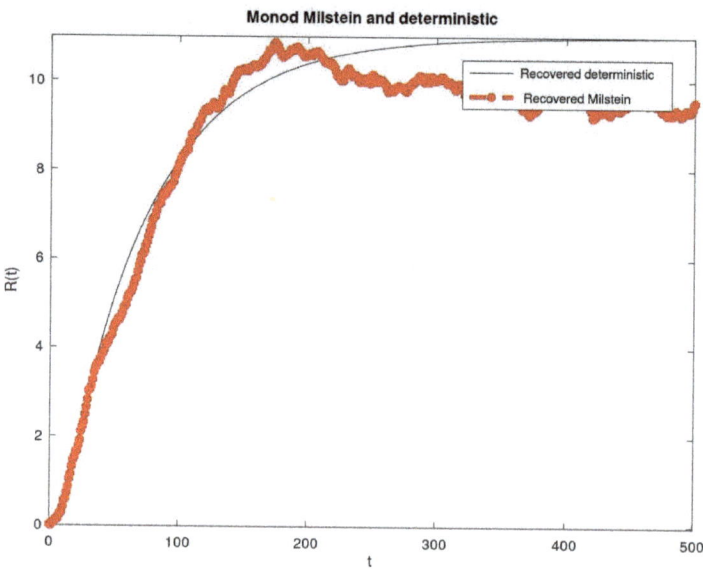

Figure 4. Plot of the dynamic behaviors of the susceptible and infected populations under the intervals $\widetilde{\beta} = 0.03$ and $\widetilde{\gamma} = 0.015$.

11. Discussion and Conclusions

In this work, we have studied the interval-based uncertain model of a three-compartment mathematical model rigorously. The homotopy analysis approach has been employed to solve the system of nonlinear equations of SIR interval uncertainty, in particular. The results obtained were compared with the known solution and are found to be in good agreement. Hence, it was established here that the homotopy analysis method has greater advantages over other analytical methods in many different ways. The HAM is a series expansion method that is directly dependent on small or large physical parameters, and hence, it is applicable for not only weakly but also strongly nonlinear problems. It also allows for the strong convergence of the solution over larger spatial and parameter domains. It also gives excellent flexibility in the expression of the solution and how the solution is explicitly obtained. It provides a simple way to ensure the convergence of the solution series. Comparing the stochastic and deterministic versions of the model, we saw that the population of the susceptible, infected, and recovered populations fell between the intervals obtained in the interval-based model. These suggest that the interval-based model give a very good range for the general SIR epidemic model. In the future, we plan to use fuzzy differential equations to capture the dynamics, and we also plan to look into a more practical problem that may be grounded with epidemiological data.

Author Contributions: Conceptualization, E.A.B., S.C. and R.P.; methodology, E.A.B. and S.C.; software, E.A.B.; validation, E.A.B., S.C. and R.P.; formal analysis, investigation, resources, and data curation, E.A.B.; writing—original draft preparation, E.A.B. and R.P.; writing—review and editing, S.C. and R.P.; funding acquisition, R.P. All authors have read and agreed to the published version of the manuscript.

Funding: This research received no external funding. The APC was funded by VAROPS granted by the Ministry of Defense of the Czech Republic.

Data Availability Statement: The data used to support the findings of this study are included within the article.

Acknowledgments: The authors thank the ICT Department of the Federal University Oye Ekiti, Ekiti State, Nigeria for the provision of space to complete this work and for their numerous support, as well as the Department of Science and Technology, Government of India and the Ministry of Defense of the Czech Republic for the support via the VAROPS grant.

Conflicts of Interest: The authors declare no conflict of interest.

Appendix A

The solutions obtained by the homotopy analysis method and the Runge–Kutta of the fourth order method for various intervals $\tilde{\beta} = \left[\underline{\beta}, \overline{\beta}\right]$ and $\tilde{\gamma} = \left[\underline{\gamma}, \overline{\gamma}\right]$ and for various constant values of β and γ are stated in Tables A1–A4. Further, Figures A1–A3 are plotted with the maximum, center, and minimum of susceptible, infected and recovered humans under the intervals $\tilde{\beta} = [0.01, 0.03]$ and $\tilde{\gamma} = [0.01, 0.015]$.

Table A1. The solutions obtained by the homotopy analysis method with the interval $\tilde{\beta} = [0.01, 0.03]$ and the constant value of $\gamma = 0.01$.

	S			I			R		
Time (t)	[min, max]	Midpoint of [min, max]	[β = 0.02, γ = 0.01]	[min, max]	Midpoint of [min, max]	[β = 0.02, γ = 0.01]	[min, max]	Midpoint of [min, max]	[β = 0.02, γ = 0.01]
0.1	[19.095, 19.700]	19.396	19.397	[15.270, 15.875]	15.573	15.587	[10.015, 10.015]	10.015	10.015
0.2	[18.183, 19.398]	18.791	18.791	[15.542, 16.757]	16.150	16.178	[10.031, 10.032]	10.032	10.031
0.3	[17.268, 19.097]	18.183	18.182	[15.813, 17.642]	16.728	16.770	[10.046, 10.049]	10.048	10.048
0.4	[16.357, 18.794]	17.576	17.572	[16.086, 18.524]	17.305	17.363	[10.062, 10.067]	10.065	10.065
0.5	[15.452, 18.492]	16.972	16.962	[16.358, 19.398]	17.878	17.955	[10.078, 10.086]	10.082	10.082
0.6	[14.560, 18.190]	16.375	16.354	[16.631, 20.260]	18.446	18.545	[10.095, 10.106]	10.101	10.101
0.7	[13.684, 17.887]	15.786	15.749	[16.903, 21.107]	19.418	19.131	[10.112, 10.127]	10.120	10.119
0.8	[12.828, 17.585]	15.207	15.149	[17.176, 21.932]	19.554	19.712	[10.129, 10.148]	10.139	10.139
0.9	[11.997, 17.283]	14.640	14.555	[17.447, 22.734]	20.091	20.286	[10.146, 10.170]	10.158	10.159
1.0	[11.193, 16.982]	14.088	13.969	[17.719, 23.508]	20.614	20.852	[10.164, 10.193]	10.179	10.179

Table A2. The solutions obtained by the homotopy analysis method with the interval $\tilde{\gamma} = [0.01, 0.015]$ and the constant value of $\beta = 0.01$.

	S			I			R		
Time (t)	[min, max]	Midpoint of [min, max]	[β = 0.02, γ = 0.01]	[min, max]	Midpoint of [min, max]	[β = 0.02, γ = 0.01]	[min, max]	Midpoint of [min, max]	[β = 0.02, γ = 0.01]
0.1	[19.699, 19.700]	19.700	19.700	[15.255, 15.286]	15.270	15.270	[10.015, 10.015]	10.015	10.015
0.2	[19.398, 19.399]	19.399	19.398	[15.511, 15.572]	15.542	15.542	[10.031, 10.032]	10.032	10.031
0.3	[19.095, 19.098]	19.097	19.097	[15.767, 15.860]	15.814	15.813	[10.046, 10.049]	10.048	10.048
0.4	[18.792, 18.797]	18.795	18.794	[16.023, 16.148]	16.086	16.086	[10.062, 10.067]	10.065	10.065

Table A2. Cont.

		S			I			R	
Time (t)	[min, max]	Midpoint of [min, max]	[β = 0.02, γ = 0.01]	[min, max]	Midpoint of [min, max]	[β = 0.02, γ = 0.01]	[min, max]	Midpoint of [min, max]	[β = 0.02, γ = 0.01]
0.5	[18.489, 18.496]	18.493	18.492	[16.280, 16.437]	16.359	16.358	[10.078, 10.086]	10.082	10.082
0.6	[18.184, 18.195]	18.190	18.190	[16.536, 16.726]	16.631	16.631	[10.095, 10.106]	10.101	10.101
0.7	[17.880, 17.894]	17.887	17.887	[16.792, 17.015]	16.904	16.903	[10.112, 10.127]	10.120	10.119
0.8	[17.576, 17.594]	17.585	17.585	[17.047, 17.304]	17.176	17.176	[10.129, 10.148]	10.139	10.139
0.9	[17.272, 17.294]	17.283	17.283	[17.302, 17.593]	17.448	17.447	[10.146, 10.170]	10.158	10.159
1.0	[16.968, 16.995]	16.982	16.982	[17.556, 17.881]	17.719	17.719	[10.164, 10.193]	10.179	10.179

Table A3. The solutions obtained by the Runge–Kutta of the fourth order method with intervals $\widetilde{\beta}$ = [0.01, 0.03] and $\widetilde{\gamma}$ = [0.01, 0.015].

		S			I			R	
Time (t)	[min, max]	Midpoint of [min, max]	[β = 0.02, γ = 0.01]	[min, max]	Midpoint of [min, max]	[β = 0.02, γ = 0.01]	[min, max]	Midpoint of [min, max]	[β = 0.02, γ = 0.01]
0.1	[19.095, 19.699]	19.397	19.397	[15.278, 15.898]	15.588	15.587	[10.008, 10.023]	10.015	10.015
0.2	[18.180, 19.398]	18.789	18.791	[15.556, 16.804]	16.179	16.178	[10.015, 10.048]	10.031	10.031
0.3	[17.263, 19.095]	18.179	18.182	[15.835, 17.713]	16.774	16.770	[10.023, 10.073]	10.048	10.048
0.4	[16.347, 18.793]	17.570	17.572	[16.113, 18.619]	17.366	17.363	[10.031, 10.101]	10.066	10.065
0.5	[15.438, 18.490]	16.964	16.962	[16.392, 19.519]	17.956	17.956	[10.039, 10.129]	10.084	10.082
0.6	14.541, 18.187]	16.364	16.354	[16.670, 20.406]	18.538	18.545	[10.048, 10.159]	10.103	10.101
0.7	[13.659, 17.884]	15.772	15.749	[16.948, 21.277]	19.113	19.131	[10.056, 10.189]	10.123	10.119
0.8	[12.822, 17.581]	15.202	15.149	[17.225, 22.127]	19.676	19.712	[10.065, 10.222]	10.143	10.139
0.9	[11.962, 17.279]	14.620	14.555	[17.502, 22.953]	20.227	20.286	[10.073, 10.256]	10.165	10.159
1.0	[11.152, 16.976]	14.064	13.969	[17.778, 23.750]	20.764	20.852	[10.083, 10.291]	10.186	10.179

Table A4. The solutions obtained by the homotopy analysis method with intervals $\widetilde{\beta}$ = [0.01, 0.03] and $\widetilde{\gamma}$ = [0.01, 0.015].

		S			I			R	
Time (t)	[min, max]	Midpoint of [min, max]	[β = 0.02, γ = 0.01]	[min, max]	Midpoint of [min, max]	[β = 0.02, γ = 0.01]	[min, max]	Midpoint of [min, max]	[β = 0.02, γ = 0.01]
0.1	[19.094, 19.699]	19.397	19.399	[15.255, 15.891]	15.573	15.572	[10.008, 10.023]	10.015	10.015
0.2	[18.181, 19.399]	18.790	18.792	[15.511, 16.789]	16.149	16.148	[10.015, 10.048]	10.031	10.031
0.3	[17.265, 19.098]	18.181	18.184	[15.767, 17.690]	16.729	16.726	[10.023, 10.073]	10.048	10.048

Table A4. Cont.

		S			I			R	
Time (t)	[min, max]	Midpoint of [min, max]	[β = 0.02, γ = 0.01]	[min, max]	Midpoint of [min, max]	[β = 0.02, γ = 0.01]	[min, max]	Midpoint of [min, max]	[β = 0.02, γ = 0.01]
0.4	[16.350, 18.797]	17.574	17.576	[16.024, 18.589]	17.307	17.304	[10.031, 10.100]	10.066	10.065
0.5	[15.443, 18.496]	16.969	16.968	[16.279, 19.482]	17.881	17.882	[10.039, 10.129]	10.084	10.082
0.6	[14.547, 18.195]	16.371	16.363	[16.536, 20.363]	18.500	18.457	[10.048, 10.158]	10.103	10.100
0.7	[13.667, 17.894]	15.780	15.761	[16.792, 21.228]	19.010	19.029	[10.056, 10.189]	10.123	10.119
0.8	[12.807, 17.594]	15.201	15.164	[17.047, 22.073]	19.560	19.597	[10.065, 10.221]	10.143	10.138
0.9	[11.972, 17.294]	14.633	14.573	[17.302, 22.893]	20.098	20.158	[10.073, 10.255]	10.164	10.164
1.0	[11.164, 16.968]	14.066	13.989	[17.556, 23.686]	20.621	20.712	[10.082, 10.289]	10.186	10.179

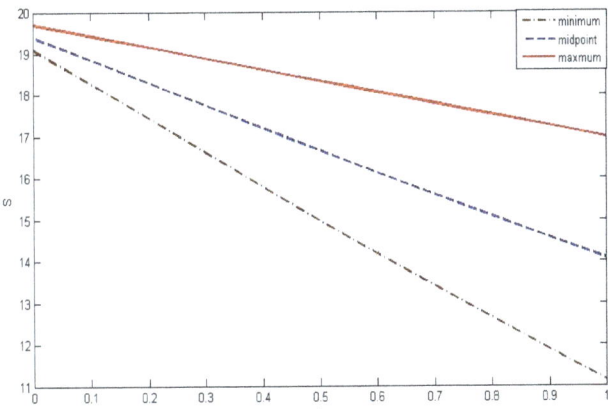

Figure A1. Plot of the maximum, center, and minimum of susceptible humana under the intervals $\tilde{\beta} = [0.01, 0.03]$ and $\tilde{\gamma} = [0.01, 0.015]$.

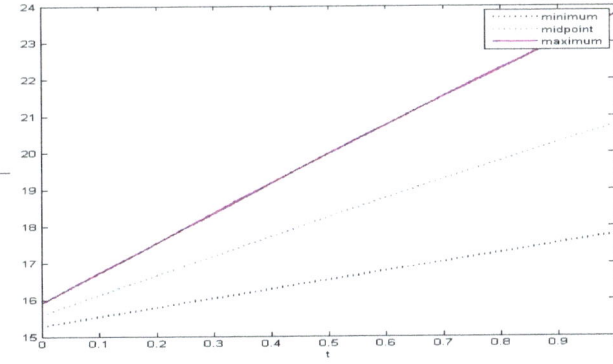

Figure A2. Plot of the maximum, center, and minimum of infected humans under the intervals $\tilde{\beta} = [0.01, 0.03]$ and $\tilde{\gamma} = [0.01, 0.015]$.

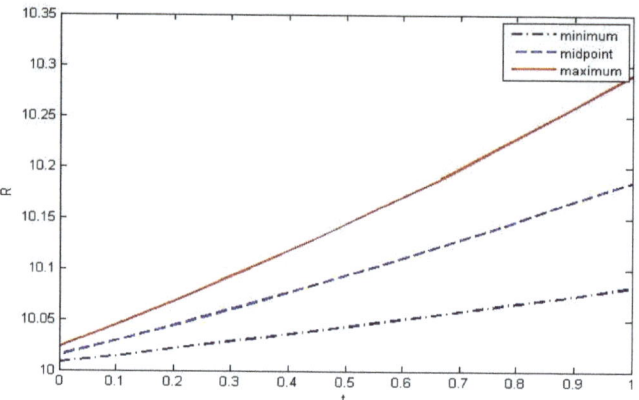

Figure A3. Plot of the maximum, center, and minimum of recovered humans under the intervals $\widetilde{\beta}$ = [0.01, 0.03] and $\widetilde{\gamma}$ [0.01, 0.015].

References

1. Moore, R.E.R.; Baker, K.; Michael, J.C. *Introduction to Interval Analysis. Society for Industrial and Applied Mathematics*; Society for Industrial and Applied Mathematics: Philadelphia, PA, USA, 2009; 235p, ISBN 978-0-898716-69-6. Available online: https://1lib.cz/book/673847/09274f (accessed on 10 December 2017).
2. Tapaswini, S.; Chakraverty, S. A New Approach to Fuzzy Initial Value Problem by Improved Euler Method. *Fuzzy Inf. Eng.* **2012**, *4*, 293–312. [CrossRef]
3. Tapaswini, S.; Chakraverty, S. Numerical Solution of n-th Order Fuzzy Linear Differential Equations by Homotopy Perturbation Method. *Int. J. Comput. Appl.* **2013**, *64*. [CrossRef]
4. Alefeld, G.; Mayer, G. Interval analysis: Theory and applications. *J. Comput. Appl. Math.* **2000**, *121*, 421–464. [CrossRef]
5. Hickey, T.; Ju, Q.; van Emden, M.H. Interval Arithmetic: From Principles to Implementation. *J. ACM* **2001**, *48*, 1038–1068. [CrossRef]
6. Liao, S.J. Homotopy Analysis Method. Available online: https://numericaltank.sjtu.edu.cn/IntroductionHAM.htm (accessed on 27 May 2021).
7. Hayat, T.; Sajid, M. Homotopy analysis of MHD boundary layer flow of an upper-convected Maxwell fluid. *Int. J. Eng. Sci.* **2007**, *45*, 393–401. [CrossRef]
8. Khan, H.; Mohapatra, R.N.; Vajravelu, K.; Liao, S. The explicit series solution of SIR and SIS epidemic models. *Appl. Math. Comput.* **2009**, *215*, 653–669. [CrossRef]
9. Kermack, W.O.; McKendrick, A.G. A contribution to the mathematical theory of epidemics. *Proc. R. Soc. Lond. Ser. A Math. Phys. Sci.* **1927**, *115*, 700–721. [CrossRef]
10. Momoh, A.A.; Ibrahim, M.O.; Tahir, A.; Adamu, I.I. Application of homotopy analysis method for solving the SEIR models of epidemics. *Nonlinear Anal. Differ. Equ.* **2015**, *3*, 53–68. [CrossRef]
11. Motsa, S.S. The homotopy analysis method solution of SIR epidemic model. *J. Adv. Res. Differ. Equ.* **2010**, *2*, 1–9.
12. Bataineh, A.S.; Noorani, M.S.M.; Hashim, I. Series Solution of the Multispecies Lotka-Volterra Equations by Means of the Homotopy Analysis Method. *Differ. Equ. Nonlinear Mech.* **2008**, *2008*, 816787. [CrossRef]
13. Ghotbi, A.R. Homotopy analysis method for solving the MHD flow over a non-linear stretching sheet. *Commun. Nonlinear Sci. Numer. Simul.* **2009**, *14*, 2653–2663. [CrossRef]
14. Odibat, Z.; Bataineh, A.S. An adaptation of homotopy analysis method for reliable treatment of strongly nonlinear problems: Construction of homotopy polynomials. *Math. Methods Appl. Sci.* **2015**, *38*, 991–1000. [CrossRef]
15. Shivanian, E.; Abbasbandy, S. Predictor homotopy analysis method: Two points second order boundary value problems. *Nonlinear Anal. Real World Appl.* **2014**, *15*, 89–99. [CrossRef]
16. Jafari, H.; Firoozjaee, M.A. Multistage Homotopy Analysis Method for Solving Nonlinear Integral Equations. *Appl. Math. Int. J.* **2010**, 34–45. Available online: https://www.pvamu.edu/aam/special-issues/august-2010/ (accessed on 10 December 2017).
17. Sadaf, M.; Akram, G. An improved adaptation of homotopy analysis method. *Math. Sci.* **2017**, *11*, 55–62. [CrossRef]
18. Liao, S. Homotopy analysis method—A new analytic approach for highly nonlinear problems. *AIP Conf. Proc.* **2015**, *1648*, 020011. [CrossRef]
19. Liao, S.J. *Beyond Perturbation: Introduction to Homotopy Analysis Method*; Chapman and Hall/CRC Press: Washington, DC, USA, 2003.

20. Noeiaghdam, S.; Araghi, M.A.F. Application of the CESTAC Method to Find the Optimal Iteration of the Homotopy Analysis Method for Solving Fuzzy Integral Equations. In *Progress in Intelligent Decision Science*; Springer: Cham, Switzerland, 2020. [CrossRef]
21. Soltani, L.A.; Shivanian, E.; Ezzati, R. Convection–radiation heat transfer in solar heat exchangers filled with a porous medium: Exact and shooting homotopy analysis solution. *Appl. Therm. Eng.* **2016**, *103*, 537–542. [CrossRef]
22. Rachah, A.; Torres, D.F.M. Predicting and controlling the Ebola infection. *Math. Methods Appl. Sci.* **2016**, *40*, 6155–6164. [CrossRef]
23. Mengxin, H.; Zhong, L.; Fengde, C.; Zhenliang, Z. Dynamic Behaviors of an N-species Lotka-Volterra Model with Nonlinear Impulses. *IAENG Int. J. Appl. Math.* **2020**, *50*, 22–30.
24. Ayub, M.; Zaman, H.; Sajid, M.; Hayat, T. Analytical solution of stagnation-point flow of a viscoelastic fluid towards a stretching surface. *Commun. Nonlinear Sci. Numer. Simul.* **2008**, *13*, 1822–1835. [CrossRef]
25. Holub, M.; Bradac, F.; Pokorny, Z.; Jelinek, A. Application of a Ballbar fordiagnostics of CNC machine tools. *MM Sci. J.* **2018**, 2601–2605. [CrossRef]
26. Biazar, J. Solution of the epidemic model by Adomian decomposition method. *Appl. Math. Comput.* **2006**, *173*, 1101–1106. [CrossRef]
27. Hethcote, H.W. The Mathematics of Infectious Diseases. *SIAM Rev.* **2000**, *42*, 599–653. [CrossRef]
28. Bakare, E.; Hoskova-Mayerova, S. Optimal Control Analysis of Cholera Dynamics in the Presence of Asymptotic Transmission. *Axioms* **2021**, *10*, 60. [CrossRef]
29. Vagaska, A.; Gombar, M. Mathematical Optimization and Application of Nonlinear Programming. In *Algorithms as a Basis of Modern Applied Mathematics*; Springer Nature: Cham, Switzerland, 2021; pp. 461–486. ISBN 978-3-030-61333-4. [CrossRef]
30. Efimov, D.; Ushirobira, R. On an interval prediction of COVID-19 development based on a SEIR epidemic model. *Annu. Rev. Control* **2021**. preprint. Available online: https://hal.inria.fr/hal-02517866v4 (accessed on 27 May 2021). [CrossRef] [PubMed]
31. Li, C.; Huang, J.; Chen, Y.-H.; Zhao, H. A Fuzzy Susceptible-Exposed-Infected-Recovered Model Based on the Confidence Index. *Int. J. Fuzzy Syst.* **2021**, 1–11. [CrossRef]
32. Bao, K.; Rang, L.; Zhang, Q. Analysis of a stochastic SIRS model with interval parameter. *Am. Inst. Math. Sci. (AIMS)* **2019**, *24*, 4827–4849. [CrossRef]
33. Park, S.W.; Champredon, D.; Weitz, J.S.; Dushoff, J. A Practical Generation Interval-Based Approach to Inferring the Strength of Epidemics from Their Speed. 2018. Available online: https://www.biorxiv.org/content/10.1101/312397v2.full.pdf (accessed on 28 May 2021).
34. Bracher, J.; Ray, E.L.; Gneiting, T.; Reich, N.G. Evaluating epidemic forecasts in an interval format. *PLoS Comput. Biol.* **2021**, *17*, e1008618. [CrossRef]
35. Bhuju, G.; Phaijoo, G.R.; Gurung, D.B. Fuzzy Approach Analyzing SEIR-SEI Dengue Dynamics. *BioMed Res. Int.* **2020**, *2020*, 1508613. [CrossRef]
36. Dhandapani, P.B.; Baleanu, D.; Thippan, J.; Sivakumar, V. On stiff, fuzzy IRD-14 day average transmission model of COVID-19 pandemic disease. *AIMS Bioeng.* **2020**, *7*, 208–223. [CrossRef]

 *axioms*

Article

Unveiling the Dynamics of the European Entrepreneurial Framework Conditions over the Last Two Decades: A Cluster Analysis

Eliana Costa e Silva *, Aldina Correia and Ana Borges

CIICESI, ESTG, Politécnico do Porto, 4610-156 Felgueiras, Portugal; aic@estg.ipp.pt (A.C.); aib@estg.ipp.pt (A.B.)
* Correspondence: eos@estg.ipp.pt

Abstract: Entrepreneurship is a theme of global interest, and it is the subject of investigations conducted by many researchers and projects. In particular, the Global Entrepreneurship Monitor project is a global project that involves several countries and years of surveys on entrepreneurship indicators. This study focuses on the 12 indicators of the entrepreneurial ecosystem defined by the Entrepreneurial Framework Conditions (EFCs). The EFCs are specifically related to the quality of the entrepreneurial ecosystem. Using clustering techniques, the present study analyzes how European experts' perceptions on the EFCs of their home country have changed between 2000 and 2019. The main finding is the existence of significant differences between the clusters obtained over the years and between countries. Therefore, in theoretical terms, this dynamical behavior in relation to the entrepreneurial conditions of economies should be considered in future works, namely, those concerning the definition of the number of clusters, which, according to the internal validation measures computed in this work, should be two.

Keywords: entrepreneurship; clustering; longitudinal analysis

MSC: 62H30; 62P20; 91-10

Citation: Costa e Silva, E.; Correia, A.; Borges, A. Unveiling the Dynamics of the European Entrepreneurial Framework Conditions over the Last Two Decades: A Cluster Analysis. *Axioms* **2021**, *10*, 149. https://doi.org/10.3390/axioms10030149

Academic Editors: Jesús Martín Vaquero, Deolinda M. L. Dias Rasteiro, Araceli Queiruga-Dios and Fatih Yilmaz

Received: 6 May 2021
Accepted: 30 June 2021
Published: 6 July 2021

Publisher's Note: MDPI stays neutral with regard to jurisdictional claims in published maps and institutional affiliations.

Copyright: © 2021 by the authors. Licensee MDPI, Basel, Switzerland. This article is an open access article distributed under the terms and conditions of the Creative Commons Attribution (CC BY) license (https://creativecommons.org/licenses/by/4.0/).

1. Introduction

In the last decades, the topic of entrepreneurship has gained increasing attention. Political leaders viewed entrepreneurial activity as a source of innovation, competitiveness and economic development, and academics set about deepening the knowledge about this core topic, resulting in it now representing a hybrid field comprised of different perspectives and theories [1]. Entrepreneurship is explained as an individual's ability to place ideas into practice; articulate project planning and management; take calculated risks; innovate; and creative with the purpose of achieving previously defined goals [2]. Thus, is it suggested that entrepreneurship may be a catalyst for economic growth and national competitiveness. In fact, as [3] explain in their extensive systematic literature review on entrepreneurial ecosystems, the growing interest in this topic is being guided largely by the interest demonstrated by policy makers in increasing entrepreneurial activity via the creation of new companies and promotion of self-employment.

The Global Entrepreneurship Monitor (GEM) research project, funded in 1997, is the largest ongoing study of entrepreneurial dynamics in the world [4]. The first report of this project was launched in 1999 and encompassed 10 developed economies—eight from the OECD (Canada, Denmark, Finland, France, Germany, Israel, Italy and the United Kingdom) as well as Japan and the United States of America [4], and it has grown to include a wide amount of economies over the world [5]. According to the GEM 2019/2020 Global Report, fifty economies participated in the GEM 2019 adult population survey, including 21 European countries.

The GEM survey is based on collecting primary data through an adult population survey (APS) of at least 2000 randomly selected adults (18–64 years of age) in each economy. Additionally, national teams collect experts' opinions about components of the entrepreneurship ecosystem through a national expert survey (NES) [4].

The present study focus on the 12 indicators compiled by the NES survey data concerning the entrepreneurial ecosystem defined by GEM, i.e., the Entrepreneurial Framework Conditions (EFCs), detailed in Table A1 in Appendix A.

Although the original GEM model expects national business activity to change with general national framework conditions, studies show that entrepreneurial activity varies according to the EFCs [2]. In line with that result, the aim of the present work is to study the changes that have occurred in the European experts' perceptions over the last two decades (between 2000 and 2019) in different countries.

There are already several studies that use GEM data in their research. Recently, [2], explained the entrepreneurial performance of economies taking into account the variables present in the EFCs combining factorial analysis with cluster analysis to group economies (countries). In addition, Pilar et al. [6] analyzed entrepreneurs' perceptions about conditions to create new and growing firms and their significance in the economic development level (EDL) of countries, using NES 2013. Braga et al. [7] analyzed GEM data in order to understand what leads certain countries' individuals to display higher levels of initiative to manage or create a high-growth business. In [8], NES datasets for 2011 until 2013 were analyzed to study the effects of different types of entrepreneurship expert specialization on the perceptions about the EFCs. Furthermore, the work of Autio et. al [9] also contributed to the understanding of the theoretical, managerial and policy implications of entrepreneurial innovation using GEM data.

Based on the similarities in economic performance across European countries, this study is mainly concerned with the evolution of experts' perceptions on the entrepreneurial framework in Europe, grouping countries in different clusters and analyzing how this grouping differs throughout the years. To achieve this goal, the present study uses multivariate cluster analysis to group all European economies according to the experts' perceptions on the EFCs of their home country (similarly to the methodology adopted by [2]). In the next section, the dataset, methods and results are presented, and in the last section, the discussion is given and future research directions are suggested.

2. Materials and Methods

2.1. Dataset

For citizens to become entrepreneurs, the conditions for entrepreneurship in their countries must be favorable. The GEM conceptual framework is based on the assumption that national economic growth is the result of the inter-dependencies between the EFCs and the personal traits and capabilities of individuals to identify and seize opportunities [10]. Thus, the behavior of these GEM indicators over the last two decades in Europe (between 2000 and 2019) are studied in this work. Although they do not directly measure the real conditions of the country, they measure them indirectly through the European experts' perceptions.

The two main sources of primary data of the GEM project are as follows:

- The adult population survey (APS), which provides standardized data on entrepreneurial activities and attitudes within each country—at least 2000 randomly selected adults (18–64 years of age)in each economy.
- The national expert survey (NES) investigates the national framework conditions for entrepreneurship by means of standardized questionnaires; national teams collect experts' opinions about components of the entrepreneurship ecosystem through a national expert survey.

In a previous study [11], the period from 2010 to 2016 was analyzed. Substantial changes in the clusters of European economies through these years were observed. In particular, it was found that despite the economic and financial similarities between Portugal, Italy, Greece and Spain, countries that all faced a dramatic period between 2010–2012, Portugal took off from the remaining countries after 2012, and only in 2016 was it caught up by Spain.

The present study aims at extending that work by considering the period before the crisis and after 2016 in order to obtain a wider view on European entrepreneurs' perceptions. For such purpose, multivariate cluster analysis techniques are used to group all of the European economies according to the experts' perceptions on the EFCs of their home country.

Therefore, the present study considers the 12 indicators of the entrepreneurial ecosystem, i.e., the EFCs, defined by the GEM project, for the whole the period of available data, namely from 2000 until 2019. The description of the EFCs is given in Table A1 in Appendix A.

The number of economies that participated in the NES survey between 2000 and 2019 ranges from a minimum of 11 countries in 2000 to a maximum of 29 countries in 2014 (see Figure 1).

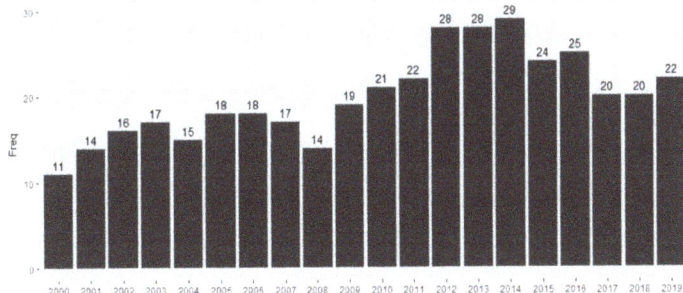

Figure 1. Number of economies in NES survey between 2000 and 2019.

Figure 2 illustrates the variation of each EFCs throughout the years and between countries. In general, large amplitudes, as observed for EFCs 2, 3, 4 and 11, reflect the differences in intra-country perceptions. The longitudinal volatility of the median, easily observed in EFCs 1, 8, 11 and 12, illustrates the annual differences in perceptions. This means that an intra-annual and intra-country difference is to be expected. The purpose of this study is to detect these differences by analyzing how countries are grouped, according to similar perceptions, over the years in the last two decades.

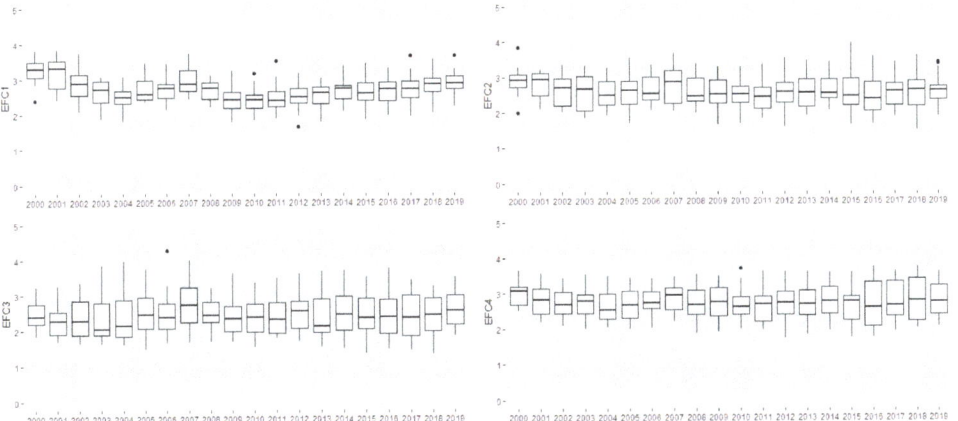

Figure 2. Cont.

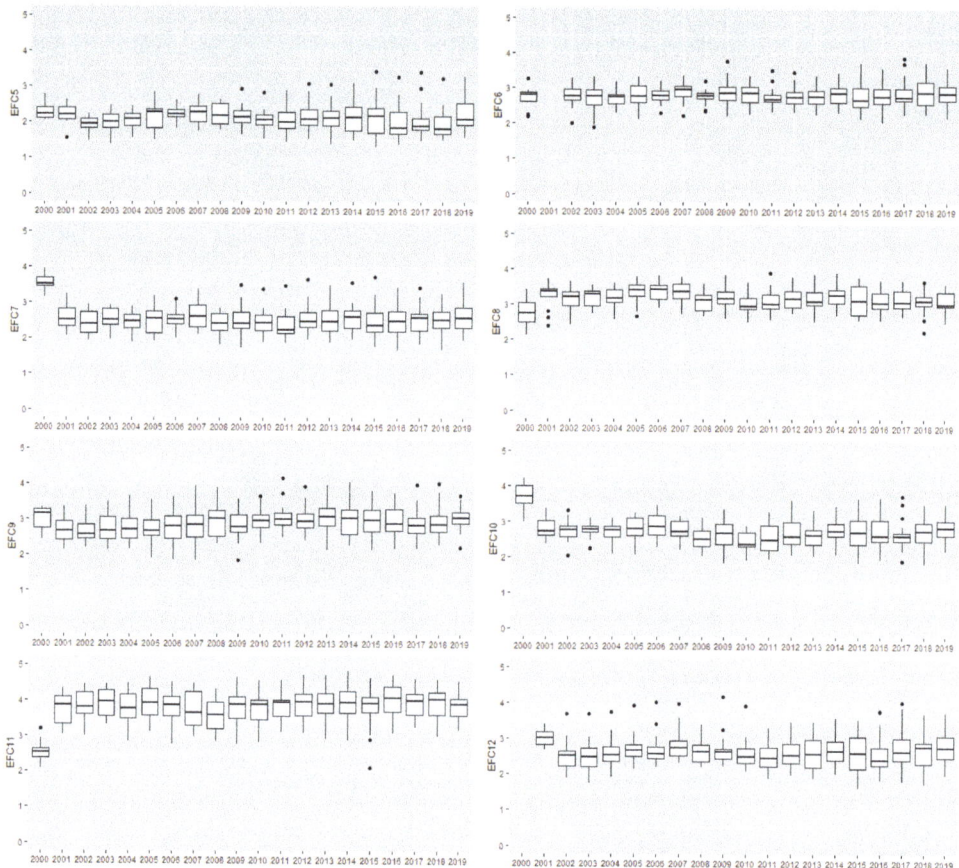

Figure 2. Box plots of the 12 EFCs in the period of 2010–2019.

2.2. Methodology

Cluster analysis includes several multivariate statistical procedures that can be used to classify objects or individuals into relatively homogeneous groups (clusters), taking into account similarities or dissimilarities between them. Sokal and Sneath presented the most popular application of these methodologies in the book [12] as early as 1963 for biological classification of species. From then on, the use of classification techniques became common practice in the most diverse of areas: in medicine to classify diseases, in the social sciences to define homogeneous cultural and scientific areas [13–15] and in marketing for segmenting markets and customers [16,17], among others.

Given a set of n individuals for whom there is information on the form of p variables, a method of cluster analysis proceeds to group individuals according to the existing information in such a way that individuals belonging to the same group are as similar as possible and always more similar to the elements of the same group than to elements of the other groups [18].

An initial difficulty in cluster analysis is that there is no single criterion, similarity measure or technique for defining the groups. The literature on the subject, as well as the available statistical packages, presents us with a very wide range of criteria, always aiming to obtain coherent groups that are significantly different from each other.

The choice of clustering technique depends on the type of variables to be considered (continuous, ratios, ordinal, nominal or binary) and must take into account different scales of measurement of the variables. In this case, it is common practice to standardize the variables, because any measure of similarity/dissimilarity will reflect the weight of the variables that have higher values and dispersion; thus, it is advisable that the variables have the same unit of measure.

Cluster analysis methods can be grouped into four types [18]:

- Optimization techniques—based on the early choice of a number of clusters, k, and a division of all cases is made by the pre-established k groups. Next, the optimization of the chosen criterion is performed. In general, it is intended that within each group, the elements are as similar as possible and as different as possible from elements in other groups;
- Hierarchical techniques—based on a matrix of similarities (or differences) in which each element of the matrix describes the degree of similarity (or difference) between each two cases, based on the chosen variables. These techniques can be agglomerative or divisive. In the first case, the procedure starts with n groups including one individual that are grouped successively until only one group is obtained including all n individuals. In the divisive, 0 the reverse process is applied: one starts from a group with all of the individuals and successive divisions are applied until obtaining n groups;
- Density or mode-seeking techniques—groups are formed by looking for regions that contain a relatively dense concentration of cases.
- Other techniques—these include those that allow groups to overlap (fuzzy clusters), additive partitive methods (kmeans and hill climbing), those that do not use a similarity matrix but that can be directly applied to the original data and others that are not included in the previous types;

Furthermore, there are several measures that can be used as measures of distance or dissimilarity between the elements of a data matrix. The most used distances are as follows:

- Euclidean distance between two cases (i and j) is the square root of the sum of the squares of the differences between values of i and j for all variables ($v = 1, 2, \ldots, p$), that is,

$$d_{ij} = \sqrt{\sum_{v=1}^{p} (X_{iv} - X_{jv})^2}; \qquad (1)$$

- Minkowski distance can be considered as a generalization of Euclidean distance (coincide when $r = 2$):

$$d_{ij} = \left(\sum_{v=1}^{p} |X_{iv} - X_{jv}|^r \right)^{1/r}; \qquad (2)$$

- Mahalanobis distance considers the covariance matrix Σ for the calculation of distances

$$d_{ij} = (X_i - X_j)^T \Sigma^{-1} (X_i - X_j) \qquad (3)$$

where X_i and X_j are the vectors of variable values for individuals i and j, respectively.

Considering the matrix of observed data $X = (x_{ui}) = \begin{bmatrix} x_{11} & x_{12} & \cdots & x_{1p} \\ x_{21} & x_{22} & \cdots & x_{2p} \\ \cdots & \cdots & \cdots & \cdots \\ x_{n1} & x_{n2} & \cdots & x_{np} \end{bmatrix}$,

where x_{ui} is the value of variable i ($i = 1, \ldots, p$) for individual u ($u = 1, 2, \ldots, n$). For a population of dimension N, the covariance matrix Σ is given as

$$\Sigma = \frac{1}{N} \sum_{u=1}^{N} \left[(X_u - \mu)(X_u - \mu)^T \right]$$

where row u, for individual u, of the matrix X is the vector of the p variables under study, i.e.,

$$X_u = \begin{bmatrix} x_{u1} \\ x_{u1} \\ \ldots \\ x_{up} \end{bmatrix}$$

and $\mu = \frac{1}{N} \sum_{u=1}^{N} X_u$ is the vector of the population means.

Other similarity indices can also be used, as long as they respect the following metric properties: symmetry, triangular inequality, differentiability of non-identicals and indifferentiability of identicals.

The indices used include, in addition to distances, correlation coefficients, association coefficients and probabilistic similarity measures, according to [19]. The correlation coefficients are more suitable if the variables have different scales and dispersion, the association coefficients are particularly useful when the variables are binary qualitative, and the probabilistic similarity measures are only used if the similarity index is to be the probability gaining information based on the initial variables.

Therefore, different definitions of distances may result in different final solutions for grouping individuals.

At each step of the agglomerative process, the similarity/distances matrix is recalculated, and the recurrence (Equation (4)) must be satisfied:

$$d_{k(i,j)} = \alpha_i \cdot d_{ki} + \alpha_j \cdot d_{kj} + \beta \cdot d_{ii} + \gamma \left| d_{ki} - d_{kj} \right| \tag{4}$$

where $d_{k(i,j)}$ is the distance between the group k and the group (i,j) formed by the junction of the groups (or elements) i and j.

Although the recurrence equation is always the same, the coefficients $\alpha_i, \alpha_j, \beta$ and γ differ according to the agglomerative method or criterion. The agglomerative method or criterion can be the following:

- Single linkage or criterion of the nearest neighbor, for which the similarity between two groups is the maximum similarity between any two cases belonging to those groups. That is, for the two groups (i,j) and (k), the distance between the two is given by Equation (5).

$$d_{(i,j)k} = \min\{d_{ik}; d_{jk}\}. \tag{5}$$

In this case, the coefficients in recurrence Equation (4) are

$$\alpha_i = \alpha_j = \tfrac{1}{2}; \beta = 0 \text{ and } \gamma = -\tfrac{1}{2}.$$

- Complete linkage or the criterion of the furthest neighbor uses the process inverse to the previous one; that is, given two groups, the distance between the two is given by Equation (6).

$$d_{(i,j)k} = \max\{d_{ik}; d_{jk}\}. \tag{6}$$

In this case, the coefficients in recurrence Equation (4) are

$$\alpha_i = \alpha_j = \tfrac{1}{2}; \beta = 0 \text{ and } \gamma = \tfrac{1}{2}.$$

- Average defines the distance as the average of the distances between all pairs of individuals constituted by elements of the two groups. This strategy is, in a way, intermediate in relation to the first two described.

In this case, the coefficients in recurrence Equation (4) are

$$\alpha_i = \tfrac{n_i}{n_i+n_j}; \alpha_j = \tfrac{n_j}{n_i+n_j}, \beta = 0 \text{ and } \gamma = 0.$$

- Centroid defines the distance between two groups as the distance between their centroids, points defined by the means of the variables that characterize the individuals in each group.
  In this case, the coefficients in recurrence Equation (4) are

$$\alpha_i = \frac{n_i}{n_i+n_j};\ \alpha_j = \frac{n_j}{n_i+n_j},\ \beta = -\alpha_i \cdot \alpha_j \text{ and } \gamma = 0.$$

- Ward method [20] is based on the loss of information resulting from the grouping of individuals and measured by adding the squares of the deviations from individual observations relative to the averages of the groups in which they are classified.
  In this case, the coefficients in recurrence Equation (4) are

$$\alpha_i = \frac{n_k+n_i}{n_k+n_i+n_j};\ \alpha_j = \frac{n_k+n_j}{n_k+n_i+n_j},\ \beta = -\frac{n_k}{n_k+n_i+n_j} \text{ and } \gamma = 0.$$

There is no better criterion for (dis)aggregation of cases in cluster analysis. It is common practice to use several criteria and to compare the results. If these are similar, it is possible to conclude that the results have been obtained with a high degree of stability and, therefore, that they are reliable [18].

Another problem with cluster analysis is the adequate number of clusters to consider. Sometimes, there is prior knowledge, on the part of the researcher, of the number of groups in which the study population should be divided; in which case, this information can be used.

Other criteria for defining the number of clusters that can be used are major changes in the fusion coefficient, the co-phenetic correlation values, the comparison of the application of different numbers of clusters and the comparison of the similarity of the results obtained, the degree of convergence of methods and internal and external validation measures.

The connectivity measure, proposed by Handl et al. in [21], the Dunn index [22] and Silhouette Width [23] are the main internal validation measures.

Given a set of n individuals for whom there is information on the form of p variables, the is defined by Equation (7):

$$Conn(\mathcal{C}) = \sum_{i=1}^{n}\sum_{j=1}^{l} x_{i,n_{ij}}. \qquad (7)$$

where n_{ij} is the jth nearest neighbor of observation i,

$$x_{i,n_{ij}} = \begin{cases} 0 & \text{if } i \text{ and } j \text{ are in the same cluster} \\ \frac{1}{j} & \text{if otherwise} \end{cases},$$

$\mathcal{C} = \{C_1, C_2, \ldots, C_k\}$ is a partition of the n observations into k disjoint clusters and l is a parameter giving the number of nearest neighbors to use, [21]. This measure has values between 0 and ∞ and should be minimized.

The Dunn Index [22] is given by Equation (8),

$$D(\mathcal{C}) = \frac{\min\limits_{C_k, C_l \in \mathcal{C}, C_k \neq C_l} \left(\min\limits_{i \in C_k, j \in C_l} dist(i,j) \right)}{\max\limits_{C_m \in \mathcal{C}} diam(C_m)}, \qquad (8)$$

where $diam(C_m)$ is the maximum distance between observations in cluster C_m. This measure has values between 0 and ∞ and should be maximized.

Silhouette Width [23] is given by Equation (9):

$$S(i) = \frac{b_i - a_i}{\max(b_i, a_i)}, \qquad (9)$$

where a_i is the average distance between i and all other observations, such as

$$b_i = \min_{C_k \in \mathcal{C} \setminus C(i)} \sum_{j \in C_k} \frac{dist(i,j)}{n(C_k)}$$

where $C(i)$ is the cluster containing observation i, $dist(i,j)$ is the considered distance between observations i and j, and $n(C)$ is the cardinality of cluster C. This measure has values between -1 and 1 and should be maximized.

These measures are implemented by Brock et al. [24] in the package clValid. This package comprises the internal validation measures and, in addition, the stability and biological validation measures. Internal validation measures take only the dataset and the clustering partition as input and use intrinsic information in the data to assess the quality of the clustering. The stability measures are a special version of internal measures. They evaluate the consistency of a clustering result by comparing it with the clusters obtained after each column is removed, one at a time. Biological validation evaluates the ability of a clustering algorithm to produce biologically meaningful clusters.

There are several cluster validation measures defined in the literature [25–28]. It is not possible to obtain the best result always with the same validation measure. Thus, several authors have proposed merging several validation measures, such as the Davies–Bouldin index, the Calinski–Harabasz index and the Dunn index, which allow for comparisons of several solutions and the selection of the internal optimal solution [26–28]. However, these validation measures focus on internal validation, but it is also important to take into account the external ones. For this reason, hybrid validation measures that combine these two types of validation have been emerging and are described by Gajawada and Toshniwal (2012) [29]. Improved measures have also been proposed based on the most common ones already mentioned; for example, since the numerical procedure to calculate the Silhouette Width criterion is rather demanding, the Simplified Silhouette Width Criterion (SSWC)—which instead of the average value, uses the distance between the elements and the clusters centroids, thus deeming the partition with the largest SSWC index to be the most appropriate partition—is usually applied[28].

3. Results and Discussion

In order to study the European countries based on the EFCs experts' perceptions during the period of 2000-2019, cluster analysis was used to group the countries into homogeneous groups. As discussed in Section 2, several measures and methods can be used for grouping countries.

In [11], the hierarchical cluster technique, Euclidean distance and the Ward method were used in order to analyze, for the period of 2010–2016, European entrepreneurs' perceptions. The present study considers the whole period of available data (between 2000 and 2019), extending that work. In that previous work, the statistical software R version 3.4.0 was used, and three clusters were considered, justified by GEM project's definition of economic development level, which considers three types of economies: (i) economies driven by factors of production; (ii) efficiency-oriented economies; and (iii) innovation-oriented economies. It was found that for each year, the countries that constitute each of the clusters observe substantial changes in the clusters throughout the years. In particular, while in 2010 and 2011 Portugal was in clusters with the second-best overall average EFCs perceptions, in 2012, Portugal was in the group with the lowest EFCs perceptions. However, from 2013 to 2016, Portugal recovered in terms of experts' perceptions and moved into the group with the second-best overall average. The behavior of Portugal was compared with that of Italy, Greece and Spain.

Considering the complete set of data, the present work intends to study the behavior of the European Expert's perceptions about their economies' entrepreneur conditions.

In order to determine the best number of clusters, internal validation measures were computed for all of the years and for hierarchical, pam, kmeans and fanny methods, as illustrated in Figure 3 (for the year 2019) and summarized in Table 1.

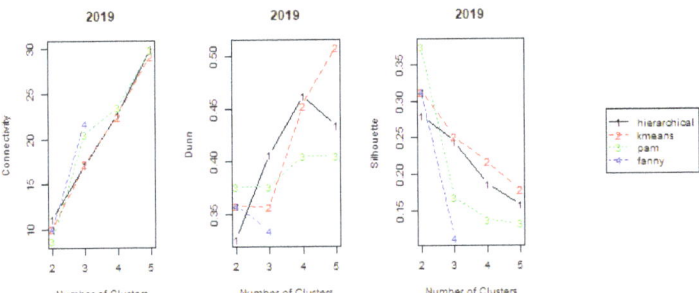

Figure 3. Clusters' internal validation measures for the year 2019.

Table 1. Optimal cluster number (k) and method for internal measures.

Year	Connectivity k	Method	Dunn Index k	Method	Silhouette Width k	Method
2000	2	hierarchical	5	hierarchical	2	hierarchical
2001	2	hierarchical	4	pam	2	hierarchical
2002	2	hierarchical	2	hierarchical	2	hierarchical
2003	2	hierarchical	5	kmeans	3	hierarchical
2004	2	hierarchical	3	hierarchical	2	hierarchical
2005	2	hierarchical	5	hierarchical	2	hierarchical
2006	2	hierarchical	4	hierarchical	2	hierarchical
2007	2	hierarchical	5	pam	2	hierarchical
2008	2	pam	5	hierarchical	2	pam
2009	2	hierarchical	4	kmeans	2	hierarchical
2010	2	hierarchical	4	hierarchical	2	kmeans
2011	2	kmeans	5	pam	2	kmeans
2012	2	hierarchical	4	hierarchical	2	kmeans
2013	2	hierarchical	5	kmeans	2	kmeans
2014	2	hierarchical	5	pam	2	hierarchical
2015	2	hierarchical	5	pam	2	hierarchical
2016	2	fanny	5	kmeans	2	kmeans
2017	2	hierarchical	3	kmeans	2	hierarchical
2018	2	pam	5	hierarchical	2	pam
2019	2	pam	5	kmeans	2	pam

The connectivity measure, Equation (7), varies between 0 and ∞ and should be minimized. Thus, looking at Figure 3 and Table 1, the optimal score for this measure, and for the year 2019, is obtained using the pam method and $k = 2$ clusters. Observing the results for all the years, for most, the optimal connectivity value is found for $k = 2$ and for the hierarchical method. The Dunn index, Equation (8), presents values between 0 and ∞ and should be maximized. It can be observed in Figure 3 and Table 1, that the best values of this measure are obtained for larger number of clusters. Silhouette Width, Equation (9), has values between −1 and 1 and should be maximized. This is achieved mostly when $k = 2$ clusters are considered and by using the hierarchical method.

Table 2 shows that the optimal validation measures are obtained mostly for two clusters and for hierarchical methods. Furthermore, observing the dendrograms in Figure 4 for the years 2000 and 2009, and considering the cutting line at height = 7, the same conclusion is reached.

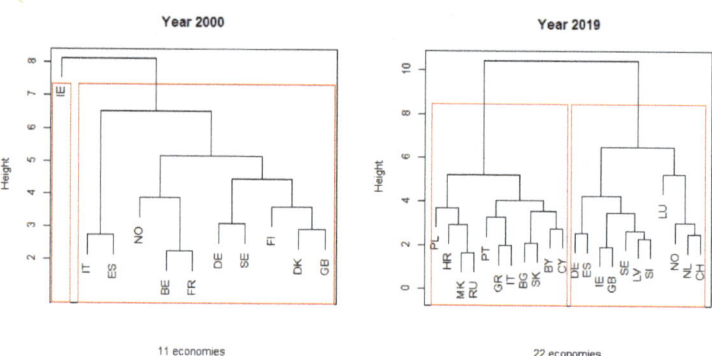

Figure 4. Dendrograms for the years 2000 and 2019.

The software R, version 3.4.0, was used, and $k = 2$ clusters were considered, as suggested in Table 1. The agglomeration of countries obtained for each year is presented in Table 2. For each year, the average of all the EFCs is shown in brackets for all countries (first and fourth columns), countries in Cluster 1 (second and fifth columns) and those in Cluster 2 (third and sixth columns). Note that Cluster 1 has an average below the global average and Cluster 2 has an average above the global average.

Analyzing the results inn Table 2, apart from Italy (IT) and Slovakia (SK), which remain in cluster 1, and Ireland (IE), Iceland (IS), Netherlands (NL) and Switzerland (CH), which maintain the allocation to cluster 2 throughout the two decades, the remaining countries' allocations vary between the two clusters.

The agglomerations of the economies present different numbers of economies and also somewhat different averages and variability. Table 3 shows, for each year, the number of economies in each cluster and for all of the economies. This table also shows the average, standard deviation and coefficient of variation (CV) in %, of the average of the 12 EFCs. The average of the EFCs for all economies varies from 2.67 in 2010 to 2.92 in 2000, while larger variability is observed in 2015 (CV = 12.8%). Since 2009, when the number of economies started to significantly increase, the CV has been larger than 9.5%, reflecting the diversity of the economies participating in the survey. When analyzing each of the clusters, it can be seen that for Cluster 1, the lowest average was 2.37, observed in 2015, and the maximum was 2.88 in 2000. For Cluster 2, the minimum average was 2.78, observed in 2004, and the maximum was 3.4 in 2016. In 2016, only three of the 25 economies (i.e., 12%) were agglomerated in Cluster 2, while the other 22 economies were in Cluster 1, which had a CV of 8.8%, the largest observed in Cluster 1. In 2011 and 2011, Cluster 2 agglomerated only 9% and 14%, respectively, of the economies, leading to large averages—3.24 and 3.18, respectively.

Some particular cases that are worthy of discussion are as follows: Denmark (DK), which was allocated to the cluster with the lowest average only in 2000, while for the other 11 years for which there are data, it was always in Cluster 2. In fact, for 2000 as well as for 2011, 2013 and 2016, the economies allocated to Cluster 1 represent more than 85% of the economies for which there were data. This could explain why economies such as Germany (DE), Finland (FI), France (FR) Belgium (BE) and the United Kingdom (GB), which for the majority of the years were allocated to Cluster 2, were in most cases in 2000, 2011, 2013 and 2016 allocated to the cluster with the lowest average EFCs. Other countries, such as Portugal (PT), Greece (GR) and Spain (ES)m present more variability in the allocation to the two clusters.

Table 2. Clusters of European Economies from 2000 until 2019.

Year	Cluster 1	Cluster 2	Year	Cluster 1	Cluster 2
2000 (2.92)	BE, DK, FI, FR, DE, IT, NO, ES, SE, GB (2.88)	IE (3.34)	2010 (2.67)	BA, HR, FR, GR, HU, IT, MK, ME, NO, PT, RU, SI, ES, SE, GB (2.55)	FI, DE, IS, IE, LV, CH (2.99)
2001 (2.85)	HU, IT, NO, PT, ES, SE (2.61)	BE, DK, FI, FR, DE, IE, NL, GB (3.04)	2011 (2.68)	BA, HR, CZ, FI, FR, DE, GR, HU, IE, LV, LT, NO, PL, PT, RU, SK, SI, ES, SE, GB (2.63)	NL, CH (3.24)
2002 (2.72)	BE, HR, HU, NO, SI, SE (2.50)	DK, FI, FR, DE, IS, IE, NL, ES, CH, GB (2.85)	2012 (2.76)	BA, HR, GR, HU, IT, LT, PL, PT, RO, RU, SK, SI, ES, SE (2.53)	AT, BE, CH, DK, EE, FI, FR, DE, IE, LV, MK, NL, NO, GB (2.99)
2003 (2.71)	HR, GR, IT, NO, SI, SE (2.49)	BE, DK, FI, FR, DE, IS, IE, NL, ES, CH, GB (2.84)	2013 (2.74)	BA, BE, CZ, DE, EE, ES, FR, GB, GR, HR, HU, IE, IT, LT, LU, MK, NO, PL, PT, RO, RU, SK, SE, SI (2.67)	CH, FI, LV, NL (3.18)
2004 (2.70)	HR, HU, PL, SI (2.47)	BE, DK, FI, DE, GR, IS, IE, NL, NO, PT, ES (2.78)	2014 (2.81)	BA, HR, GR, HU, IT, NA, PL, RO, RU, SK, SI, ES, GB (2.58)	AT, BE, DK, EE, FI, FR, DE, IE, LV, LT, LU, NL, NO, PT, SE, CH (3.00)
2005 (2.79)	HR, HU, IT, SI (2.41)	AT, BE, DK, FI, DE, GR, IS, IE, LV, NL, NO, ES, CH, GB (2.90)	2015 (2.76)	BG, ES, GR, HR, HU, IT, RO, SK, (2.37)	BE, CH, DE, EE, FI, GB, IE, LU, LV, MK, NL, NO, PL, PT, SE, SI (2.95)
2006 (2.81)	HR, CZ, HU, IT, LV, RU, SI (2.60)	BE, DK, FI, DE, GR, IS, IE, NL, NO, ES, GB (2.94)	2016 (2.73)	AT, FI, FR, DE, IE, LV, LU, PT, ES, BG, HR, CY, GR, HU, IT, MK, PL, RU, SK, SI, SE, GB (2.64)	CH, EE, NL (3.40)
2007 (2.88)	HR, GR, IT, RO, RU, RS, SI, ES (2.64)	AT, BE, DK, FI, IS, IE, NO, CH, GB (3.09)	2017 (2.78)	BA, BG, HR, CY, GR, IT, PL, SK, ES (2.52)	EE, NL, FR, DE, IE, LV, LU, SI, SE, CH, GB (2.99)
2008 (2.73)	BA, HR, GR, IT, MK, RU, RS, SI, ES (2.59)	DK, FI, DE, IE, NO (2.98)	2018 (2.78)	BG, HR, GR, IT, PL, RU, SK (2.46)	AT, CY, FR, DE, IE, LV, LU, NL, SI, ES, SE, CH, GB (2.96)
2009 (2.73)	BA, HR, GR, HU, IT, LV, RU, RS, SI, ES, GB (2.52)	BE, DK, FI, DE, IS, NL, NO, CH (3.02)	2019 (2.85)	BY, BG, HR, CY, GR, IT, MK, PL, PT, RU, SK (2.61)	DE, IE, LV, LU, NL, NO, SI, ES, SE, CH, GB (3.10)

AT—Austria, BA—Bosnia and Herzegovina, BE—Belgium, BG—Bulgaria, BY—Belarus, CH—Switzerland, CY—Cyprus, CZ—Czech Republic, DE—Germany, DK—Denmark, EE—Estonia, ES—Spain, FI—Finland, FR—France, GB—United Kingdom, GR—Greece, HR—Croatia, HU—Hungary, IE—Ireland, IS—Iceland, IT—Italy, LT—Lithuania, LU—Luxembourg, LV—Latvia, ME—Montenegro, MK—North Macedonia, NA—Kosovo, NL—Netherlands, NO—Norway, PL—Poland, PT—Portugal, RO—Romania, RS—Serbia, RU—Russia, RU—Russia, SE—Sweden, SI—Slovenia, SK—Slovakia.

To understand the pattern and exemplify differences in the cluster agglomeration over the years, we compared the allocations of the top European Economies with the best three and the three worst total early-stage entrepreneurial activity (TEA) values. TEA is a GEM indicator that represents the percentage of the 18–64-year-old population who are either a nascent entrepreneur or owner-manager of a new business.

Italy (TEA = 2.79), Poland (TEA = 5.39) and Belarus (TEA = 3.78) are the three countries with lower TEA values, and, in fact, Italy remains in Cluster 1 throughout the two decades, Poland, besides being allocated to Cluster 2 in 2015, is allocated to Cluster 1 in the remaining years. Belarus has only information in 2019, and it is allocated to Cluster 1, as expected.

On the other hand, the allocation of Latvia, which registers a higher TEA value for 2019 (TEA = 15.43), changes between Cluster 1 and Cluster 2, throughout the years. Slovakia, with the second-highest TEA value (TEA = 13.33), contrary to what was excepted, maintains its allocation to Cluster 1 in all years with information. Portugal (TEA = 12.89), the country with the third-highest TEA value, also presents differences in its allocation between Cluster 1 and cluster 2 throughout the years.

Table 3. Characterization of the clusters in Table 2.

Year	Cluster 1				Cluster 2				All Economies			
	#	Av.	StD	CV	#	Av.	StD	CV	#	Av.	StD	CV
2000	10	2.88	0.175	6.1	1	3.34	—	—	11	2.92	0.216	7.4
2001	6	2.61	0.153	5.9	8	3.04	0.07	2.3	14	2.85	0.244	8.6
2002	6	2.50	0.118	4.7	10	2.85	0.099	3.5	16	2.72	0.203	7.5
2003	5	2.49	0.065	2.6	11	2.84	0.123	4.3	16	2.71	0.203	7.5
2004	4	2.47	0.048	1.9	11	2.78	0.187	6.7	15	2.70	0.221	8.2
2005	4	2.41	0.057	2.4	14	2.90	0.156	5.4	18	2.79	0.250	9.0
2006	7	2.60	0.071	2.7	11	2.94	0.144	4.9	18	2.81	0.209	7.4
2007	8	2.64	0.088	3.3	9	3.09	0.074	2.4	17	2.88	0.243	8.4
2008	9	2.59	0.161	6.2	5	2.98	0.049	1.6	14	2.73	0.234	8.6
2009	11	2.52	0.131	5.2	8	3.02	0.173	5.7	19	2.73	0.294	10.8
2010	15	2.55	0.181	7.1	6	2.99	0.118	3.9	21	2.67	0.261	9.8
2011	20	2.63	0.210	8.8	2	3.24	0.061	1.9	22	2.68	0.272	10.1
2012	14	2.53	0.145	5.7	14	2.99	0.178	6.0	28	2.76	0.286	10.4
2013	24	2.67	0.229	8.6	4	3.18	0.133	4.2	28	2.74	0.283	10.3
2014	13	2.58	0.158	6.1	16	3.00	0.166	5.5	29	2.81	0.267	9.5
2015	8	2.37	0.126	5.3	16	2.95	0.253	8.6	24	2.76	0.352	12.8
2016	22	2.64	0.233	8.8	3	3.40	0.095	2.8	25	2.73	0.333	12.2
2017	9	2.52	0.137	5.4	11	2.99	0.278	9.3	20	2.78	0.322	11.6
2018	7	2.46	0.200	8.1	13	2.96	0.206	7.0	20	2.78	0.319	11.5
2019	11	2.61	0.115	4.4	11	3.10	0.202	7.1	22	2.85	0.298	10.5

The obtained results indicate the need to consider annual and intra-country dynamics in studies on the topic of entrepreneurship, especially if they analyze data from GEM. Most studies (for example, the recent study of [2,11]) perform cross-sectional studies combining information from GEM with group economies. However, neglecting to consider a longitudinal dynamic may result in biased results.

4. Conclusions

In order to understand the dynamics of the European entrepreneurial framework conditions over the last two decades, cluster analysis was used to group the countries in homogeneous groups based on the EFCs experts' perceptions during the period of 2000–2019.

The cluster analysis revealed that there are significant differences between the clusters obtained over the years and also that the distribution of the countries in each cluster considerably varies.

This study contributes to the existing literature in the sense that it clarifies the existence of a dynamic, entrepreneurial behavior of economies regarding entrepreneurial framework conditions, which should be considered in future works.

In the future, as a result of the differences encountered in countries' agglomerations through time, a longitudinal clustering approach will be performed to compare results instead of the desegregated cross-sectional approach for each year. Furthermore, we intend to analyze the impact of the EFCs on entrepreneurship intentions and on total early-stage entrepreneurial activity (TEA) in Europe, making use of dynamic longitudinal models, in particular the system GMM procedure, to capture the intra-year and intra-country variability.

Author Contributions: This work was conducted by the three authors in collaboration through joint and distributed tasks. Joint tasks included conceptualization, writing—original draft preparation and writing—review and editing. Major contribution in software implementation, validation and visualization was given by E.C.e.S.; A.B. mostly contributed with state-of-the-art investigations, formal analysis of the results and the finalization of the conclusions. A.C. defined the methodology, collected the resources and contributed to the interpretation and organization of the paper. All authors have read and agreed to the published version of the manuscript.

Funding: This work has been supported by national funds from FCT—Fundação para a Ciência e Tecnologia through project UIDB/04728/2020.

Institutional Review Board Statement: Not applicable.

Informed Consent Statement: Not applicable.

Data Availability Statement: Data used in this work is available on the GEM project website at https://www.gemconsortium.org/data and accessed 29 June 2020.

Conflicts of Interest: The authors declare no conflict of interest.

Abbreviations

The following abbreviations are used in this manuscript:

APS	adult population survey
EDL	economic development level
EFCs	Entrepreneurial Framework Conditions
GEM	Global Entrepreneurship Monitor
NES	national expert survey
TEA	total early-stage entrepreneurial activity

Appendix A

Table A1. Description of Entrepreneurial Framework Conditions (EFCs). Source: [11].

EFC	Description	Indicator
1	The availability of financial resources—equity and debt—for small and medium enterprises (SMEs) (including grants and subsidies)	Financing for entrepreneurs
2	The extent to which public policies support entrepreneurship—entrepreneurship as a relevant economic issue	Governmental support and policies
3	The extent to which public policies support entrepreneurship—taxes or regulations are either size neutral or encourage new and SMEs	Taxes and bureaucracy
4	The presence and quality of programs directly assisting SMEs at all levels of government (national, regional, municipal)	Governmental programs
5	The extent to which training in creating or managing SMEs is incorporated within the education and training system at primary and secondary levels	Basic school entrepreneurial education and training
6	The extent to which training in creating or managing SMEs is incorporated within the education and training system in higher education, such as vocational education, college, business schools, etc.	Post-school entrepreneurial education and training
7	The extent to which national research and development will lead to new commercial opportunities and is available to SMEs	R&D transfer
8	The presence of property rights, commercial, accounting and other legal and assessment services and institutions that support or promote SMEs	Commercial and professional infrastructure
9	The level of change in markets from year to year	Internal market dynamics
10	The extent to which new firms are free to enter existing markets	Internal market openness
11	Ease of access to physical resources—communication, utilities, transportation, land or space—at a price that does not discriminate against SMEs	Physical and services infrastructure
12	The extent to which social and cultural norms encourage or allow actions leading to new business methods or activities that can potentially increase personal wealth and income	Cultural and social norms

References

1. Fernandes, A.J.; Ferreira, J.J. Entrepreneurial ecosystems and networks: A literature review and research agenda. *Rev. Manag. Sci.* **2021**, 1–59. [CrossRef]
2. Farinha, L.; Lopes, J.; Bagchi-Sen, S.; Sebastião, J.R.; Oliveira, J. Entrepreneurial dynamics and government policies to boost entrepreneurship performance. *Socio Econ. Plan. Sci.* **2020**, *72*, 100950. [CrossRef]
3. De Brito, S.; Leitão, J. Mapping and defining entrepreneurial ecosystems: A systematic literature review. *Knowl. Manag. Res. Pract.* **2020**, *19*, 1–22. [CrossRef]
4. Herrington, M.; Kew, P.K. *Global Entrepreneurship Monitor: 2016/17 Global Report*; Technical Report; Global Entrepreneurship Research Association (GERA): London, UK, 2017.
5. Kelley, D.; Bosma, N.; Amorós, J.E. *Global Entrepreneurship Monitor 2010 Global Report*; Technical Report; Global Entrepreneurship Research Association (GERA): London, UK, 2011.
6. Pilar, M.D.F.; Marques, M.; Correia, A. New and growing firms entrepreneurs' perceptions and their discriminant power in edl countries. *Glob. Bus. Econ. Rev.* **2018**, *21*, 474–499. [CrossRef]
7. Braga, V.; Queirós, M.; Correia, A.; Braga, A. High-Growth Business Creation and Management: A Multivariate Quantitative Approach Using GEM Data. *J. Knowl. Econ.* **2017**, *9*, 424–445. [CrossRef]
8. Correia, A.; Costa e Silva, E.; Lopes, I.C.; Braga, A.; Braga, V. Experts' perceptions on the entrepreneurial framework conditions. In *AIP Conference Proceedings*; AIP Publishing: New York, NY, USA, 2017; Volume 1906, p. 110004.
9. Autio, E.; Kenney, M.; Mustar, P.; Siegel, D.; Wright, M. Entrepreneurial innovation: The importance of context. *Res. Policy* **2014**, *43*, 1097–1108. [CrossRef]
10. Singer, S.; Herrington, M.; Menipaz, E. *Global Entrepreneurship Monitor: Global Report 2017/18*; Technical Report; Global Entrepreneurship Research Association (GERA): London, UK, 2018.
11. Costa e Silva, E.; Correia, A.; Duarte, F. How Portuguese experts' perceptions on the entrepreneurial framework conditions have changed over the years: A benchmarking analysis. In *AIP Conference Proceedings*; AIP Publishing LLC: New York, NY, USA, 2018; Volume 2040, p. 110005.
12. Sokal, R.R. The principles and practice of numerical taxonomy. *Taxon* **1963**, *12*, 190–199. [CrossRef]
13. Driver, H.E. Survey of numerical classification in anthropology. In *The Use of Computers in Anthropology*; De Gruyter Mouton: Berlin, Germany, 2011; pp. 301–344.
14. Johnson, M.E. *Multivariate Statistical Simulation: A Guide to Selecting and Generating Continuous Multivariate Distributions*; John Wiley & Sons: Hoboken, NJ, USA, 1987; Volume 192.
15. Walter, G.A.; Barney, J.B. Management objectives in Mergers and Acquisitions. *Strateg. Manag. J. II(I)* **1990**, *11*, 79–86. [CrossRef]
16. Doyle, P.; Saunders, J. Market segmentation and positioning in specialized industrial markets. *J. Mark.* **1985**, *49*, 24–32. [CrossRef]
17. Green, P.E.; Schaffer, C.; Patterson, K. A reduced space approach to the clustering of categorical data in market segmentation. *J. Mark.* **1991**, *55*, 20–31. [CrossRef]
18. Reis, E. *Estatística Multivariada Aplicada*, 2nd ed.; Edições Sílabo: Lisboa, Portugal, 2001; ISBN 972-618-247-6.
19. Aldenderfer, M.S.; Blashfield, R.K. Cluster analysis software and the literature on clustering. In *Cluster Analysis*; SAGE Publications Inc.: Thousand Oaks, CA, USA, 1984; pp. 75–81. [CrossRef]
20. Ward, J.H., Jr. Hierarchical grouping to optimize an objective function. *J. Am. Stat. Assoc.* **1963**, *58*, 236–244. [CrossRef]
21. Handl, J.; Knowles, J.; Kell, D.B. Computational cluster validation in post-genomic data analysis. *Bioinformatics* **2005**, *21*, 3201–3212. [CrossRef] [PubMed]
22. Dunn, J.C. Well-separated clusters and optimal fuzzy partitions. *J. Cybern.* **1974**, *4*, 95–104. [CrossRef]
23. Rousseeuw, P.J. Silhouettes: A graphical aid to the interpretation and validation of cluster analysis. *J. Comput. Appl. Math.* **1987**, *20*, 53–65. [CrossRef]
24. Brock, G.; Pihur, V.; Datta, S.; Datta, S. clValid, an R package for cluster validation. *J. Stat. Softw.* **2008**, *5*, 1–22. [CrossRef]
25. Bezdek, J.C.; Keller, J.; Krisnapuram, R.; Pal, N. *Fuzzy Models and Algorithms for Pattern Recognition and Image Processing*; Springer Science & Business Media: Berlin, Germany, 1999; Volume 4.
26. Scitovski, R.; Sabo, K.; Martínez Álvarez, F.; Ungar, S. *Cluster Analysis and Applications*; Springer International Publishing: Berlin, Germany, 2021.
27. Theodoridis, S.; Koutroumbas, K. *Pattern Recognition*, 4th ed.; Academic Press: Cambridge, MA, USA, 2009.
28. Vendramin, L.; Campello, R.J.; Hruschka, E.R. On the comparison of relative clustering validity criteria. In Proceedings of the 2009 SIAM International Conference on Data Mining, SIAM, Sparks, NV, USA, 30 April–2 May 2009; pp. 733–744.
29. Gajawada, S.; Toshniwal, D. Hybrid cluster validation techniques. In *Advances in Computer Science, Engineering & Applications*; Springer: Berlin, Germany, 2012; pp. 267–273.

Article

Gauss–Newton–Secant Method for Solving Nonlinear Least Squares Problems under Generalized Lipschitz Conditions

Ioannis K. Argyros [1], Stepan Shakhno [2,*], Roman Iakymchuk [3,4], Halyna Yarmola [5] and Michael I. Argyros [6]

1. Department of Mathematical Sciences, Cameron University, Lawton, OK 73905, USA; iargyros@cameron.edu
2. Department of Theory of Optimal Processes, Ivan Franko National University of Lviv, Universytetska Str. 1, 79000 Lviv, Ukraine
3. PEQUAN, LIP6, Sorbonne Université, 4 Place Jussieu, 75252 Paris, France; roman.iakymchuk@sorbonne-universite.fr
4. Fraunhofer ITWM, Fraunhofer-Platz 1, 67663 Kaiserslautern, Germany
5. Department of Computational Mathematics, Ivan Franko National University of Lviv, Universytetska Str. 1, 79000 Lviv, Ukraine; halyna.yarmola@lnu.edu.ua
6. Department of Computer Science, University of Oklahoma, Norman, OK 73071, USA; michael.i.argyros-1@ou.edu
* Correspondence: stepan.shakhno@lnu.edu.ua; Tel.: +38-0322-394-791

Abstract: We develop a local convergence of an iterative method for solving nonlinear least squares problems with operator decomposition under the classical and generalized Lipschitz conditions. We consider the case of both zero and nonzero residuals and determine their convergence orders. We use two types of Lipschitz conditions (center and restricted region conditions) to study the convergence of the method. Moreover, we obtain a larger radius of convergence and tighter error estimates than in previous works. Hence, we extend the applicability of this method under the same computational effort.

Keywords: nonlinear least squares problem; differential-difference method; divided differences; radius of convergence; residual; error estimates

MSC: 65J15

1. Introduction

Nonlinear least squares problems often arise while solving overdetermined systems of nonlinear equations, estimating parameters of physical processes by measurement results, constructing nonlinear regression models for solving engineering problems, etc. The most used method for solving nonlinear least squares problems is the Gauss–Newton method [1]. In the case when the derivative can not be calculated, difference methods are used [2,3].

Some nonlinear functions have a differentiable and a nondifferentiable part. In this case, a good idea is to use a sum of the derivative of the differentiable part of the operator and the divided difference of the nondifferentiable part instead of the Jacobian [4–6]. Numerical study shows that these methods converge faster than Gauss–Newton type's method or difference methods.

In this paper, we study the local convergence of the Gauss–Newton–Secant method under the classical and generalized Lipschitz conditions for first-order Fréchet derivative and divided differences.

Let us consider the nonlinear least squares problem:

$$\min_{x \in R^p} \frac{1}{2}(F(x)+G(x))^T(F(x)+G(x)), \qquad (1)$$

147

where residual function $F + G : R^p \to R^m$ ($m \geq p$) is nonlinear in x, F is a continuously differentiable function, and G is a continuous function, the differentiability of which, in general, is not required.

We propose the following modification of the Gauss–Newton method combined with the Secant-type method [4,6] for finding the solution to problem (1):

$$x_{n+1} = x_n - (A_n^T A_n)^{-1} A_n^T (F(x_n) + G(x_n)), n = 0, 1, \ldots, \tag{2}$$

where $A_n = F'(x_n) + G(x_n, x_{n-1})$, $F'(x_n)$ is a Fréchet derivative of $F(x)$; $G(x_n, x_{n-1})$ is a divided difference of the first order of function $G(x)$ [7] at points x_n, x_{n-1}; and x_0, x_{-1} are given.

Setting $A_n = F'(x_n)$, for solving problem (1), from (2) we obtain an iterative Gauss–Newton-type method:

$$x_{n+1} = x_n - (F'(x_n)^T F'(x_n))^{-1} F'(x_n)^T (F(x_n) + G(x_n)), \quad n = 0, 1, \ldots. \tag{3}$$

For $m = p$, problem (1) turns into a system of nonlinear equations:

$$F(x) + G(x) = 0. \tag{4}$$

In this case, method (2) is transformed into the combined Newton–Secant method [8–10]:

$$x_{n+1} = x_n - (F'(x_n) + G(x_n, x_{n-1}))^{-1}(F(x_n) + G(x_n)), n = 0, 1, \ldots, \tag{5}$$

and method (3) into the Newtons-type method for solving nonlinear equations [11]:

$$x_{n+1} = x_n - (F'(x_n))^{-1}(F(x_n) + G(x_n)), n = 0, 1, \ldots. \tag{6}$$

The convergence domain is small (in general), and error estimates are pessimistic. These problems restrict the applicability of these methods. The novelty of our work is in the claim that these problems can be addressed without adding hypotheses. In particular, our idea is to use a center and restricted radius Lipschitz conditions. Such an approach to the study of the convergence of methods allows for extending the convergence ball of the method and improving error estimates.

The remainder of the paper is organized as follows: Section 2 deals with the local convergence analysis. The numerical experiments appear in Section 3. Section 4 contains the concluding remarks and ideas about future works.

2. Local Convergence Analysis

Let us consider, at first, some auxiliary lemmas needed to obtain the main results. Let D be an open subset of R^p.

Lemma 1 ([4]). *Let $e(t) = \int_0^t E(u)du$, where E is an integrable and positive nondecreasing function on $[0, T]$. Then, $e(t)$ is monotonically increasing with respect to t on $[0, T]$.*

Lemma 2 ([1,12]). *Let $h(t) = \frac{1}{t} \int_0^t H(u)du$, where H is an integrable and positive nondecreasing function on $[0, T]$. Then, $h(t)$ is nondecreasing with respect to t on $(0, T]$.*

Additionally, $h(t)$ at $t = 0$ is defined as $h(0) = \lim_{t \to 0} \left(\frac{1}{t} \int_0^t H(u)du \right)$.

Lemma 3 ([13]). *Let $s(t) = \frac{1}{t^2} \int_0^t S(u)u\,du$, where S is an integrable and positive nondecreasing function on $[0, T]$. Then, $s(t)$ is nondecreasing with respect to t on $(0, T]$.*

Definition 1. *The Fréchet derivative F' satisfies the center Lipschitz condition on D with L_0 average if*

$$\|F'(x) - F'(x^*)\| \leq \int_0^{\rho(x)} L_0(u) du, \quad \text{for each } x \in D \subset R^p, \tag{7}$$

where $\rho(x) = \|x - x^*\|$, $x^* \in D$ is a solution of problem (1), and L_0 is an integrable, positive, and nondecreasing function on $[0, T]$.

The functions M_0, L, M, L_1 and M_1 introduced next are as the function L_0: integrable, positive, and nondecreasing functions defined on $[0, 2R]$.

Definition 2. *The first order divided difference $G(x, y)$ satisfies the center Lipschitz condition on $D \times D$ with M_0 average if*

$$\|G(x, y) - G(x^*, x^*)\| \leq \int_0^{\rho(x) + \rho(y)} M_0(u) du, \quad \text{for each } x, y \in D. \tag{8}$$

Let $B > 0$ and $\alpha > 0$. We define function φ on $[0, +\infty)$ by

$$\varphi(t) = B\left[2\alpha + \int_0^t L_0(u) du + \int_0^{2t} M_0(u) du\right]\left[\int_0^t L_0(u) du + \int_0^{2t} M_0(u) du\right]. \tag{9}$$

Suppose that equation

$$\varphi(t) = 1 \tag{10}$$

has at least one positive solution. Denote by γ the minimal such solution. Then, we can define $\Omega_0 = D \cap \Omega(x^*, \gamma)$, where $\Omega(x^*, \gamma) = \{x : \|x - x^*\| < \gamma\}$.

Definition 3. *The Fréchet derivative F' satisfies the restricted radius Lipschitz condition on Ω_0 with L average if*

$$\|F'(x) - F'(x^\tau)\| \leq \int_{\tau\rho(x)}^{\rho(x)} L(u) du, \ x^\tau = x^* + \tau(x - x^*), 0 \leq \tau \leq 1, \text{ for each } x \in \Omega_0. \tag{11}$$

Definition 4. *The first order divided difference $G(x, y)$ satisfies the restricted radius Lipschitz condition on Ω_0 with M average if*

$$\|G(x, y) - G(u, v)\| \leq \int_0^{\|x-u\|+\|y-v\|} M(u) du, \quad \text{for each } x, y, u, v \in \Omega_0. \tag{12}$$

Definition 5. *The Fréchet derivative F' satisfies the radius Lipschitz condition on D with L_1 average if*

$$\|F'(x) - F'(x^\tau)\| \leq \int_{\tau\rho(x)}^{\rho(x)} L_1(u) du, \quad \text{for each } x \in D. \tag{13}$$

Definition 6. *The first order divided difference $G(x, y)$ satisfies the radius Lipschitz condition on D with M_1 average if*

$$\|G(x, y) - G(u, v)\| \leq \int_0^{\|x-u\|+\|y-v\|} M_1(u) du, \quad \text{for each } x, y, u, v \in D. \tag{14}$$

Remark 1. *It follows from the preceding definitions that $L = L(L_0, M_0)$, $M = M(L_0, M_0)$, and for each $t \in [0, \gamma]$*

$$L_0(t) \leq L_1(t), \tag{15}$$
$$L(t) \leq L_1(t), \tag{16}$$
$$M(t) \leq M_1(t), \tag{17}$$

since $\Omega_0 \subseteq D$. By $L(L_0, M_0)$, we mean that L (or M) depends on L_0 and M_0 by the definition of Ω_0. In case any of (15)–(17) are strict inequalities, the following benefits are obtained over the work in [4] using L_1, M_1 instead of the new functions:

(a1) An at least as large convergence region leading to at least as many initial choices;
(a2) At least as tight upper bounds on the distances $\|x_n - x^*\|$, so at least as few iterations are needed to obtain a desired error tolerance.

These benefits are obtained under the same computational effort as in [4], since the new functions L_0, M_0, L, and M are special cases of the functions L_1 and M_1. This technique of using the center Lipschitz condition in combination with the restricted convergence region has been used by us on Newton's, Secant, Newton-like methods [14,15], and can be used on other methods, too, with the same benefits.

The proof of the next result follows as the corresponding one in [4], but there are crucial differences, where we use (L_0, L) instead of L_1 and (M_0, M) instead of M_1 used in [4].

We use the Euclidean norm. Note that the following equality is satisfied for the Euclidean norm $\|A - B\| = \|A^T - B^T\|$, where $A, B \in R^{m \times p}$.

Theorem 1. *Let $F + G : R^p \to R^m$ be continuous on an open convex subset $D \subset R^p$, F be a continuously differentiable function, and G be a continuous function. Suppose that problem (1) has a solution $x^* \in D$; the inverse operation*

$$(A_*^T A_*)^{-1} = [(F'(x^*) + G(x^*, x^*))^T (F'(x^*) + G(x^*, x^*))]^{-1} \tag{18}$$

exists, such that $\|(A_*^T A_*)^{-1}\| \leq B$; (7), (8), (11), and (12) hold, and γ given in (10) exists. Furthermore,

$$\|F(x^*) + G(x^*)\| \leq \eta, \qquad \|F'(x^*) + G(x^*, x^*)\| \leq \alpha; \tag{19}$$

$$\frac{B}{R}\left(\int_0^R L_0(u)du + \int_0^{2R} M_0(u)du\right)\eta < 1 \tag{20}$$

and $\Omega = \Omega(x^*, r_*) \subseteq D$, where r_* is the unique positive zero of the function q given by

$$\begin{aligned} q(r) = & B\bigg[\bigg(\alpha + \int_0^r L_0(u)du + \int_0^{2r} M_0(u)du\bigg)\bigg(\int_0^r L(u)udu + \int_0^r M(u)du\bigg) \\ & + \bigg(2\alpha + \int_0^r L_0(u)du + \int_0^{2r} M_0(u)du\bigg)\bigg(\int_0^r L_0(u)du + \int_0^{2r} M_0(u)du\bigg) \\ & + \bigg(\frac{1}{r}\int_0^r L_0(u)du + \frac{1}{r}\int_0^{2r} M_0(u)du\bigg)\eta\bigg] - 1. \end{aligned} \tag{21}$$

Then, for $x_0, x_{-1} \in \Omega$, the iterative sequence $\{x_n\}$, $n = 0, 1, \ldots$, generated by (2), is well defined, remains in Ω, and converges to x^*. Moreover, the following error estimates hold for each $n = 0, 1, 2, \ldots$:

$$\begin{aligned} \|x_{n+1} - x^*\| \leq & C_1 \|x_{n-1} - x^*\| + C_2 \|x_n - x^*\| + C_3 \|x_{n-1} - x^*\| \|x_n - x^*\| \\ & + C_4 \|x_n - x^*\|^2, \end{aligned} \tag{22}$$

where

$$g(r) = \frac{B}{1-\varphi(r)}; \quad C_1 = g(r_*)\frac{1}{2r_*}\int_0^{2r_*} M_0(u)du\,\eta; \quad (23)$$

$$C_2 = g(r_*)\left(\frac{1}{r_*}\int_0^{r_*} L_0(u)du + \frac{1}{2r_*}\int_0^{2r_*} M_0(u)du\right)\eta; \quad (24)$$

$$C_3 = g(r_*)\left(\alpha + \int_0^{r_*} L_0(u)du + \int_0^{2r_*} M_0(u)du\right)\frac{1}{r_*}\int_0^{r_*} M(u)du; \quad (25)$$

$$C_4 = g(r_*)\left(\alpha + \int_0^{r_*} L_0(u)du + \int_0^{2r_*} M_0(u)du\right)\frac{1}{r_*}\int_0^{r_*} L(u)u\,du. \quad (26)$$

Proof. We obtain

$$\lim_{r\to 0^+}\frac{1}{r}\int_0^r L_0(u)du \le \lim_{r\to 0^+}\frac{L_0(r)r}{r} \le L_0(0), \quad (27)$$

$$\lim_{r\to 0^+}\frac{1}{r}\int_0^{2r} M_0(u)du \le \lim_{r\to 0^+}\frac{M_0(2r)2r}{r} \le 2M_0(0), \quad (28)$$

since L_0 and M_0 are positive and nondecreasing functions on $[0,R]$, and $[0,2R]$, respectively. Taking into account Lemma 1 for a sufficiently small η, $q(0) = B(L_0(0) + 2M_0(0))\eta - 1 < 0$. With a sufficiently large R, the inequality $q(R) > 0$ holds. By the intermediate value theorem, the function q has a positive zero on $(0,R)$ denoted by r_*. Moreover, this zero is the only one on $(0,R)$. Indeed, according to Lemma 2, the function $\left(\frac{1}{r}\int_0^r L_0(u)du + \frac{1}{r}\int_0^{2r} M_0(u)du\right)\eta$ is non-decreasing with respect to r on $(0,R]$. By Lemma 1, functions $\int_0^r L(u)du, \int_0^r M(u)du$, and $\int_0^{2r} M(u)du$ are monotonically increasing on $[0,R]$. Furthermore, by Lemma 3, the function $\int_0^r L(u)u\,du = r^2\left(\frac{1}{r^2}\int_0^r L(u)u\,du\right)$ is monotonically increasing with respect to r on $(0,R]$. Therefore, $q(r)$ is monotonically increasing on $(0,R]$. Thus, the graph of function $q(r)$ crosses the positive r-axis only once on $(0,R)$. Finally, from the monotonicity of q and since $q(\gamma) > 0$, we obtain $r_* < \gamma$, so $\Omega(x^*, r_*) \subset \Omega_0$.

We denote $A_n = F'(x_n) + G(x_n, x_{n-1})$. Let $n = 0$. By the assumption $x_0, x_{-1} \in \Omega$, we obtain the following estimation:

$$\left\|I - (A_*^T A_*)^{-1}A_*^T A_0\right\| = \left\|(A_*^T A_*)^{-1}(A_*^T A_* - A_0^T A_0)\right\|$$

$$= \left\|(A_*^T A_*)^{-1}\left[A_*^T(A_* - A_0) + (A_*^T - A_0^T)(A_0 - A_*) + (A_*^T - A_0^T)A_*\right]\right\|$$

$$\le \left\|(A_*^T A_*)^{-1}\right\|\left[\|A_*^T\|\|A_* - A_0\| + \|A_*^T - A_0^T\|\|A_0 - A_*\| + \|A_*^T - A_0^T\|\|A_*\|\right]$$

$$\le B\left[\alpha\|A_* - A_0\| + \|A_*^T - A_0^T\|\|A_0 - A_*\| + \alpha\|A_*^T - A_0^T\|\right]. \quad (29)$$

Using conditions (11) and (12), we obtain

$$\|A_0 - A_*\| = \|(F'(x_0) + G(x_0, x_{-1})) - (F'(x^*) + G(x^*, x^*))\|$$

$$= \|F'(x_0) - F'(x_*) + G(x_0, x_{-1}) - G(x^*, x^*)\|$$

$$\le \|F'(x_0) - F'(x^*)\| + \|G(x_0, x_{-1}) - G(x^*, x^*)\|$$

$$\le \int_0^{\rho_0} L_0(u)du + \int_0^{\rho_0+\rho_{-1}} M_0(u)du, \quad (30)$$

where $\rho_k = \rho(x_k)$. Then, from inequality (29) and the equation $q(r) = 0$, we obtain by (10)

$$\|I - (A_*^T A_*)^{-1} A_0^T A_0\| \leq B\left[2\alpha + \int_0^{\rho_0} L_0(u)du + \int_0^{\rho_0 + \rho_{-1}} M_0(u)du\right]$$
$$\times \left[\int_0^{\rho_0} L_0(u)du + \int_0^{\rho_0 + \rho_{-1}} M_0(u)du\right] \leq B\left[2\alpha + \int_0^{r_*} L_0(u)du\right.$$
$$\left. + \int_0^{2r_*} M_0(u)du\right]\left[\int_0^{r_*} L_0(u)du + \int_0^{2r_*} M_0(u)du\right] < 1. \quad (31)$$

Next, from (29)–(31) and the Banach lemma [16], it follows that $(A_0^T A_0)^{-1}$ exists, and

$$\left\|(A_0^T A_0)^{-1}\right\| \leq g_0 = B\left\{1 - B\left[2\alpha + \int_0^{\rho_0} L_0(u)du + \int_0^{\rho_0 + \rho_{-1}} M_0(u)du\right]\right.$$
$$\times \left[\int_0^{\rho_0} L_0(u)du + \int_0^{\rho_0 + \rho_{-1}} M_0(u)du\right]\right\}^{-1}$$
$$\leq g(r_*) = B\left\{1 - B\left[2\alpha + \int_0^{r_*} L_0(u)du + \int_0^{2r_*} M_0(u)du\right]\right.$$
$$\times \left[\int_0^{r_*} L_0(u)du + \int_0^{2r_*} M_0(u)du\right]\right\}^{-1}. \quad (32)$$

Hence, x_1 is correctly defined. Next, we will show that $x_1 \in \Omega(x^*, r_*)$. Using the fact

$$A_*^T(F(x^*) + G(x^*)) = (F'(x^*) + G(x^*, x^*))^T(F(x^*) + G(x^*)) = 0, \quad (33)$$

$x_0, x_{-1} \in \Omega(x^*, r_*)$ and the choice of r_*, we obtain the estimate

$$\|x_1 - x^*\| = \left\|x_0 - x^* - (A_0^T A_0)^{-1}[A_0^T(F(x_0) + G(x_0)) - A_*^T(F(x^*) + G(x^*))]\right\|$$
$$\leq \left\|-(A_0^T A_0)^{-1}\right\|\left\| - A_0^T\left[A_0 - \int_0^1 F'(x_* + t(x_0 - x^*))dt\right.\right.$$
$$\left.\left. - G(x_0, x^*)\right](x_0 - x^*) + (A_0^T - A_*^T)(F(x^*) + G(x^*))\right\|. \quad (34)$$

So, considering the inequalities

$$\left\|A_0 - \int_0^1 F'(x^* + t(x_0 - x^*))dt - G(x_0, x^*)\right\|$$
$$= \left\|F'(x_0) - \int_0^1 F'(x^* + t(x_0 - x^*))dt + G(x_0, x_{-1}) - G(x_0, x^*)\right\|$$
$$= \left\|\int_0^1 [F'(x_0) - F'(x^* + t(x_0 - x^*))]dt + G(x_0, x_{-1}) - G(x_0, x^*)\right\|$$
$$= \left\|\int_0^1 [F'(x_0) - F'(x_0^t)]dt + G(x_0, x_{-1}) - G(x_0, x^*)\right\|$$
$$\leq \int_0^1 \int_{t\rho_0}^{\rho_0} L(u)dudt + \int_0^{\rho_{-1}} M(u)du = \int_0^{\rho_0} L(u)udu + \int_0^{\rho_{-1}} M(u)du$$
$$\leq \frac{1}{r_*^2}\int_0^{r_*} L(u)udu \, \rho_0^2 + \frac{1}{r_*}\int_0^{r_*} M(u)du \, \rho_{-1}, \quad (35)$$

$$\|A_0\| \leq \|A_*\| + \|A_0 - A_*\| \leq \alpha + \int_0^{\rho_0} L_0(u)du + \int_0^{\rho_0 + \rho_{-1}} M_0(u)du, \quad (36)$$

we obtain

$$\|x_1 - x^*\| \leq g_0 \Big\{ \Big[\alpha + \int_0^{\rho_0} L_0(u)du + \int_0^{\rho_0+\rho_{-1}} M_0(u)du\Big]$$
$$\times \Big[\int_0^{\rho_0} L(u)udu + \int_0^{\rho_{-1}} M(u)du\Big] \|x_0 - x^*\| + \eta \Big[\int_0^{\rho_0} L_0(u)du$$
$$+ \int_0^{\rho_0+\rho_{-1}} M_0(u)du\Big] \Big\} \leq g_0 \Big\{ \Big[\alpha + \int_0^{r_*} L_0(u)du + \int_0^{2r_*} M_0(u)du\Big]$$
$$\times \Big[\frac{1}{r_*^2} \int_0^{r_*} L(u)udu \rho_0^2 + \frac{1}{r_*} \int_0^{r_*} M(u)du \rho_{-1}\Big] \|x_0 - x^*\|$$
$$+ \eta \Big[\frac{1}{r_*} \int_0^{r_*} L_0(u)du \rho_0 + \frac{1}{2r_*} \int_0^{2r_*} M_0(u)du(\rho_0 + \rho_{-1})\Big] \Big\} \qquad (37)$$
$$< g(r_*) \Big\{ \Big[\alpha + \int_0^{r_*} L_0(u)du + \int_0^{2r_*} M_0(u)du\Big] \Big[\int_0^{r_*} L(u)udu + \int_0^{r_*} M(u)du\Big]$$
$$+ \frac{1}{r_*} \Big[\int_0^{r_*} L_0(u)du + \int_0^{2r_*} M_0(u)du\Big] \eta \Big\} r_* = p(r_*)r_* = r_*,$$

where

$$p(r) = g(r) \Big\{ \Big[\alpha + \int_0^r L_0(u)du + \int_0^{2r} M_0(u)du\Big] \Big[\int_0^r L(u)udu + \int_0^r M(u)du\Big]$$
$$+ \frac{1}{r} \Big[\int_0^r L_0(u)du + \int_0^{2r} M_0(u)du\Big] \eta \Big\}. \qquad (38)$$

Therefore, $x_1 \in \Omega(x^*, r_*)$, and estimate (22) holds for $n = 0$.

Let us assume that $x_n \in \Omega(x^*, r_*)$ for $n = 0, 1, ..., k$ and estimate (22) holds for $n = 0, 1, ..., k-1$, where $k \geq 1$ is an integer. We shall show $x_{n+1} \in \Omega$ and that the estimate (22) holds for $n = k$.

We can write

$$\|I - (A_*^T A_*)^{-1} A_k^T A_k\| = \|(A_*^T A_*)^{-1} (A_*^T A_* - A_k^T A_k)\|$$
$$= \|(A_*^T A_*)^{-1} (A_*^T (A_* - A_k) + (A_*^T - A_k^T)(A_k - A_*) + (A_*^T - A_k^T) A_*)\|$$
$$\leq B \Big(\alpha \|A_* - A_k\| + \|A_*^T - A_k^T\| \|A_k - A_*\| + \alpha \|A_*^T - A_k^T\|\Big)$$
$$\leq B \Big[2\alpha + \int_0^{\rho_k} L_0(u)du + \int_0^{\rho_k + \rho_{k-1}} M_0(u)du\Big] \Big[\int_0^{\rho_k} L_0(u)du \qquad (39)$$
$$+ \int_0^{\rho_k + \rho_{k-1}} M_0(u)du\Big] \leq B \Big[2\alpha + \int_0^{r_*} L_0(u)du + \int_0^{2r_*} M_0(u)du\Big]$$
$$\times \Big[\int_0^{r_*} L_0(u)du + \int_0^{2r_*} M_0(u)du\Big] < 1.$$

Consequently, $\left(A_k^T A_k\right)^{-1}$ exists, and

$$\|(A_{k+1}^T A_{k+1})^{-1}\| \leq g_k = B \Big\{ 1 - B \Big[2\alpha + \int_0^{\rho_k} L_0(u)du + \int_0^{\rho_k + \rho_{k-1}} M_0(u)du\Big]$$
$$\times \Big[\int_0^{\rho_k} L_0(u)du + \int_0^{\rho_k + \rho_{k-1}} M_0(u)du\Big] \Big\}^{-1} \leq g(r_*). \qquad (40)$$

Therefore, x_{k+1} is correctly defined, and the following estimate holds:

$$\|x_{k+1} - x_*\| = \|x_k - x^* - (A_k^T A_k)^{-1}[A_k^T(F(x_k) + G(x_k)) - A_*^T(F(x^*)$$
$$+ G(x^*))]\| \leq \| - (A_k^T A_k)^{-1}\| \left\| - A_k^T \left[A_k - \int_0^1 F'(x^* + t(x_k - x^*))dt \right. \right.$$
$$\left. - G(x_k, x_*) \right] (x_k - x^*) + (A_k^T - A_*^T)(F(x^*) + G(x^*)) \right\|$$
$$\leq \| - (A_k^T A_k)^{-1}\| \left\| - A_k^T \left[A_k - \int_0^1 F'(x^* + t(x_k - x^*))dt \right. \right.$$
$$\left. - G(x_k, x_*) \right] (x_k - x^*) + (A_k^T - A_*^T)(F(x^*) + G(x^*)) \right\| \tag{41}$$
$$\leq g_k \left\{ \left[\alpha + \int_0^{\rho_k} L_0(u)du + \int_0^{\rho_k + \rho_{k-1}} M_0(u)du \right] \left[\int_0^{\rho_k} L(u)udu \right. \right.$$
$$\left. + \int_0^{\rho_{k-1}} M(u)du \right] \|x_k - x^*\| + \eta \left[\int_0^{\rho_k} L_0(u)du + \int_0^{\rho_k + \rho_{k-1}} M_0(u)du \right] \right\}$$
$$\leq g(r_*) \left\{ \left[\alpha + \int_0^{r_*} L_0(u)du + \int_0^{2r_*} M_0(u)du \right] \right.$$
$$\times \left[\frac{1}{r_*^2} \int_0^{r_*} L(u)udu \rho_k^2 + \frac{1}{r_*} \int_0^{r_*} M(u)du \rho_{k-1} \right] \|x_k - x^*\|$$
$$\left. + \eta \left[\frac{1}{r_*} \int_0^{r_*} L_0(u)du \rho_k + \frac{1}{2r_*} \int_0^{2r_*} M_0(u)du(\rho_k + \rho_{k-1}) \right] \right\} < p(r_*)r_* = r_*.$$

This proves that $x_{k+1} \in \Omega(x^*, r_*)$ and estimate (22) for $n = k$.

Thus, by the induction method, (2) is correctly defined, $x_n \in \Omega(x^*, r_*)$, and estimate (22) holds for each $n = 0, 1, 2, \ldots$.

It remains to be proven that $x_n \to x^*$ for $n \to \infty$.

Let us define functions a and b on $[0, r_*]$ as

$$a(r) = g(r) \left\{ \left[\alpha + \int_0^r L_0(u)du + \int_0^{2r} M_0(u)du \right] \left[\int_0^r L(u)udu + \int_0^r M(u)du \right] \right.$$
$$\left. + \left[\frac{1}{r} \int_0^r L_0(u)du + \frac{1}{2r} \int_0^{2r} M_0(u)du \right] \eta \right\}; \tag{42}$$

$$b(r) = g(r) \frac{1}{2r} \int_0^{2r} M(u)du \, \eta. \tag{43}$$

According to the choice of r_*, we obtain

$$a(r_*) \geq 0, \quad b(r_*) \geq 0, \quad a(r_*) + b(r_*) = 1. \tag{44}$$

Using estimate (22), the definition of functions a, b and constants C_i ($i = 1, 2, 3, 4$), we have

$$\|x_{n+1} - x^*\| \leq C_1 \|x_{n-1} - x^*\| + (C_2 + C_3 r_* + C_4 r_*) \|x_n - x^*\|$$
$$= a(r_*) \|x_n - x^*\| + b(r_*) \|x_{n-1} - x^*\|. \tag{45}$$

According to the proof in [17], under the conditions (42)–(45), the sequence $\{x_n\}$ converges to x^* for $n \to \infty$. □

Corollary 1 ([4]). *The convergence order of method (2) for the problem (1) with zero residual is equal to $\dfrac{1 + \sqrt{5}}{2}$.*

If $\eta = 0$, we have the nonlinear least squares problem with zero residual. Then, the constants $C_1 = 0$ and $C_2 = 0$, and estimate (22) takes the form

$$\|x_{n+1} - x^*\| \leq C_3 \|x_{n-1} - x^*\| \, \|x_n - x^*\| + C_4 \|x_n - x^*\|^2. \tag{46}$$

This inequality can be written as

$$\|x_{n+1} - x^*\| \leq (C_3 + C_4) \|x_{n-1} - x^*\| \, \|x_n - x^*\|. \tag{47}$$

Then, we can write an equation for determining the convergence order as follows:

$$t^2 - t - 1 = 0. \tag{48}$$

Therefore, the positive root, $t^* = \dfrac{1 + \sqrt{5}}{2}$ of the latter equation is the order of convergence of method (2).

In case $G(x) \equiv 0$ in (1), we obtain the following consequences.

Corollary 2 ([4]). *The convergence order of method (2) for problem (1) with zero residual is quadratic.*

Indeed, if $G(x) \equiv 0$, then $C_3 = 0$, and estimate (22) takes the form

$$\|x_{n+1} - x^*\| \leq C_4 \|x_n - x^*\|^2, \tag{49}$$

which indicates the quadratic convergence rate of method (2).

Remark 2. *If $L_0 = L = L_1$ and $M_0 = M = M_1$, our results specialize to the corresponding ones in [4]. Otherwise, they constitute an improvement as already noted in Remark 1. As an example, let $q_1, g_1, C_1^1, C_2^1, C_3^1, C_4^1, r_*^1$ denote the functions and parameters where L_0, L, M_0, M are replaced by L_1, L_1, M_1, M_1, respectively. Then, we have in view of (15)–(17) that*

$$q(r) \leq q_1(r), \tag{50}$$
$$g(r) \leq g_1(r), \tag{51}$$
$$C_1 \leq C_1^1, \tag{52}$$
$$C_2 \leq C_2^1, \tag{53}$$
$$C_3 \leq C_3^1, \tag{54}$$

and

$$C_4 \leq C_4^1. \tag{55}$$

Hence, we have

$$r_*^1 \leq r_*, \tag{56}$$

the new error bounds (22) being tighter than the corresponding (6) in [4], and the rest of the advantages (already mentioned in Remark 1) holding true.

Next, we study the convergence of method (2) if L_0, L, M_0, M are constants, as a consequence of Theorem 1.

Corollary 3. *Let $F + G : R^p \to R^m$ be continuous on an open convex subset $D \subset R^p$, F be a continuously differentiable, and G be a continuous function on D. Suppose that problem (1) has a solution $x^* \in D$, and the inverse operation*

$$(A_*^T A_*)^{-1} = [(F'(x^*) + G(x^*, x^*))^T (F'(x^*) + G(x^*, x^*))]^{-1} \tag{57}$$

exists, such that $\|(A_*^T A_*)^{-1}\| \le B$.

Suppose that the Fréchet derivative F' satisfies the classic Lipschitz conditions

$$\|F'(x) - F'(x^*)\| \le L_0 \|x - x^*\|, \text{ for each } x \in D, \quad (58)$$
$$\|F'(x) - F'(y)\| \le L \|x - y\|, \text{ for each } x, y \in \Omega_0 \quad (59)$$

and the function G has a first order divided difference $G(x,y)$ that satisfies

$$\|G(x,y) - G(x^*, x^*)\| \le M_0(\|x - x^*\| + \|y - x^*\|), \text{ for each } x, y \in D, \quad (60)$$
$$\|G(x,y) - G(u,v)\| \le M(\|x - u\| + \|y - v\|), \text{ for each } x, y, u, v \in \Omega_0, \quad (61)$$

where $\Omega_0 = D \cap \Omega\left(x^*, \dfrac{\sqrt{B^2\alpha^2 + B} - B\alpha}{B(L_0 + 2M_0)}\right)$.

Furthermore,

$$\|F(x^*) + G(x^*)\| \le \eta, \|F'(x^*) + G(x^*, x^*)\| \le \alpha, B(L_0 + 2M_0)\eta < 1 \quad (62)$$

and $\Omega = \Omega(x^*, r_*) \subseteq D$, where

$$r_* = \dfrac{4(1 - BT_0\eta)}{B\alpha(4T_0 + T) + \sqrt{B^2\alpha^2(4T_0 + T)^2 + 8BT_0(2T_0 + T)(1 - BT_0\eta)}}, \quad (63)$$

$T_0 = L_0 + 2M_0$, $T = L + 2M$. Then, for each $x_0, x_{-1} \in \Omega$, the iterative sequence $\{x_n\}$, $n = 0, 1, \ldots$, generated by (2) is well defined, remains in Ω, and converges to x^*, such that the following error estimate holds for each $n = 0, 1, 2, \ldots$:

$$\|x_{n+1} - x^*\| \le C_1 \|x_{n-1} - x^*\| + C_2 \|x_n - x^*\|$$
$$+ C_3 \|x_{n-1} - x^*\| \|x_n - x^*\| + C_4 \|x_n - x^*\|^2, \quad (64)$$

where

$$g(r) = B[1 - B(2\alpha + (L_0 + 2M_0)r)(L_0 + 2M_0)r]^{-1}; \quad (65)$$
$$C_1 = g(r_*)M_0\eta;\ C_2 = g(r_*)(L_0 + M_0)\eta; \quad (66)$$
$$C_3 = g(r_*)(\alpha + (L_0 + 2M_0)r_*)M; \quad (67)$$
$$C_4 = g(r_*)(\alpha + (L_0 + 2M_0)r_*)\dfrac{L}{2}. \quad (68)$$

The proof of Corollary 3 is analogous to the proof of Theorem 1.

3. Numerical Examples

In this section, we give examples to show the applicability of method (2) and to confirm Remark 2. We use the norm $\|x\| = \sqrt{\sum_{i=1}^{p} x_i^2}$ for $x \in \mathbb{R}^p$.

Example 1. *Let function $F + G : \mathbb{R}^2 \to \mathbb{R}^3$ be defined by*

$$F(x) + G(x) = \begin{pmatrix} 3u^2v + v^2 - 1 + |u^2 - 1| \\ u^4 + uv^3 - 1 + |v| \\ v - 0.3 + |u - 1| \end{pmatrix}, \quad (69)$$

$$F(x) = \begin{pmatrix} 3u^2v + v^2 - 1 \\ u^4 + uv^3 - 1 \\ v - 0.3 \end{pmatrix},\ G(x) = \begin{pmatrix} |u^2 - 1| \\ |v| \\ |u - 1| \end{pmatrix}, \quad (70)$$

where $x = (u,v)$. The solution of this problem $x^* \approx (0.917889, 0.288314)$ and $\eta \approx 0.079411$.

Let us give the number of iterations needed to obtain an approximate solution of this problem. We test method (2) for the different initial points $x_0 = \delta(1.1, 0.5)^T$, where $\delta \in R$, and use the stopping criterion $\|x_{n+1} - x_n\| \leq \varepsilon$. The additional point $x_{-1} = x_0 + 10^{-4}$. The numerical results are shown in Table 1.

Table 1. Results for Example 1, $\varepsilon = 10^{-8}$.

	$\delta = 0.1$	$\delta = 1$	$\delta = 5$	$\delta = 10$	$\delta = 100$
Number of iterations	12	8	15	17	25

In Table 2, we give values of x_{n+1}, $\|x_{n+1} - x_n\|$ and the norm of residual at each iteration.

Table 2. Iterative sequence, norm of growth, and residual for Example 1, $x_0 = (0.8, 0.2)^T$, $\varepsilon = 10^{-6}$.

n	x_{n+1}	$\|x_{n+1} - x_n\|$	$\|F(x_{n+1}) + G(x_{n+1})\|$
0	(0.937901, 0.312602)	0.178033	0.143759
1	(0.918455, 0.290216)	2.965298×10^{-2}	7.973496×10^{-2}
2	(0.917850, 0.288333)	1.977741×10^{-3}	7.941104×10^{-2}
3	(0.917888, 0.288313)	4.346993×10^{-5}	7.941092×10^{-2}
4	(0.917889, 0.288314)	7.873833×10^{-7}	7.941092×10^{-2}

Example 2. Let function $F + G : D \subseteq R \to R^3$ be defined by [5]:

$$F(x) + G(x) = \begin{pmatrix} x + \mu \\ \lambda x^3 + x - \mu \\ \lambda |x^2 - 1| - \lambda \end{pmatrix}, \tag{71}$$

$$F(x) = \begin{pmatrix} x + \mu \\ \lambda x^3 + x - \mu \\ 0 \end{pmatrix}, \quad G(x) = \begin{pmatrix} 0 \\ 0 \\ \lambda |x^2 - 1| - \lambda \end{pmatrix}, \tag{72}$$

where $\lambda, \mu \in R$ are two parameters. Here $x^* = 0$ and $\eta = \sqrt{2}|\mu|$. Thus, if $\mu = 0$, then we have a problem with zero residual.

Let us consider Example 2 and show that $r_*^1 \leq r_*$ and the new error estimates (64) are tighter than the corresponding ones in [4]. We consider the case of the classical Lipschitz conditions (Corollary 3). Error estimates from [4] are as follows:

$$\|x_{n+1} - x^*\| \leq C_1^1 \|x_{n-1} - x^*\| + C_2^1 \|x_n - x^*\| + C_3^1 \|x_{n-1} - x^*\| \|x_n - x^*\| + C_4^1 \|x_n - x^*\|^2, \tag{73}$$

where

$$g^1(r) = B[1 - B(2\alpha + (L_1 + 2M_1)r)(L_1 + 2M_1)r]^{-1}; \tag{74}$$

$$C_1^1 = g^1(r_*^1)M_1\eta; \quad C_2^1 = g^1(r_*^1)(L_1 + M_1)\eta; \tag{75}$$

$$C_3^1 = g^1(r_*^1)(\alpha + (L_1 + 2M_1)r_*^1)M_1; \tag{76}$$

$$C_4^1 = g^1(r_*^1)(\alpha + (L_1 + 2M_1)r_*^1)\frac{L_1}{2}. \tag{77}$$

They can be obtained from (64) by replacing r_*, L_0, L, M_0, M in $g(r), C_1, C_2, C_3, C_4$ by $r_*^1, L_1, L_1, M_1, M_1$, respectively. Similarly,

$$r_*^1 = \frac{4(1 - BT_1\eta)}{5B\alpha T_1 + \sqrt{25B^2\alpha^2 T_1^2 + 24BT_1^2(1 - BT_1\eta)}}, \quad T_1 = L_1 + 2M_1. \tag{78}$$

Let us choose $D = (-0.5; 0.5)$. Thus, we have $B = 0.5$, $\eta = \sqrt{2}|\mu|$, $\alpha = \sqrt{2}$, $L_0 = \max\limits_{x \in D} 3|\lambda|\|x\|$, $L = \max\limits_{x,y \in \Omega_0} 3|\lambda|\|x + y\|$, $L_1 = \max\limits_{x,y \in D} 3|\lambda|\|x + y\|$, $M_0 = M = M_1 = |\lambda|$. Radii are written in Table 3.

Table 3. Radii of convergence domains.

λ	μ	L_0	L	L_1	M	r_*	r_*^1
0.4	0	0.6	1.004205	1.2	0.4	0.319259	0.235702
0.1	0.2	0.15	0.3	0.3	0.1	1.192633	0.885163

Tables 4 and 5 report the left and right side of error estimates (64) and (73). We obtained these results for $\varepsilon = 10^{-8}$ and starting approximations $x_{-1} = 0.2001$, $x_0 = 0.2$. We see that the new error bounds (64) are tighter than the corresponding (73) from [4].

Table 4. Results for $\lambda = 0.4$, $\mu = 0$.

n	$|x_{n+1} - x^*|$	The Right Side of (64)	The Right Side of (73)
0	4.364164×10^{-3}	0.125318	0.169740
1	1.425535×10^{-5}	1.245455×10^{-3}	1.529729×10^{-3}
2	2.179258×10^{-11}	8.675961×10^{-8}	1.060957×10^{-7}
3	3.542853×10^{-22}	4.314684×10^{-16}	5.272102×10^{-16}

Table 5. Results for $\lambda = 0.1$, $\mu = 0.2$.

n	$|x_{n+1} - x^*|$	The Right Side of (64)	The Right Side of (73)
0	2.063103×10^{-3}	5.909333×10^{-2}	8.484100×10^{-2}
1	5.453349×10^{-7}	9.113893×10^{-3}	1.080560×10^{-2}
2	2.054057×10^{-14}	9.051468×10^{-5}	1.057648×10^{-4}
3	1.447579×10^{-18}	2.390964×10^{-8}	2.792694×10^{-8}

4. Conclusions

We developed an improved local convergence analysis of the Gauss–Newton–Secant method for solving nonlinear least squares problems with nondifferentiable operator. We use a center and restricted radius Lipschitz conditions to study the method. As a consequence, we obtain a larger radius of convergence and tighter error estimates under the same computational effort as in earlier papers. This idea can be used to extend the usage of other methods with inverses, such as Newton-type, Secant-type, single-step, or multi-step, to mention a few. This should be our future work. Finally, it is worth mentioning that except for the methods used in this paper, some of the most representative computational intelligence algorithms can be used to solve the problems, such as monarch butterfly optimization (MBO) [18], the earthworm optimization algorithm (EWA) [19], elephant herding optimization (EHO) [20], the moth search (MS) algorithm [21], the slime mould algorithm (SMA), and Harris hawks optimization (HHO) [22].

Author Contributions: Editing, I.K.A.; Conceptualization S.S.; Investigation I.K.A., S.S., R.I., H.Y. and M.I.A. All authors have read and agreed to the published version of the manuscript.

Funding: This research received no external funding.

Institutional Review Board Statement: Not applicable.

Informed Consent Statement: Not applicable.

Data Availability Statement: Not applicable.

Conflicts of Interest: The authors declare no conflict of interest.

References

1. Li, C.; Zhang, W.; Jin, X. Convergence and uniqueness properties of Gauss-Newton's method. *Comput. Math. Appl.* **2004**, *47*, 1057–1067. [CrossRef]
2. Argyros, I.K.; Ren, H. A derivative free iterative method for solving least squares problems. *Numer. Algorithms* **2011**, *58*, 555–571.
3. Shakhno, S.M.; Gnatyshyn, O.P. On an iterative algorithm of order 1.839... for solving the nonlinear least squares problems. *Appl. Math. Comput.* **2005**, *161*, 253–264. [CrossRef]
4. Shakhno, S.M.; Iakymchuk, R.P.; Yarmola, H.P. An iterative method for solving nonlinear least squares problems with nondifferentiable operator. *Mat. Stud.* **2017**, *48*, 97–107. [CrossRef]
5. Shakhno, S.M.; Iakymchuk, R.P.; Yarmola, H.P. Convergence analysis of a two-step method for the nonlinear least squares problem with decomposition of operator. *J. Numer. Appl. Math.* **2018**, *128*, 82–95.
6. Shakhno, S.; Shunkin, Yu. One combined method for solving nonlinear least squares problems. *Visnyk Lviv Univ. Ser. Appl. Math. Comp. Sci.* **2017**, *25*, 38–48. (In Ukrainian)
7. Ulm, S. On generalized divided differences. *Izv. ESSR Ser. Phys. Math.* **1967**, *16*, 13–26. (In Russian)
8. Cătinaş, E. On some iterative methods for solving nonlinear equations. *Rev. Anal. Numér. Théor. Approx.* **1994**, *23*, 47–53.
9. Shakhno, S.M.; Mel'nyk, I.V.; Yarmola, H.P. Convergence analysis of combined method for solving nonlinear equations. *J. Math. Sci.* **2016**, *212*, 16–26. [CrossRef]
10. Shakhno, S.M. Convergence of combined Newton-Secant method and uniqueness of the solution of nonlinear equations. *Sci. J. Tntu* **2013**, *1*, 243–252. (In Ukrainian)
11. Zabrejko, P.P.; Nguen, D.F. The majorant method in the theory of Newton-Kantorovich approximations and the Pták error estimates. *Numer. Funct. Anal. Optim.* **1987**, *9*, 671–686. [CrossRef]
12. Wang, X.; Li, C. Convergence of Newton's method and uniqueness of the solution of equations in Banach space II. *Acta Math. Sin.* **2003**, *19*, 405–412. [CrossRef]
13. Wang, X. Convergence of Newton's method and uniqueness of the solution of equations in Banach space. *IMA J. Numer. Anal.* **2000**, *20*, 123–134. [CrossRef]
14. Argyros, I.K.; Hilout, S. On an improved convergence analysis of Newton's method. *Appl. Math. Comput.* **2013**, *225*, 372–386. [CrossRef]
15. Argyros, I.K.; Magreñán, A.A. *Iterative Methods and Their Dynamics with Applications: A Contemporary Study*; CRC Press: Boca Raton, FL, USA, 2017.
16. Dennis, J.E.; Schnabel, R.B. *Numerical Methods for Unconstrained Optimization and Nonlinear Equations*; SIAM: Philadelphia, PA, USA, 1996.
17. Ren, H.; Argyros, I.K. Local convergence of a secant type method for solving least squares problems. *Appl. Math. Comput.* **2010**, *217*, 3816–3824. [CrossRef]
18. Wang, G.G.; Deb, S.; Cui, Z. Monarch butterfly optimization. *Neural Comput. Appl.* **2019**, *31*, 1995–2014. [CrossRef]
19. Wang, G.G.; Deb, S.; Dos, L.; Coelho, L.D.S. Earthworm optimization algorithm: A bio-inspired metaheuristic algorithm for global optimization problems. *Int. J. Bio-Inspired Comput.* **2018**, *12*, 1–22. [CrossRef]
20. Wang, G.G.; Deb, S.; Coelho, L.D.S. Elephant Herding Optimization. In Proceedings of the 3rd International Symposium on Computational and Business Intelligence (ISCBI 2015), Bali, Indonesia, 7–9 December 2015; pp. 1–5.
21. Mirjalili, S. Moth-flame optimization algorithm: A novel nature-inspired heuristic paradigm. *Knowl.-Based Syst.* **2015**, *89*, 228–249 [CrossRef]
22. Zhao, J.; Gao, Z.-M. The hybridized Harris hawk optimization and slime mould algorithm. *J. Phys. Conf. Ser.* **2020**, *1682*, 012029. [CrossRef]

Article

Numerical Algorithms for Computing an Arbitrary Singular Value of a Tensor Sum

Asuka Ohashi [1,*] and Tomohiro Sogabe [2]

1. National Institute of Technology, Kagawa College, Takuma Campus, Mitoyo 769-1192, Japan
2. Department of Applied Physics, Nagoya University, Furo-cho, Chikusa-ku, Nagoya 464-8603, Japan; sogabe@na.nuap.nagoya-u.ac.jp
* Correspondence: ohashi-a@dg.kagawa-nct.ac.jp
† This paper is an extended version of our paper published in International Conference on Mathematics and Its Applications in Science and Engineering (ICMASE2020), Ankara Haci Bayram Veli University, Turkey (Online), 9–10 July 2020.

Abstract: We consider computing an arbitrary singular value of a tensor sum: $T := I_n \otimes I_m \otimes A + I_n \otimes B \otimes I_\ell + C \otimes I_m \otimes I_\ell \in \mathbb{R}^{\ell mn \times \ell mn}$, where $A \in \mathbb{R}^{\ell \times \ell}$, $B \in \mathbb{R}^{m \times m}$, $C \in \mathbb{R}^{n \times n}$. We focus on the shift-and-invert Lanczos method, which solves a shift-and-invert eigenvalue problem of $(T^T T - \tilde{\sigma}^2 I_{\ell mn})^{-1}$, where $\tilde{\sigma}$ is set to a scalar value close to the desired singular value. The desired singular value is computed by the maximum eigenvalue of the eigenvalue problem. This shift-and-invert Lanczos method needs to solve large-scale linear systems with the coefficient matrix $T^T T - \tilde{\sigma}^2 I_{\ell mn}$. The preconditioned conjugate gradient (PCG) method is applied since the direct methods cannot be applied due to the nonzero structure of the coefficient matrix. However, it is difficult in terms of memory requirements to simply implement the shift-and-invert Lanczos and the PCG methods since the size of T grows rapidly by the sizes of A, B, and C. In this paper, we present the following two techniques: (1) efficient implementations of the shift-and-invert Lanczos method for the eigenvalue problem of $T^T T$ and the PCG method for $T^T T - \tilde{\sigma}^2 I_{\ell mn}$ using three-dimensional arrays (third-order tensors) and the n-mode products, and (2) preconditioning matrices of the PCG method based on the eigenvalue and the Schur decomposition of T. Finally, we show the effectiveness of the proposed methods through numerical experiments.

Keywords: tensor sum; singular value; shift-and-invert Lanczos method; preconditioned conjugate gradient method

MSC: 65F15; 65F08

1. Introduction

We consider computing an arbitrary singular value of a tensor sum:

$$T := I_n \otimes I_m \otimes A + I_n \otimes B \otimes I_\ell + C \otimes I_m \otimes I_\ell \in \mathbb{R}^{\ell mn \times \ell mn}, \quad (1)$$

where $A \in \mathbb{R}^{\ell \times \ell}$, $B \in \mathbb{R}^{m \times m}$, $C \in \mathbb{R}^{n \times n}$, I_n is the $n \times n$ identity matrix, and the symbol "$\otimes$" denotes the Kronecker product. The tensor sum T arises from a finite difference discretization of three-dimensional constant coefficient partial differential equations (PDE) defined as follows:

$$[-\boldsymbol{a} \cdot (\nabla * \nabla) + \boldsymbol{b} \cdot \nabla + c]u(x,y,z) = g(x,y,z) \text{ in } \Omega, \quad u(x,y,z) = 0 \text{ on } \partial\Omega, \quad (2)$$

where $\Omega = (0,1) \times (0,1) \times (0,1)$, $\boldsymbol{a}, \boldsymbol{b} \in \mathbb{R}^3$, $c \in \mathbb{R}$, and the symbol "$*$" denotes element-wise products. If $\boldsymbol{a} = (1,1,1)$, then $\boldsymbol{a} \cdot (\nabla * \nabla) = \Delta$. Matrix T tends to be too large even if A, B and C are not. Hence it is difficult to compute singular values of T with regard to the memory requirement.

Previous studies [1,2] provided methods to compute the maximum and minimum singular values of T. By the previous studies, one can compute only the maximum and minimum singular values of T without shift. On the other hand, one can compute arbitrary singular values of T with the shift by this work. The previous studies are based on the Lanczos bidiagonalization method (see, e.g., [3]), which computes the maximum and minimum singular values of a matrix. For insights on Lanczos bidiagonalization method, see, e.g., [4–6]. The Lanczos bidiagonalization method for T was implemented using tensors and their operations to reduce the memory requirement.

The Lanczos method with the shift-and-invert technique, see, e.g., [3], is widely known for computing an arbitrary eigenvalue λ of a symmetric matrix $M \in \mathbb{R}^{n \times n}$. This method solves the shift-and-invert eigenvalue problem: $(M - \tilde{\sigma} I_n)^{-1} x = (\lambda - \tilde{\sigma})^{-1} x$, where x is the eigenvector of M corresponding to λ, and $\tilde{\sigma}$ is a shift point which is set to the nearby λ ($\tilde{\sigma} \neq \lambda$). Since the eigenvalue problem has the eigenvalue $(\lambda - \tilde{\sigma})^{-1}$ as the maximum eigenvalue, the method is effective for computing the desired eigenvalue λ near $\tilde{\sigma}$. For successful work using the shift-and-invert technique, see, e.g., [7–13].

Therefore, we obtain a computing method for an arbitrary singular value of T based on the shift-and-invert Lanczos method. The method solves the following shift-and-invert eigenvalue problem: $(T^T T - \tilde{\sigma}^2 I_{\ell m n})^{-1} x = (\sigma^2 - \tilde{\sigma}^2)^{-1} x$, where σ is the desired singular value of T, x is the corresponding right-singular vector, and $\tilde{\sigma}$ is close to σ ($\tilde{\sigma} \neq \sigma$). This shift-and-invert Lanczos method requires the solution of large-scale linear systems with the coefficient matrix $T^T T - \tilde{\sigma}^2 I_{\ell m n}$. Here, $T^T T - \tilde{\sigma}^2 I_{\ell m n}$ can be a dense matrix whose number of elements is $O(n^6)$ even if T is a sparse matrix whose number of elements is $O(n^4)$ when $A, B, C \in \mathbb{R}^{n \times n}$ are dense.

Since it is difficult regarding the memory requirement to apply the direct method, e.g., the Cholesky decomposition, which needs generating matrix $T^T T - \tilde{\sigma}^2 I_{\ell m n}$, the preconditioned conjugate gradient (PCG) method, see, e.g., [14], is applied, even though it is difficult in terms of memory requirements to simply implement this shift-and-invert Lanczos method and the PCG method since the size of T grows rapidly by the sizes of A, B, and C.

We propose the following two techniques in this paper: (1) Efficient implementations of the shift-and-invert Lanczos method for the eigenvalue problem of $T^T T$ and the PCG method for $T^T T - \tilde{\sigma}^2 I_{\ell m n}$ using three-dimensional arrays (third-order tensors) and the n-mode products, see, e.g., [15]. (2) Preconditioning matrices based on the eigenvalue decomposition and the Schur decomposition of T for faster convergence of the PCG method. Finally, we show the effectiveness of the proposed method through numerical experiments.

2. Preliminaries of Tensor Operations

A tensor means a multidimensional array. Particularly, the third-order tensor $\mathcal{X} \in \mathbb{R}^{I \times J \times K}$ plays an important role. In the rest of this section, the definitions of some tensor operations are shown. For more details, see, e.g., [15].

Firstly, a summation, a subtraction, an inner product, and a norm for $\mathcal{X}, \mathcal{Y} \in \mathbb{R}^{I \times J \times K}$ are defined as follows:

$$(\mathcal{X} \pm \mathcal{Y})_{ijk} := \mathcal{X}_{ijk} \pm \mathcal{Y}_{ijk}, \quad (\mathcal{X}, \mathcal{Y}) := \sum_{i=1}^{I} \sum_{j=1}^{J} \sum_{k=1}^{K} \mathcal{X}_{ijk} \mathcal{Y}_{ijk}, \quad \|\mathcal{X}\| = \sqrt{(\mathcal{X}, \mathcal{X})},$$

where $\mathcal{X}_{ijk}$ denotes the (i, j, k) element of $\mathcal{X}$. Secondly, the n-mode product of a tensor $\mathcal{X} \in \mathbb{R}^{I_1 \times I_2 \times \cdots \times I_N}$ and a matrix $M \in \mathbb{R}^{J \times I_n}$ is defined as

$$(\mathcal{X} \times_n M)_{i_1 \ldots i_{n-1} j i_{n+1} \ldots i_N} = \sum_{i_n=1}^{I_n} \mathcal{X}_{i_1 i_2 \ldots i_N} M_{j i_n},$$

where $n \in \{1, 2, \ldots N\}$, $i_k \in \{1, 2, \ldots, I_k\}$ for $k = 1, 2, \ldots N$, and $j \in \{1, 2, \ldots, J\}$. Finally, vec and vec^{-1} operators are the following maps between a vector space $\mathbb{R}^{IJK}$ and a tensor space $\mathbb{R}^{I \times J \times K}$: vec : $\mathbb{R}^{I \times J \times K} \to \mathbb{R}^{IJK}$ and vec^{-1} : $\mathbb{R}^{IJK} \to \mathbb{R}^{I \times J \times K}$. vec operator can vectorize a tensor by combining all column vectors of the tensor into one long vector. Conversely, vec^{-1} operator can reshape a tensor from one long vector.

3. Shift-and-Invert Lanczos Method for an Arbitrary Singular Value over Tensor Space

This section gives an algorithm for computing an arbitrary singular value of the tensor sum T. Let σ and x be a desired singular value of T and the corresponding right singular vectors, respectively. Then, the eigenvalue problem of T is written by $T^T T x = \sigma^2 x$. Here, introducing a shift $\tilde{\sigma} \approx \sigma$, the shift-and-invert eigenvalue problem is

$$(T^T T - \tilde{\sigma}^2 I_{\ell mn})^{-1} x = \frac{1}{\sigma^2 - \tilde{\sigma}^2} x. \tag{3}$$

The shift-and-invert Lanczos method (see, e.g., [3]) computes the nearest singular value σ based on Equation (3). Reconstructing this method over the $\ell \times m \times n$ tensor space, we obtain Algorithm 1 whose memory requirement is of $O(n^3)$ when $n = m = \ell$.

Algorithm 1: Shift-and-invert Lanczos method for an arbitrary singular value over tensor space

---

1: Choose an initial tensor $\mathcal{Q}_0 \in \mathbb{R}^{\ell \times m \times n}$;
2: $\mathcal{V} := \mathcal{Q}_0, \beta_0 := \|\mathcal{V}\|$;
3: **for** $k = 1, 2, \ldots$, until convergence **do**
4:     $\mathcal{Q}_k := \mathcal{V}/\beta_{k-1}$;
5:     $\mathcal{V} := \text{vec}^{-1}\left\{ \left(T^T T - \tilde{\sigma}^2 I_{\ell mn} \right)^{-1} \text{vec}(\mathcal{Q}_k) \right\}$;
    (Computed by Algorithms 3 or 4 in Section 4)
6:     $\mathcal{V} := \mathcal{V} - \beta_{k-1} \mathcal{Q}_{k-1}$;
7:     $\alpha_k := (\mathcal{Q}_k, \mathcal{V})$;
8:     $\mathcal{V} := \mathcal{V} - \alpha_k \mathcal{Q}_k$;
9:     $\beta_k := \|\mathcal{V}\|$;
10: **end for**
11: Approximate singular value $\sigma = \sqrt{\tilde{\sigma}^2 + \frac{1}{\tilde{\lambda}^{(k)}}}$, where $\tilde{\lambda}^{(k)}$ is the maximum eigenvalue of $\tilde{T}_k$.

At step k, we have the following $\tilde{T}_k$ by Algorithm 1:

$$\tilde{T}_k := \begin{pmatrix} \alpha_1 & \beta_1 & & & \\ \beta_1 & \alpha_2 & \beta_2 & & \\ & \ddots & \ddots & \ddots & \\ & & \beta_{k-2} & \alpha_{k-1} & \beta_{k-1} \\ & & & \beta_{k-1} & \alpha_k \end{pmatrix} \in \mathbb{R}^{k \times k}.$$

To implement Algorithm 1, we need to iteratively solve the linear system

$$\mathcal{V} := \text{vec}^{-1}\left\{ \left(T^T T - \tilde{\sigma} I_{\ell mn} \right)^{-1} \text{vec}(\mathcal{Q}_k) \right\}, \tag{4}$$

whose coefficient matrix is $\ell mn \times \ell mn$, that is, the memory requirement is $O(n^6)$ when $n = m = \ell$. Here, the convergence rate of the shift and invert Lanczos method depends on the ratio of gaps between the maximum, the second maximum, and the minimum singular values $\sigma_1, \sigma_2, \sigma_m$ of $(T^T T - \tilde{\sigma}^2 I_{\ell mn})^{-1}$ as follows: $(\sigma_1^2 - \sigma_2^2)/(\sigma_1^2 - \sigma_m^2)$.

In the next section, we consider solving the linear systems with memory requirement of $O(n^3)$ when $n = m = \ell$.

4. Preconditioned Conjugate Gradient (PCG) Method over Tensor Space

This section provides an efficient solver of Equation (4) using tensors. This linear system is rewritten by $v = \left(T^\mathrm{T} T - \tilde{\sigma}^2 I_{\ell mn}\right)^{-1} q_k$, where $v := \mathrm{vec}(\mathcal{V})$ and $q_k := \mathrm{vec}(\mathcal{Q}_k)$. Then we solve $\left(T^\mathrm{T} T - \tilde{\sigma}^2 I_{\ell mn}\right) v = q_k$, where v and q_k are unknown and known vectors. Since the coefficient matrix is symmetric positive definite, we can use the preconditioned conjugate gradient method (PCG method, see, e.g., [14]), which is one of the widely used solvers. However, it is difficult to simply apply the method due to the complex nonzero structure of the coefficient matrix $T^\mathrm{T} T - \tilde{\sigma}^2 I_{\ell mn}$. For applying the PCG method, we consider transforming the linear system $\left(T^\mathrm{T} T - \tilde{\sigma}^2 I_{\ell mn}\right) v = q_k$ by the eigendecomposition and the complex Schur decomposition as shown in the next subsections.

4.1. PCG Method by the Eigendecomposition

Firstly, T is decomposed into $T := XDX^{-1}$, where X and D are a matrix whose column vectors are eigenvectors and a diagonal matrix with eigenvalues, respectively. Then, it follows that

$$\left(T^\mathrm{T} T - \tilde{\sigma}^2 I_{\ell mn}\right) v = q_k \Leftrightarrow \left((XDX^{-1})^\mathrm{H}(XDX^{-1}) - \tilde{\sigma}^2 I_{\ell mn}\right) v = q_k$$

$$\Leftrightarrow \left(\overline{D} X^\mathrm{H} X D - \tilde{\sigma}^2 X^\mathrm{H} X\right)\left(X^{-1} v\right) = X^\mathrm{H} q_k,$$

where $\overline{D}$ is the complex conjugate of D. We rewrite the above linear system into $\tilde{A} \tilde{y} = \tilde{b}$, where $\tilde{A} := \overline{D} X^\mathrm{H} X D - \tilde{\sigma}^2 X^\mathrm{H} X$, $\tilde{y} := X^{-1} v$, and $\tilde{b} := X^\mathrm{H} q_k$. Here, X is easily computed by small matrices X_A, X_B, and X_C whose column vectors are eigenvectors of A, B, and C as follows: $X = X_C \otimes X_B \otimes X_A$. Moreover, eigenvalues of T in D are obtained by summations of each eigenvalue of A, B, and C.

The PCG method for solving $\tilde{A}\tilde{y} = \tilde{b}$ is shown in Algorithm 2. Since this algorithm computes $\tilde{y}$, we need to compute $v = X\tilde{y}$. Section 4.1.1 proposes a preconditioning matrix and Section 4.1.2 provides efficient computations using tensors.

Algorithm 2: PCG method over vector space for $\tilde{A}\tilde{y} = \tilde{b}$

1: Choose an initial vector $x_0 \in \mathbb{R}^{\ell mn}$ and $p_0 = 0 \in \mathbb{R}^{\ell mn}$, and an initial scalar $\beta_0 = 0$;
2: $r_0 = \tilde{b} - \tilde{A} x_0$;
3: $z_0 = M^{-1} r_0$;
4: **for** $k' = 1, 2, \ldots,$ until convergence **do**
5: $\quad p_{k'} = z_{k'-1} + \beta_{k'-1} p_{k'-1}$;
6: $\quad \hat{p}_{k'} = \tilde{A} p_{k'}$;
7: $\quad \alpha_{k'} = (z_{k'-1}, r_{k'-1}) / (p_{k'-1}, \hat{p}_{k'})$;
8: $\quad x_{k'} = x_{k'-1} + \alpha_{k'} p_{k'}$;
9: $\quad r_{k'} = r_{k'-1} - \alpha_{k'} \hat{p}_{k'}$;
10: $\quad z_{k'} = M^{-1} r_{k'}$;
11: $\quad \beta_{k'} = (z_{k'}, r_{k'}) / (z_{k'-1}, r_{k'-1})$;
12: **end for**
13: Obtain an approximate solution $\tilde{y} \approx x_{k'}$;

4.1.1. Preconditioning Matrix

Algorithm 2 solves

$$\left(M^{-1} \tilde{A} M^{-\mathrm{H}}\right)\left(M^\mathrm{H} \tilde{y}\right) = M^{-1} \tilde{b},$$

where $M \in \mathbb{R}^{\ell mn \times \ell mn}$ is a preconditioning matrix. M must satisfy the following two conditions: (1) a condition number of $M^{-1} \tilde{A}$ is close to 1; (2) the matrix-vector multiplication of M^{-1} is easily computed.

Therefore, we propose a preconditioning matrix based on the eigendecomposition of T

$$M := \overline{D}D - \tilde{\sigma}^2 I_{\ell mn}. \tag{5}$$

Since M is the diagonal matrix, the second condition of the preconditioning matrix is satisfied. Moreover, if T is symmetric, X is the unitary matrix, that is, $X^H X = I_{\ell mn}$. In the case of the symmetric matrix T, we obtain $M = \tilde{A}$. Namely, the proposed matrix satisfies the first conditions when T is symmetric. So, even if T is not exactly symmetric, if T is almost symmetric, then we can expect the preconditioning matrix M to be effective.

4.1.2. Efficient Implementation of Algorithm 2 by the Eigendecomposition

Similarly to obtaining Algorithm 1, to improve an implementation of Algorithm 2, we reconstruct ℓmn dimensional vectors into $\ell \times m \times n$ tensors via vec^{-1} operator as follows: $\mathcal{X}_{k'} := \text{vec}^{-1}(x_{k'})$, $\mathcal{R}_{k'} := \text{vec}^{-1}(r_{k'})$, $\mathcal{P}_{k'} := \text{vec}^{-1}(p_{k'})$, $\mathcal{Z}_{k'} := \text{vec}^{-1}(z_{k'})$, and $\hat{\mathcal{P}}_{k'} := \text{vec}^{-1}(\hat{p}_{k'})$. Most computations of vectors are simply transformed into computations of tensors because of the linearity of vec^{-1} operator.

In the rest of this section, we show the computations of $\text{vec}^{-1}(\tilde{A}\text{vec}(\mathcal{P}_{k'}))$ and $\text{vec}^{-1}(M^{-1}\text{vec}(\mathcal{R}_{k'}))$, which are required in the PCG method, using the 1, 2, and 3-mode products for tensors and the definition of T. First, from the definitions of $\tilde{A}$ and X, $\text{vec}^{-1}(\tilde{A}\text{vec}(\mathcal{P}_{k'})) = \text{vec}^{-1}(\overline{D}X^H X D \text{vec}(\mathcal{P}_{k'})) - \tilde{\sigma}^2 \text{vec}^{-1}(X^H X \text{vec}(\mathcal{P}_{k'}))$ holds. Let $\mathcal{D} = \text{vec}^{-1}(\text{diag}(D))$, where $\text{diag}(D)$ returns an ℓmn-dimensional column vector with diagonals of D. Then, $\mathcal{D}_{ijk} := \lambda_i^{(A)} + \lambda_j^{(B)} + \lambda_k^{(C)}$, where $\lambda_i^{(A)}, \lambda_j^{(B)}$, and $\lambda_k^{(C)}$ denote the eigenvalues of A, B, and C. Note that $(\text{vec}^{-1}(D\text{vec}(\mathcal{P}_{k'})))_{ijk} = \mathcal{D}_{ijk}(\mathcal{P}_{k'})_{ijk}$ since we compute $(D p_{k'})_i = D_{ii}(p_{k'})_i$ for $i = 1, 2, \ldots, \ell mn$. Using the relation between the Kronecker product and the mode products via vec^{-1} operator, we compute

$$\begin{aligned}
&\text{vec}^{-1}(\tilde{A}\text{vec}(\mathcal{P})) \\
&= \overline{\mathcal{D}} * \left\{ (\mathcal{D} * \mathcal{P}_{k'}) \times_1 X_A^H X_A + (\mathcal{D} * \mathcal{P}_{k'}) \times_2 X_B^H X_B + (\mathcal{D} * \mathcal{P}_{k'}) \times_3 X_C^H X_C \right\} \\
&\quad - \tilde{\sigma}^2 \left(\mathcal{P}_{k'} \times_1 X_A^H X_A + \mathcal{P}_{k'} \times_2 X_B^H X_B + \mathcal{P}_{k'} \times_3 X_C^H X_C \right),
\end{aligned} \tag{6}$$

where "$*$" denotes elementwise product.

Next, from the definition of the diagonal matrix M in Equation (5), we easily obtain

$$\left(M^{-1}\right)_{ii} = \frac{1}{(\overline{D})_{ii}(D)_{ii} - \tilde{\sigma}^2}, \quad i = 1, 2, \ldots, \ell mn.$$

Here, let $\mathcal{M} = \text{vec}^{-1}(\text{diag}(M^{-1}))$. Then it follows that $\mathcal{M}_{ijk} = 1/(\overline{\mathcal{D}}_{ijk}\mathcal{D}_{ijk} - \tilde{\sigma}^2)$. $\text{vec}^{-1}(M^{-1}\text{vec}(\mathcal{R}_{k'}))$ is computed by

$$\text{vec}^{-1}\left(M^{-1}\text{vec}(\mathcal{R}_{k'})\right) = \mathcal{M} * \mathcal{R}_{k'}. \tag{7}$$

As shown in Algorithm 3, the PCG method can be implemented using the preconditioning matrix M and the aforementioned computations, where the linear system $\tilde{A}\tilde{y} = \tilde{b}$ is transformed into $\tilde{A} \text{vec}(\tilde{\mathcal{Y}}) = \text{vec}(\tilde{\mathcal{B}})$, where $\text{vec}(\tilde{\mathcal{B}}) := \tilde{b} = \text{vec}(\mathcal{Q}_k \times_1 X_A^H + \mathcal{Q}_k \times_2 X_B^H + \mathcal{Q}_k \times_3 X_C^H)$ and $\text{vec}(\tilde{\mathcal{Y}}) := \tilde{y}$. Algorithm 3 requires only small matrices A, B, and C and $\ell \times m \times n$ tensors $\mathcal{X}_{k'}, \mathcal{R}_{k'}, \mathcal{P}_{k'}$, and $\mathcal{Z}_{k'}$. Therefore the memory requirement is of $O(n^3)$ in the case of $n = m = \ell$.

4.2. PCG Method by the Schur Decomposition

Firstly, the Schur decomposition of T is $T := QRQ^H$, where R and Q are upper triangular and unitary matrices, respectively. Then,

$$\left(T^T T - \tilde{\sigma}^2 I_{\ell mn}\right) v = q_k \Leftrightarrow \left((QRQ^H)^H (QRQ^H) - \tilde{\sigma}^2 I_{\ell mn}\right) v = q_k$$

$$\Leftrightarrow \left(R^H R - \tilde{\sigma}^2 I_{\ell mn}\right)(Q^H v) = Q^H q_k.$$

This linear system denotes $\tilde{A}\tilde{y} = \tilde{b}$, where $\tilde{A} := R^H R - \tilde{\sigma}^2 I_{\ell mn}$, $\tilde{y} := Q^H v$, and $\tilde{b} := Q^H q_k$. The PCG method for $\tilde{A}\tilde{y} = \tilde{b}$ is shown in Algorithm 2. R and Q are obtained from the complex Schur decomposition of $A, B,$ and C as follows: $R = I_n \otimes I_m \otimes R_A + I_n \otimes R_B \otimes I_\ell + R_C \otimes I_m \otimes I_\ell$ and $Q = Q_C \otimes Q_B \otimes Q_A$ from the definition of T, where $A = Q_A R_A Q_A^H$, $B = Q_B R_B Q_B^H$, and $C = Q_C R_C Q_C^H$ by the Schur decomposition of $A, B,$ and C.

Algorithm 3: PCG method over tensor space for the 5-th line of Algorithm 1 [Proposed inner algorithm using the eigendecomposition]

1: Choose an initial tensor $\mathcal{X}_0 \in \mathbb{R}^{\ell \times m \times n}$ and $\mathcal{P}_0 = O_{\ell \times m \times n}$, and an initial scalar $\beta_0 = 0$;
2: $\mathcal{R}_0 = (\mathcal{Q}_k \times_1 X_A^H + \mathcal{Q}_k \times_2 X_B^H + \mathcal{Q}_k \times_3 X_C^H)$
    $- [\overline{\mathcal{D}} * \{(\mathcal{D} * \mathcal{X}_0) \times_1 X_A^H X_A + (\mathcal{D} * \mathcal{X}_0) \times_2 X_B^H X_B + (\mathcal{D} * \mathcal{X}_0) \times_3 X_C^H X_C\}$
    $- \tilde{\sigma}^2 (\mathcal{X}_0 \times_1 X_A^H X_A + \mathcal{X}_0 \times_2 X_B^H X_B + \mathcal{X}_0 \times_3 X_C^H X_C)]$;
3: $\mathcal{Z}_0 = \mathcal{M} * \mathcal{R}_0$;
4: **for** $k' = 1, 2, \ldots,$ until convergence **do**
5: $\quad \mathcal{P}_{k'} = \mathcal{Z}_{k'-1} + \beta_{k'-1} \mathcal{P}_{k'-1}$;
6: $\quad \hat{\mathcal{P}}_{k'} = \overline{\mathcal{D}} * \{(\mathcal{D} * \mathcal{P}_{k'}) \times_1 X_A^H X_A + (\mathcal{D} * \mathcal{P}_{k'}) \times_2 X_B^H X_B + (\mathcal{D} * \mathcal{P}_{k'}) \times_3 X_C^H X_C\}$
    $- \tilde{\sigma}^2 (\mathcal{P}_{k'} \times_1 X_A^H X_A + \mathcal{P}_{k'} \times_2 X_B^H X_B + \mathcal{P}_{k'} \times_3 X_C^H X_C)$;
7: $\quad \alpha_{k'} = (\mathcal{Z}_{k'-1}, \mathcal{R}_{k'-1}) / (\mathcal{P}_{k'-1}, \hat{\mathcal{P}}_{k'})$;
8: $\quad \mathcal{X}_{k'} = \mathcal{X}_{k'-1} + \alpha_{k'} \mathcal{P}_{k'}$;
9: $\quad \mathcal{R}_{k'} = \mathcal{R}_{k'-1} - \alpha_{k'} \hat{\mathcal{P}}_{k'}$;
10: $\quad \mathcal{Z}_{k'} = \mathcal{M} * \mathcal{R}_{k'-1}$;
11: $\quad \beta_{k'} = (\mathcal{Z}_{k'}, \mathcal{R}_{k'}) / (\mathcal{Z}_{k'-1}, \mathcal{R}_{k'-1})$;
12: **end for**
13: Obtain an approximate solution $\tilde{y} \approx \mathcal{X}_{k'}$;
14: $\mathcal{V} = \tilde{\mathcal{Y}} \times_1 X_A + \tilde{\mathcal{Y}} \times_2 X_B + \tilde{\mathcal{Y}} \times_3 X_C$;

4.2.1. Preconditioning Matrix

A preconditioning matrix for $\tilde{A}\tilde{y} = \tilde{b}$ satisfies the conditions in Section 4.1.1. Therefore, we propose the preconditioning matrix based on the Schur decomposition

$$M := \overline{D}_R D_R - \tilde{\sigma}^2 I_{\ell mn},$$

where D_R is a diagonal matrix with diagonals of R. Since M is also the diagonal matrix, the above second conditions are satisfied. Moreover, if T is symmetric, R is a diagonal matrix, that is, $R = D_R$. Therefore $M = \tilde{A}$ in the case of the symmetric matrix T. From this, we expect that the preconditioning matrix M is effective if T is not symmetric but almost symmetric.

4.2.2. Efficient Implementation of Algorithm 2 by the Schur Decomposition

We show the computations of $\text{vec}^{-1}(\tilde{A} \text{vec}(\mathcal{P}_{k'}))$ and $\text{vec}^{-1}(M^{-1} \text{vec}(\mathcal{R}_{k'}))$ for the PCG method over tensor space using the 1, 2, and 3-mode products for tensors and the definition of T. First, from the definitions of $\tilde{A}$ and R, we have $\text{vec}^{-1}(\tilde{A}\text{vec}(\mathcal{P}_{k'})) = \text{vec}^{-1}(R^H(R\text{vec}(\mathcal{P}_{k'})) - \tilde{\sigma}^2\text{vec}(\mathcal{P}_{k'}))$. Therefore,

$$\text{vec}^{-1}(\tilde{A}\text{vec}(\mathcal{P}_{k'})) = \mathcal{P}_{k'} \times_1 R_A^H R_A + \mathcal{P}_{k'} \times_2 R_B^H R_B + \mathcal{P}_{k'} \times_3 R_C^H R_C - \tilde{\sigma}^2 \mathcal{P}_{k'}.$$

Next, from $M = \overline{D}_R D_R - \tilde{\sigma}^2 I_{\ell mn}$, we easily obtain

$$\left(M^{-1}\right)_{ii} = \frac{1}{(\overline{D}_R)_{ii}(D_R)_{ii} - \tilde{\sigma}^2}, \quad i = 1, 2, \ldots, \ell mn.$$

Similarly to Section 4.1.2, let $\mathcal{D} = \text{vec}^{-1}(\text{diag}(D_R))$ and $\mathcal{M} = \text{vec}^{-1}(\text{diag}(M^{-1}))$. Then, we have $\mathcal{M}_{ijk} = 1/(\overline{\mathcal{D}}_{ijk}\mathcal{D}_{ijk} - \tilde{\sigma}^2)$, where $\mathcal{D}_{ijk} = (R_A)_{ii} + (R_B)_{jj} + (R_C)_{kk}$. $\text{vec}^{-1}(M^{-1}\text{vec}(\mathcal{R}_{k'}))$ is computed by (7).

As shown in Algorithm 4, the PCG method can be implemented using the preconditioning matrix M and the aforementioned computations, where the linear system $\tilde{A}\tilde{y} = \tilde{b}$ is transformed into $\tilde{A}\text{vec}(\tilde{\mathcal{Y}}) = \text{vec}(\tilde{\mathcal{B}})$, where $\text{vec}(\tilde{\mathcal{B}}) := \tilde{b} = \text{vec}(\mathcal{Q}_k \times_1 Q_A + \mathcal{Q}_k \times_2 Q_B + \mathcal{Q}_k \times_3 Q_C)$ and $\text{vec}(\tilde{\mathcal{Y}}) := \tilde{y}$. Algorithm 4 just requires small matrices A, B, and C and $\ell \times m \times n$ tensors $\mathcal{X}_{k'}, \mathcal{R}_{k'}, \mathcal{P}_{k'}$, and $\mathcal{Z}_{k'}$, namely, do not require large matrix T. Therefore the memory requirement is of $O(n^3)$ in the case of $n = m = \ell$.

Algorithm 4: PCG method over tensor space for the 5-th line of Algorithm 1 [Proposed inner algorithm using the Schur decomposition]

1: Choose an initial tensor $\mathcal{X}_0 \in \mathbb{R}^{\ell \times m \times n}$ and $\mathcal{P}_0 = O_{\ell \times m \times n}$, and an initial scalar $\beta_0 = 0$;
2: $\mathcal{R}_0 = (\mathcal{Q}_k \times_1 Q_A + \mathcal{Q}_k \times_2 Q_B + \mathcal{Q}_k \times_3 Q_C)$
    $- (\mathcal{X}_0 \times_1 R_A^H R_A + \mathcal{X}_0 \times_2 R_B^H R_B + \mathcal{X}_0 \times_3 R_C^H R_C - \tilde{\sigma}^2 \mathcal{X}_0)$;
3: $\mathcal{Z}_0 = \mathcal{M} * \mathcal{R}_0$;
4: **for** $k' = 1, 2, \ldots$, until convergence **do**
5:   $\mathcal{P}_{k'} = \mathcal{Z}_{k'-1} + \beta_{k'-1}\mathcal{P}_{k'-1}$;
6:   $\hat{\mathcal{P}}_{k'} = \mathcal{P}_{k'} \times_1 R_A^H R_A + \mathcal{P}_{k'} \times_2 R_B^H R_B + \mathcal{P}_{k'} \times_3 R_C^H R_C - \tilde{\sigma}^2 \mathcal{P}_{k'}$;
7:   $\alpha_{k'} = (\mathcal{Z}_{k'-1}, \mathcal{R}_{k'-1}) / (\mathcal{P}_{k'-1}, \hat{\mathcal{P}}_{k'})$;
8:   $\mathcal{X}_{k'} = \mathcal{X}_{k'-1} + \alpha_{k'}\mathcal{P}_{k'}$;
9:   $\mathcal{R}_{k'} = \mathcal{R}_{k'-1} - \alpha_{k'}\hat{\mathcal{P}}_{k'}$;
10: $\mathcal{Z}_{k'} = \mathcal{M} * \mathcal{R}_{k'}$;
11: $\beta_{k'} = (\mathcal{Z}_{k'}, \mathcal{R}_{k'}) / (\mathcal{Z}_{k'-1}, \mathcal{R}_{k'-1})$;
12: **end for**
13: Obtain an approximate solution $\tilde{\mathcal{Y}} \approx \mathcal{X}_{k'}$;
14: $\mathcal{V} = \tilde{\mathcal{Y}} \times_1 Q_A + \tilde{\mathcal{Y}} \times_2 Q_B + \tilde{\mathcal{Y}} \times_3 Q_C$;

---

5. Numerical Experiments

This section provides results of numerical experiments using Algorithm 1 with Algorithm 3 and Algorithm 1 with Algorithm 4. There are the two purposes of this experiments: (1) to confirm convergence to the singular value of T nearest to the shift by Algorithm 1, and (2) to confirm the effectiveness of the proposed precondition matrix in Algorithms 3 and 4. For comparison, the results using Algorithms 3 and 4 in the case of $M = I$ are also given as the results by the CG method. All the initial guesses of Algorithms 1, 3, and 4 are tensors with random numbers. The stopping criteria used in Algorithm 1 was $\beta_k \|e_k^T s_{\text{MAX}}^k\| < 10^{-8}$, where s_{MAX}^k is the eigenvector corresponding to the maximum eigenvalue of $\tilde{T}_k$ and e_k denotes the k-th canonical basis for k dimensional vector space. Algorithms 3 and 4 were stopped when either the relative residual $\|\mathcal{R}_{k'}\|/\|\tilde{\mathcal{B}}\| < 10^{-12}$ or the maximum number of iterations $k' > 20{,}000$ were satisfied.

All computations were carried out using MATLAB R2021a version on a workstation with Xeon processor 3.7 GHz and 128 GB of RAM.

In the following subsection, we show the results computing the 5-th maximum, median, and 5-th minimum singular values σ of the test matrices T. For all the cases, for the first purpose, we set the shift value in Algorithm 1 as

$$\tilde{\sigma} = \sigma - 10^{-2}, \tag{8}$$

where $\tilde{\sigma}$'s and σ's are the perturbed singular values of T and the aforementioned singular values computed by the svd function in MATLAB, respectively.

Test matrices T in Equation (1) are obtained from a seven-point central difference discretization of the PDE (2) in over an $(n+1) \times (n+1) \times (n+1)$ grid. The test matrices T in Equation (1), whose size is $n^3 \times n^3$, are generated from

$$A = B = C, \quad A := \frac{1}{h^2} a M_1 + \frac{1}{2h} b M_2 + \frac{1}{3} c I_n, \tag{9}$$

where $h = 1/(n+1)$, M_1 and M_2 are symmetric and skew-symmetric matrices given below.

$$M_1 = \begin{pmatrix} -2 & 1 & & & \\ 1 & -2 & 1 & & \\ & \ddots & \ddots & \ddots & \\ & & 1 & -2 & 1 \\ & & & 1 & -2 \end{pmatrix} \in \mathbb{R}^{n \times n}, M_2 = \begin{pmatrix} 0 & 1 & & & \\ -1 & 0 & 1 & & \\ & \ddots & \ddots & \ddots & \\ & & -1 & 0 & 1 \\ & & & -1 & 0 \end{pmatrix} \in \mathbb{R}^{n \times n}.$$

Numerical Results

In all tables, the number of iterations of the shift-and-invert Lanczos method ("the Lanczos method" hereafter) and the average of the number of iterations of the CG or PCG method based on the eigendecomposition or the Schur decomposition are summarized. "Not converged" denotes Algorithm 3 or 4 did not converge.

We show the first results in the case of almost symmetric matrix with $a = c = 1$ and $b = 0.01$ in Equation (9) for the shift (8). From Tables 1–3, the numbers of iterations of Lanczos methods using any inner algorithms were almost the same. Focusing on the effectiveness of the proposed preconditioning matrix M, the numbers of iterations of both PCG methods were less than 19 regardless of the size of T. On the other hand, the numbers of iterations of both CG methods linearly increased depending on the size of T. From these facts, the preconditioning matrix M is effective in the case of almost symmetric matrix T. Moreover, the number of iterations of the shift and invert Lanczos method for the median singular value is larger than the number for other singular values since the distance between the maximum and second maximum singular values of $(T^T T - \tilde{\sigma}^2 I_{\ell mn})^{-1}$ for the median singular value of T is closer than the cases of other singular values.

Here, the running time of Table 1 is summarized in Table 4. All the running time by the PCG method were less than the time by the CG method. Moreover, the running time by the PCG methods of Algorithms 3 and 4 were similar since the computational complexities of these algorithms are similar. Thus, the running time is strongly correlated with the number of iterations of Algorithms 3 and 4.

In addition, convergence histories of $n = 15$ in Tables 1 and 2 are shown in Figures 1 and 2. Figure 2 displays that the relative residual norms unsteadily decreased when the number of iterations of the shift and invert Lanczos method is not small.

Table 1. Number of iterations of the Lanczos method and the average of numbers of iterations of the CG/PCG method in the case of the 5-th max. singular value of almost symmetric matrix T with $a = c = 1$ and $b = 0.01$ in (9) for the shift (8).

Method		Algorithms 1 and 3 (by Eigendecompn.)				Algorithms 1 and 4 (by Schur Decompn.)		
		Lanczos	CG	Lanczos	PCG	Lanczos	CG	Lanczos	PCG
n	5	4	43.0	4	16.0	4	35.8	4	15.0
	10	4	90.0	4	17.0	4	86.5	4	15.0
	15	3	134.7	3	17.0	3	128.0	3	17.0
	20	4	180.0	4	17.0	4	169.3	4	17.0
	25	3	225.7	3	17.0	3	211.0	3	17.0
	30	3	273.0	3	17.0	3	252.3	3	17.0

Table 2. Number of iterations of the Lanczos method and the average of numbers of iterations of the CG/PCG method in the case of the median of singular value of almost symmetric matrix T with $a = c = 1$ and $b = 0.01$ in (9) for the shift (8).

Method		Algorithms 1 and 3 (by Eigendecompn.)				Algorithms 1 and 4 (by Schur Decompn.)		
		Lanczos	CG	Lanczos	PCG	Lanczos	CG	Lanczos	PCG
n	5	15	139.3	15	13.0	15	109.1	15	15.0
	10	7	1081.0	7	13.0	7	943.7	7	15.0
	15	41	5201.6	39	15.0	39	4858.1	41	17.0
	20	13	8339.5	4	16.0	4	6847.3	4	15.0
	25	(Not converged.)		48	17.0	(Not converged.)		48	17.0
	30	(Not converged.)		8	19.0	(Not converged.)		8	15.0

Table 3. Number of iterations of the Lanczos method and the average of numbers of iterations of the CG/PCG method in the case of the 5-th min. singular value of almost symmetric matrix T with $a = c = 1$ and $b = 0.01$ in (9) for the shift (8).

Method		Algorithms 1 and 3 (by Eigendecompn.)				Algorithms 1 and 4 (by Schur Decompn.)		
		Lanczos	CG	Lanczos	PCG	Lanczos	CG	Lanczos	PCG
n	5	3	124.7	3	13.0	3	97.9	3	14.6
	10	3	748.1	3	13.0	3	654.9	3	14.1
	15	3	4872.5	3	14.9	3	4531.1	3	16.7
	20	3	775.5	3	13.8	3	2358.8	3	15.0
	25	3	1494.0	3	16.8	3	1472.0	3	16.8
	30	3	2158.0	3	16.8	3	2088.0	3	15.0

Table 4. Running time (seconds) of the Lanczos method using the CG/PCG method in the case of the 5-th max. singular value of almost symmetric matrix T with $a = c = 1$ and $b = 0.01$ in (9) for the shift (8).

Method		Algorithms 1 and 3 (by Eigendecompn.)		Algorithms 1 and 4 (by Schur Decompn.)	
		Lanczos with CG	Lanczos with PCG	Lanczos with CG	Lanczos with PCG
n	5	0.114	0.071	0.095	0.061
	10	0.578	0.095	0.566	0.086
	15	1.402	0.137	1.448	0.192
	20	6.639	0.437	6.446	0.335
	25	13.632	0.558	12.686	0.432
	30	35.121	1.145	33.203	0.836

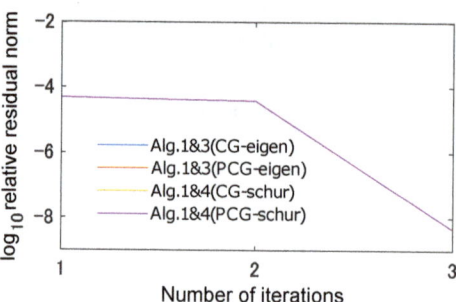

Figure 1. Convergence histories with relative residual norm of the Lanczos method for the 5-th max. singular value of the almost symmetric matrix T whose size is $n = 15$.

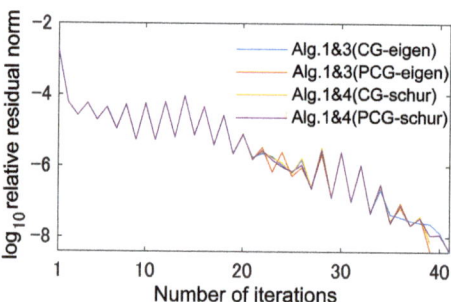

Figure 2. Convergence histories with relative residual norm of the Lanczos method for the median singular value of the almost symmetric matrix T whose size is $n = 15$.

Next, we show the second results in the case of slightly symmetric matrix with $a = c = 1$ and $b = 0.1$ in Equation (9) for the shift (8). From Table 5, both PCG methods did not converge for computing the 5-th maximum singular values of slightly symmetric matrix T. It seems that the linear system for $T^T T - \hat{\sigma}^2 I_{\ell m n}$ is ill-conditioned since 10^{-2} in the shift (8) is much less than the 5-th maximum singular values of the matrix. In Appendix A, we show the results using relative shift without the effect of the magnitude of the singular values. Table 6 shows Algorithms 1 and 4, that is, the algorithms based on Schur decomposition, was more robust than Algorithms 1 and 3, that is, the algorithms based on the eigendecomposition. From Table 7, both PCG methods converged regardless of n, and the numbers of iterations of both PCG methods were less than the number of iterations of both CG method. Namely, it seems that the preconditioning matrix M can be effective in the case of a slightly symmetric matrix T when computing the 5-th minimum and median singular values of T.

Table 5. Number of iterations of the Lanczos method and the average of numbers of iterations of the CG/PCG method in the case of the 5-th max. singular value of slightly symmetric matrix T with $a = c = 1$ and $b = 0.1$ in (9) for the shift (8).

Method		Algorithms 1 and 3 (by Eigendecompn.)			Algorithms 1 and 4 (by Schur Decompn.)		
		Lanczos	CG	PCG	Lanczos	CG	PCG
n	5	4	50.0	(Not converged.)	4	37.8	(Not converged.)
	10	4	100.0	(Not converged.)	4	87.3	(Not converged.)
	15	3	152.0	(Not converged.)	3	131.0	(Not converged.)
	20	4	205.5	(Not converged.)	4	172.0	(Not converged.)
	25	3	262.0	(Not converged.)	3	230.0	(Not converged.)
	30	3	306.7	(Not converged.)	3	259.0	(Not converged.)

Table 6. Number of iterations of the Lanczos method and the average of numbers of iterations of the CG/PCG method in the case of the median of singular value of slightly symmetric matrix T with $a = c = 1$ and $b = 0.1$ in (9) for the shift (8).

Method		Algorithms 1 and 3 (by Eigendecompn.)				Algorithms 1 and 4 (by Schur Decompn.)			
		Lanczos	CG	Lanczos	PCG	Lanczos	CG	Lanczos	PCG
n	5	13	231.8	(Not converged.)		13	193.1	13	73.0
	10	6	1582.7	6	55.0	6	1272.7	6	29.0
	15	30	12,174.6	(Not converged.)		31	11061.9	31	97.0
	20	4	13,777.8	(Not converged.)		4	8799.3	(Not converged.)	
	25	(Not converged.)		(Not converged.)		(Not converged.)		83	116.0
	30	(Not converged.)		(Not converged.)		(Not converged.)		27	45.0

Table 7. Number of iterations of the Lanczos method and the average of numbers of iterations of the CG/PCG method in the case of the 5-th min. singular value of slightly symmetric matrix T with $a = c = 1$ and $b = 0.1$ in (9) for the shift (8).

Method		Algorithms 1 and 3 (by Eigendecompn.)				Algorithms 1 and 4 (by Schur Decompn.)			
		Lanczos	CG	Lanczos	PCG	Lanczos	CG	Lanczos	PCG
n	5	3	195.5	3	59.0	3	163.9	3	68.4
	10	3	949.8	3	58.0	3	771.2	3	44.0
	15	3	616.0	3	63.0	3	597.3	3	94.5
	20	3	859.8	3	63.0	3	2999.8	3	57.0
	25	3	1695.7	3	63.0	3	1559.3	3	114.4
	30	3	2388.0	3	63.0	3	2262.0	3	46.1

Finally, we show the third results in the case of marginally symmetric matrix with $a = c = 1$ and $b = 0.2$ in Equation (9) for the shift (8). Both PCG methods did not converge for computing the 5-th maximum singular values of T as shown in Table 8, similarly to Table 5. Moreover, computing the median singular values of T sometimes did not converge from Table 9. In Table 10, all methods converged for the 5-th minimum singular value of T. The numbers of iterations by the PCG method with the proposed preconditioning matrix were less than the number of iterations by the CG method in most cases. It seems that the preconditioning matrix M can be effective in the case of the marginally symmetric matrix T when computing the 5-th minimum singular values of T.

Table 8. Number of iterations of the Lanczos method and the average of numbers of iterations of the CG/PCG method in the case of the 5-th max. singular value of marginally symmetric matrix T with $a = c = 1$ and $b = 0.2$ in (9) for the shift (8).

Method		Algorithms 1 and 3 (by Eigendecompn.)				Algorithms 1 and 4 (by Schur Decompn.)			
		Lanczos	CG	Lanczos	PCG	Lanczos	CG	Lanczos	PCG
n	5	4	52.8	(Not converged.)		4	39.8	(Not converged.)	
	10	4	107.0	(Not converged.)		4	93.0	(Not converged.)	
	15	3	162.0	(Not converged.)		3	135.7	(Not converged.)	
	20	4	217.8	(Not converged.)		4	206.3	(Not converged.)	
	25	3	271.3	(Not converged.)		3	248.0	(Not converged.)	
	30	3	334.0	(Not converged.)		3	260.0	(Not converged.)	

Table 9. Number of iterations of the Lanczos method and the average of numbers of iterations of the CG/PCG method in the case of the median of singular value of marginally symmetric matrix T with $a = c = 1$ and $b = 0.2$ in (9) for the shift (8).

Method		Algorithms 1 and 3 (by Eigendecompn.)				Algorithms 1 and 4 (by Schur Decompn.)			
		Lanczos	CG	Lanczos	PCG	Lanczos	CG	Lanczos	PCG
n	5	11	481.4	(Not converged.)		10	355.6	(Not converged.)	
	10	6	2123.8	(Not converged.)		10	1550.8	10	4262.4
	15	(Not converged.)		(Not converged.)		(Not converged.)		(Not converged.)	
	20	15	13,268.7	89	6358.9	7	10019.9	108	160.0
	25	(Not converged.)		(Not converged.)		(Not converged.)		(Not converged.)	
	30	(Not converged.)		90	1150.0	(Not converged.)		28	11,764.0

Table 10. Number of iterations of the Lanczos method and the average of numbers of iterations of the CG/PCG method in the case of the 5-th min. singular value of marginally symmetric matrix T with $a = c = 1$ and $b = 0.2$ in (9) for the shift (8).

Method		Algorithms 1 and 3 (by Eigendecompn.)				Algorithms 1 and 4 (by Schur Decompn.)			
		Lanczos	CG	Lanczos	PCG	Lanczos	CG	Lanczos	PCG
n	5	5	313.9	5	295.8	5	213.9	5	1077.8
	10	3	1247.8	3	187.0	3	1150.6	3	3048.8
	15	3	650.0	3	201.0	3	606.7	3	177.0
	20	3	929.3	3	6151.3	3	847.8	3	162.8
	25	3	1800.3	3	205.0	3	1683.7	3	249.0
	30	3	2585.7	3	1118.6	3	2371.3	3	181.0

6. Conclusions

We considered computing an arbitrary singular value of a tensor sum. The shift-and-invert Lanczos method and the PCG method reconstructed over tensor space. We proposed the preconditioning matrices which are the following two diagonal matrices: (1) whose diagonals of the eigenvalues by the eigendecomposition, and (2) whose diagonals of the upper diagonal matrix by the Schur decomposition. This preconditioning matrix can be effective if the tensor sum is almost symmetric.

From numerical results, we confirmed that the proposed method reduces memory requirements without any conditions. The numbers of iterations of the PCG method by the proposed preconditioning matrix were reduced in most cases of the almost and slightly symmetric matrix. Moreover, the numbers of iterations of the PCG method by the proposed preconditioning matrix were also reduced in certain cases of the marginally symmetric matrix.

For future work, we will consider a robust preconditioning matrix for slightly or marginally symmetric tensor sum, a suitable preconditioning matrix for non-symmetric tensor sum, parallel implementations of the proposed algorithms, and finding real-life applications.

Author Contributions: Conceptualization, A.O. and T.S.; investigation, A.O. and T.S.; software, A.O.; validation, T.S.; writing—original draft, A.O.; writing—review and editing, A.O. and T.S. All authors have read and agreed to the published version of the manuscript.

Funding: This work was supported by JSPS KAKENHI Grant Numbers: JP21K17754, JP20H00581, JP20K20397, JP17H02829.

Institutional Review Board Statement: Not applicable.

Informed Consent Statement: Not applicable.

Data Availability Statement: Not applicable.

Conflicts of Interest: The authors declare no conflict of interest.

Abbreviations

The following abbreviations are used in this manuscript:

PCG   Preconditioned Conjugate Gradient
CG   Conjugate Gradient
PDE   Partial Differential Equation

Appendix A

This appendix gives the numerical results in the case of the 5-th maximum and the median singular values of slightly and marginally symmetric matrices by the relative shift

$$\tilde{\sigma} = \sigma - 10^{-2}\sigma, \tag{A1}$$

where σ's are the singular values of T computed by the svd function in MATLAB. The condition of the numerical experiments except for the setting of the shift is the same as the experiments in Section 5.

Firstly, we show the results in the case of slightly symmetric matrix with $a = c = 1$ and $b = 0.1$ in Equation (9) for the shift (A1) in Tables A1 and A2. Computing the 5-th and the median singular values of the slightly symmetric matrix using the shift (A1), the number of iterations of both PCG methods is much less than the number of iterations of both CG methods.

Secondly, Tables A3 and A4 are the results in the case of marginally symmetric matrix with $a = c = 1$ and $b = 0.2$ in Equation (9) for the shift (A1). From Tables A3 and A4, both PCG methods converged faster than both CG method using the relative shift. Moreover, the PCG method by Algorithm 4 is more robust than the PCG method by Algorithm 3.

Consequently, when we compute the 5-th maximum and the median singular values of the slightly symmetric matrix, the numerical experiments of Section 5 and Appendix A imply that the proposed preconditioning matrix can work in the case of a suitable shift.

Table A1. Number of iterations of the Lanczos method and the average of numbers of iterations of the CG/PCG method in the case of the 5-th max. singular value of slightly symmetric matrix T with $a = c = 1$ and $b = 0.1$ in (9) for the shift (A1).

Method		Algorithms 1 and 3 (by Eigendecompn.)				Algorithms 1 and 4 (by Schur Decompn.)			
		Lanczos	CG	Lanczos	PCG	Lanczos	CG	Lanczos	PCG
n	5	5	41.0	5	17.0	5	35.8	5	15.0
	10	7	84.0	7	20.0	7	77.0	7	13.0
	15	4	162.0	4	22.0	4	154.0	4	17.0
	20	7	223.7	7	23.0	7	197.3	7	12.0
	25	5	383.6	5	24.0	5	307.4	5	15.0
	30	6	522.7	6	24.0	6	426.5	6	13.0

Table A2. Number of iterations of the Lanczos method and the average of numbers of iterations of the CG/PCG method in the case of the median of singular value of slightly symmetric matrix T with $a = c = 1$ and $b = 0.1$ in (9) for the shift (A1).

Method		Algorithms 1 and 3 (by Eigendecompn.)				Algorithms 1 and 4 (by Schur Decompn.)			
		Lanczos	CG	Lanczos	PCG	Lanczos	CG	Lanczos	PCG
n	5	23	139.3	23	38.0	23	105.9	23	14.0
	10	10	1081.0	10	21.0	10	1074.4	10	22.0
	15	21	5201.6	21	21.0	21	3470.4	21	14.0
	20	17	7333.1	(Not converged.)		17	6242.4	17	16.0
	25	11	16,034.1	11	32.0	11	14,360.6	11	14.0
	30	(Not converged.)		12	23.0	(Not converged.)		12	17.0

Table A3. Number of iterations of the Lanczos method and the average of numbers of iterations of the CG/PCG method in the case of the 5-th max. singular value of marginally symmetric matrix T with $a = c = 1$ and $b = 0.2$ in (9) for the shift (A1).

Method		Algorithms 1 and 3 (by Eigendecompn.)				Algorithms 1 and 4 (by Schur Decompn.)			
		Lanczos	CG	Lanczos	PCG	Lanczos	CG	Lanczos	PCG
n	5	5	43.0	5	25.0	5	38.4	5	17.0
	10	7	90.0	7	29.0	7	79.0	7	13.0
	15	4	174.5	4	32.0	4	152.5	4	62.0
	20	7	253.0	7	33.0	7	198.1	7	15.0
	25	5	403.0	5	35.0	5	319.0	5	19.0
	30	6	626.0	6	36.0	6	441.2	6	16.0

Table A4. Number of iterations of the Lanczos method and the average of numbers of iterations of the CG/PCG method in the case of the median of singular value of marginally symmetric matrix T with $a = c = 1$ and $b = 0.2$ in (9) for the shift (A1).

Method		Algorithms 1 and 3 (by Eigendecompn.)				Algorithms 1 and 4 (by Schur Decompn.)			
		Lanczos	CG	Lanczos	PCG	Lanczos	CG	Lanczos	PCG
n	5	24	138.3	(Not converged.)		25	115.6	23	17.0
	10	10	1479.0	(Not converged.)		10	1119.2	10	18.0
	15	21	4506.6	21	34.0	21	3787.0	21	16.0
	20	17	8603.3	(Not converged.)		17	6604.5	17	21.0
	25	11	18,267.0	(Not converged.)		11	14,991.6	11	30.0
	30	(Not converged.)		13	35.0	(Not converged.)		13	31.0

References

1. Ohashi, A.; Sogabe, T. On computing maximum/minimum singular values of a generalized tensor sum. *Electron. Trans. Numer. Anal.* **2015**, *43*, 244–254.
2. Ohashi, A.; Sogabe, T. On computing the minimum singular value of a tensor sum. *Special Matrices* **2019**, *7*, 95–106. [CrossRef]
3. Bai, Z.; Demmel, J.; Dongarra, J.; Ruhe, A.; van der Vorst, H.A. *Templates for the Solution of Algebraic Eigenvalue Problems: A Practical Guide*; SIAM: Philadelphia, PA, USA, 2000.
4. Beik, F.P.A.; Jbilou, K.; Najafi-Kalyani, M.; Reichel, L. Golub–Kahan bidiagonalization for ill-conditioned tensor equations with applications. *Numer. Algorithms* **2020**, *84*, 1535–1563. [CrossRef]
5. Rezaie, M.; Moradzadeh, A.; Kalateh, A.N. Fast 3D inversion of gravity data using solution space priorconditioned lanczos bidiagonalization. *J. Appl. Geophys.* **2017**, *136*, 42–50. [CrossRef]
6. Zhong, H.X.; Xu, H. Weighted Golub-Kahan-Lanczos bidiagonalization algorithms. *Electron. Trans. Numer. Anal.* **2017**, *47*, 153–178. [CrossRef]
7. Garber, D.; Hazan, E.; Jin, C.; Musco, C.; Netrapalli, P.; Sidford, A. Faster eigenvector computation via shift-and-invert preconditioning. In Proceedings of the 33rd International Conference on Machine Learning, New York, NY, USA, 19–24 June 2016; pp. 2626–2634.
8. Huang, W.Q.; Lin, W.W.; Lu, H.H.S.; Yau, S.T. SIRA: Integrated shift—Invert residual Arnoldi method for graph Laplacian matrices from big data. *J. Comput. Appl. Math.* **2019**, *346*, 518–531. [CrossRef]
9. Katrutsa, A.; Botchev, M.; Oseledets, I. Practical shift choice in the shift-and-invert Krylov subspace evaluations of the matrix exponential. *arXiv* **2019**, arXiv:1909.13059.
10. Xu, Z. Gradient descent meets shift-and-invert preconditioning for eigenvector computation. *Adv. Neural. Inf. Process. Syst.* **2018**, *31*, 2825–2834.
11. Yue, S.F.; Zhang, J.J. An extended shift-invert residual Arnoldi method. *Comput. Appl. Math.* **2021**, *40*, 1–15. [CrossRef]
12. Zemaityte, M.; Tisseur, F.; Kannan, R. Filtering Frequencies in a Shift-and-invert Lanczos Algorithm for the Dynamic Analysis of Structures. *SIAM J. Sci. Comput.* **2019**, *41*, B601–B624. [CrossRef]
13. Zhong, H.X.; Chen, G.L.; Shen, W.Q. Shift and invert weighted Golub-Kahan-Lanczos bidiagonalization algorithm for linear response eigenproblem. *J. Comput. Anal. Appl.* **2019**, *26*, 1169–1178.
14. Van Loan, C.F.; Golub, G.H. *Matrix Computations*; Johns Hopkins University Press: Baltimore, MD, USA, 1983.
15. Kolda T.G.; Bader B.W. Tensor Decompositions and Applications. *SIAM Rev.* **2008**, *51*, 455–500. [CrossRef]

Article

Using Free Mathematical Software in Engineering Classes

Víctor Gayoso Martínez [1,2,*,†,‡], Luis Hernández Encinas [1,‡], Agustín Martín Muñoz [1,‡] and Araceli Queiruga Dios [3,‡]

[1] Institute of Physical and Information Technologies (ITEFI), Spanish National Research Council (CSIC), 28006 Madrid, Spain; luis@iec.csic.es (L.H.E.); agustin@iec.csic.es (A.M.M.)
[2] Engineering Department, Centro Universitario U-tad 28290 Las Rozas, Spain
[3] Department of Applied Mathematics, Institute of Fundamental Physics and Mathematics, University of Salamanca, 37008 Salamanca, Spain; queirugadios@usal.es
* Correspondence: victor.gayoso@iec.csic.es or victor.gayoso@u-tad.com
† Current address: C/Serrano, 144, 28006 Madrid, Spain.
‡ All authors contributed equally to this work.

Abstract: There are many computational applications and engines used in mathematics, with some of the best-known arguably being Maple, Mathematica, MATLAB, and Mathcad. However, although they are very complete and powerful, they demand the use of commercial licences, which can be a problem for some education institutions or in cases where students desire to use the software on an unlimited number of devices or to access it from several of them simultaneously. In this contribution, we show how GeoGebra, WolframAlpha, Python, and SageMath can be applied to the teaching of different mathematical courses in engineering studies, as they are some of the most interesting representatives of free (and mostly open source) mathematical software. As the best way to show a topic in mathematics is by providing examples, this article explains how to make calculations for some of the main topics associated with Calculus, Algebra, and Coding theories. In addition to this, we provide some results associated with the usage of Mathematica in different graded activities. Moreover, the comparison between the results from students that use Mathematica and students that participate in a "traditional" course, solving problems and attending to master classes, is shown.

Keywords: coding theory; engineering; GeoGebra; mathematica; Python; SageMath; WolframAlpha

MSC: 97D10; 97D60; 97U10; 97U50; 97U70

1. Introduction

The Bologna Accord is an agreement on a common model of higher education reached in 1999 that implies the creation of a common European area of university studies. It emphasizes the creation of a European Area of Higher Education (EAHE) as a key to promoting students' mobility, aiming to simplify Europe's educational qualifications and ensuring that credentials granted by an institution in one country are comparable with those earned elsewhere [1].

There are 48 countries currently involved in the Bologna Accord. The cornerstones of such an open space are mutual recognition of degrees and other higher education qualifications, transparency (readable and comparable degrees organized in a three-cycle structure), and European cooperation in quality assurance.

Due to the Bologna Accord, the teaching of mathematics has suffered important changes, such as the necessity to enhance the traditional teaching–learning process with practical cases, the possibility to introduce some key concepts, and the reinforcement of the learning process by using technology and specific mathematical software [2,3]. Nowadays, there are many computational packages focused on mathematics, with Mathematica and MATLAB being two of the best known [4,5]. However, even though they are certainly very complete and powerful, they require the use of commercial licenses, which can be a problem for some education institutions or in cases where students desire to use the software on an unlimited number of devices or to access them from several of them simultaneously.

In this paper, our goal is to show how free mathematical software (which most of the time is also open source software, but not always) can be applied to the teaching of different engineering courses at the University of Salamanca and Centro Universitario U-tad, from first-year algebra or calculus to more specialized topics such as coding theory. The University of Salamanca was founded by King Alfonso IX of León in 1218, which makes it the oldest Spanish university in existence and one of the oldest in Europe. The university offers 81 courses in the first and second cycles spread throughout five branches of knowledge, including Science and Engineering [6]. U-tad is the acronym for Centro Universitario de Tecnología y Arte Digital (Technology and Digital Art University Centre), a private university centre founded in 2011 with a strong focus on the creation, programming, and management of digital content, products, and services [7]. U-tad is based near Madrid, and it currently offers three higher technical education courses, five undergraduate degrees and twelve postgraduate courses.

Learning a programming language is highly important for pre-university and university students. One of the goals of the Europe 2020 growth strategy [8] is the implantation of information and communication technologies at all educational levels. In this sense, Scratch and App Inventor are widely used in Spanish secondary education and high schools, and Python is also included among the technology tools used in formal education institutions [9]. However, the number of students that arrive at university with a fair programming knowledge is still low. In fact, the first contact with a formal programming language for most engineering students takes place during their first semester. At the University of Salamanca and U-tad, for example, C is the first programming language that is taught to students.

In this study, GeoGebra, WolframAlpha, Python, and SageMath have been used for providing actual examples used in class, as they are good representatives of free mathematical software. In addition, we have analysed the relationship between the use of these tools and the final grades obtained by engineering students. We have also included data about a statistical study of two academic courses in which we proposed the use of Mathematica (and as an alternative WolframAlpha) as a tool for solving mathematical problems.

The rest of this contribution is organized as follows: Section 2 describes other articles associated with this topic. Section 3 presents the most relevant information about GeoGebra, WolframAlpha, Python, and SageMath, while Section 4 provides several examples used at class. After that, Section 5 provides some statistics associated with the usage of Mathematica software in some engineering classes. Finally, in Section 6 we offer some conclusions and ideas for future work.

2. Related Work

There are several publications that analyse the use of mathematical software for teaching at different levels and from different points of view. For example, Hillmayr et al. presented a comprehensive analysis about how the use of technology can enhance learning in secondary school mathematics and science in [10]. They compared learning outcomes of students using digital tools to those of a control group taught without the use of digital tools. Their results showed that the use of digital tools had a positive effect on student learning outcomes and that the use of intelligent tutoring systems or simulations (dynamic mathematical tools) was significantly more beneficial than hypermedia systems. Moreover, in [11] a taxonomy of five categories of tool-based mathematics software is considered: (a) review and practice, (b) general, (c) specific, (d) environment, and (e) communication. A description of the affordances and constraints of such categories of software is provided, and how each one facilitates different aspects of student learning is discussed.

Other contributions study the use of different software for teaching mathematics. Among them, we highlight the following: [12–20]. In comparison to those articles, this contribution focuses on a specific set of open source engines and provides examples used in actual engineering classes.

3. Computational Engines

Engineering is considered "the application of mathematics and sciences to the building and design of projects for the use of society" [21]. Moreover, mathematical theory and practical engineering challenges are linked to computational procedures [22]. Representation of functions or surfaces, the calculation of the Taylor polynomial for a given function, solving systems of linear equations or making calculations with matrices are some of the examples that engineering students need everyday in their studies [23].

There are arguably three possibilities regarding the usage of computational engines:

- Commercial software: MATLAB, Mathematica, Maple, Mathcad, SPSS, etc.
- Free software: GeoGebra, WolframAlpha, SageMath, Maxima, Scilab, Octave, R, FreeMat, Demetra+, etc.
- Programming languages such as Python or Julia.

Each of these options has its benefits and disadvantages. Applications such as MATLAB or Mathematica are very powerful, but obviously they require commercial licences and the installation of many software packages that in some cases have to be managed manually and need to allocate several gigabytes of hard drive space. In addition to this, those applications sometimes have processor and memory requirements that cannot be satisfied by all type of students. Even though universities usually provide computing resources to students, events such as the coronavirus pandemic have shown that students cannot depend solely on the university infrastructure.

In comparison, the computational capabilities of free software engines are lower in some instances, but for introductory subjects they may be more than enough. Finally, programming languages such as Python are very versatile and allow one to perform symbolic and numeric calculations, but many first-year students are not familiar with them. Even though it could be argued that first-year students are also unfamiliar with the syntaxis of mathematical engines, it is true that many computations can be achieved with a sole command in those mathematical engines, while they would require creating a small application using a programming language, with the difficulties that that option brings (importing the proper libraries, formatting the code in a proper way, etc.).

In this paper, we have focused on GeoGebra, WolframAlpha, Python, and SageMath, not only because they are free to use, but also because as a side effect that freedom allows us to chose the best option for each topic inside a course, preventing educators from being tied to a single solution.

Several authors have analysed the benefits and disadvantages of different educational applications [24–26], and they found that all of them have similar characteristics and are suitable for classes. Sometimes, the decision on which application to use depends on the usage of the same software by the teacher in his/her own research activities [27].

It is important to mention other open source applications and libraries of mathematical software different to those considered in detail in this work, which are included in Table 1.

Table 1. Additional open source mathematical software applications.

Name	Author(s)	Web Page (accessed on 31 August 2021)	Field
Axiom	Axiom Team	http://www.axiom-developer.org/	CAS
Cadabra	K. Peeters et al.	https://cadabra.science/	CAl
CoCoA	L. Robbiano	http://cocoa.dima.unige.it/	CmA
Demetra+	Eurostat	https://github.com/jdemetra/jdemetra-app	CDm
Flint	W. Hart	http://www.flintlib.org/	ODE
FreeMat	S. Basu	http://freemat.sourceforge.net/	Alg
GAP	Araújo et al.	https://www.gap-system.org/	DAl
Gfan	A. Jensen	http://home.imf.au.dk/jensen/software/gfan/gfan.html	AlG
GiNaC	C. Bauer et al.	https://www.ginac.de/	AlC
Gnuplot	Gnuplot team	http://www.gnuplot.info/	2/3D
Gretl	A. Cottrell	http://gretl.sourceforge.net/	EcA
LiPS	M. Melnick	http://lipside.sourceforge.net/	LiP
Mathics	B. Jones et al.	https://mathics.org/	CAS
Maxima	W. Schelter	https://maxima.sourceforge.io/	CAS
Macaulay 2	D. Grayson et al.	http://www.math.uiuc.edu/Macaulay2/	AlG
MPFR	MPFR team	https://www.mpfr.org/	FPA
MPIR	B. Gladman et al.	https://www.mpir.org/	Art
MuPAD-Combinat	F. Hivert et al.	http://mupad-combinat.sourceforge.net/	CAl
NTL	V. Shoup	http://www.shoup.net/ntl//	NTh
Octave	J.B. Rawlings et al.	https://www.octave.org	SyC
PARI/GP	H. Cohen et al.	http://pari.math.u-bordeaux.fr/	NTh
R	R. Ihaka and R. Gentleman	https://www.r-project.org/	CDm
Reduce	T. Hearn	http://www.reduce-algebra.com/	CAS
Scilab	INRIA	https://www.scilab.org/	NuC
Xcas	B. Parisse	http://xcas.sourceforge.net/fr/index.php	CAS

3.1. GeoGebra

GeoGebra is an interactive geometry, algebra, statistics, and calculus application available both as an online resource and a native application in Windows, macOS, and Linux systems [28].

The GeoGebra website includes several services such as a calculator and a graphics plotter, but the most widely used option is what is called GeoGebra Classic, which puts together those individual tools.

Figure 1 shows the GeoGebra Classic interface, where it is possible to find modules for two- and three-dimensional plotting, an input bar, and the CAS (Computer Algebra System) module, among others.

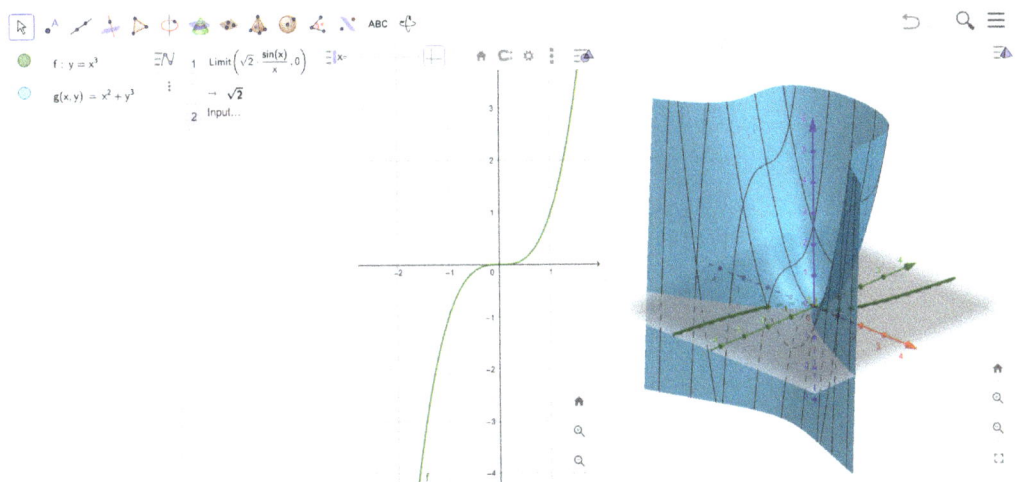

Figure 1. GeoGebra Classic screen.

GeoGebra's interface is easy to use and allows the configuration of several aspects associated with function representation, such as line width, colour, and style. These representations can be integrated into online books that can be shared with students so, for instance, they can navigate through all the examples and solutions associated with a certain topic [29].

3.2. WolframAlpha

WolframAlpha is a computational knowledge engine developed by a subsidiary of Wolfram Research, the company behind Mathematica [30]. Given that WolframAlpha is a reduced version of the Mathematica software, all options must be entered as text in the application's input box. However, the website provides access to many examples, so students can find the right expression in a relatively short time. Obviously, the advantage of using WolframAlpha instead of Mathematica is that it can be accessed by anyone as a web service free of charge.

One of the most interesting aspects of WolframAlpha is the possibility to use both natural language and Mathematica syntax for computations, so even students with little or no knowledge of the Mathematica syntax can use the engine without effort.

3.3. Python

Python is an interpreted, high-level, and general-purpose programming language that emphasizes code readability [31]. Python was first released in 1991, but it was not until the launch of versions 2.0 and 3.0 in 2000 and 2008, respectively, that Python was really popularized among programmers. Since 1 January 2020, Python 2 is no longer officially supported [32], which means that Python 3 is the only version which is active nowadays.

One of the advantages of Python over other programming languages is the number of modules and extensions that can be used [33]. From an engineering point of view, some of the most useful are NumPy (which defines types for numerical arrays and matrices together with the basic operations that can be applied to them) [34], SymPy (a library for symbolic mathematics) [35], and SciPy (which uses NumPy in order to perform advanced mathematical, signal processing, optimization, and statistics calculations) [36].

3.4. SageMath

SageMath is a computer algebra system with features covering many aspects of mathematics, including algebra, combinatorics, graph theory, numerical analysis, number theory, calculus, and statistics [37].

The first version of SageMath was released in 2005 as free and open source software under the GNU General Public License version 2, with the initial goal of becoming an open source alternative to Magma, Maple, Mathematica, and MATLAB.

Instead of developing another computational engine from scratch, SageMath integrates many already existing open source packages such as NumPy, SciPy, matplotlib, Sympy, Maxima, and R, among others, using a syntax similar to the one provided by Python.

SageMath can be installed as a stand-alone application or run in the cloud using CoCalc [38], a web-based cloud computing service (see Figure 2).

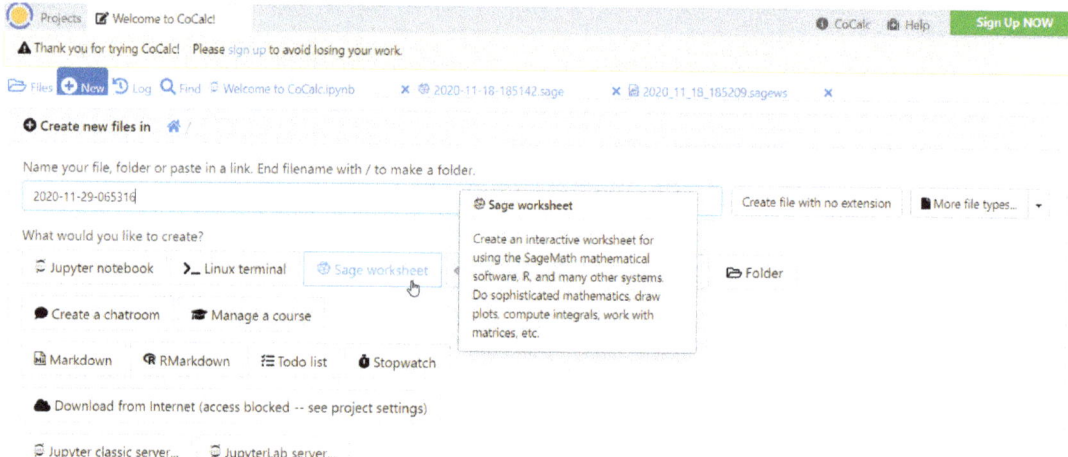

Figure 2. CoCalc website.

4. Examples

4.1. Calculus

Many Calculus key concepts can be reinforced or at least better understood by students when presented in a graphical way. Allowing students to replicate some model computations in similar problems has the benefit of providing a durable link between what is taught in class and what they study at home [39].

Figure 3 shows an example associated with the graphical representation of a function and its asymptotes. If, for instance, we intend to show how the Taylor polynomials work, we can include in the same solution the initial function and Taylor polynomials of different degrees, so students can realize that a higher degree implies a better approximation for real functions (see Figure 3).

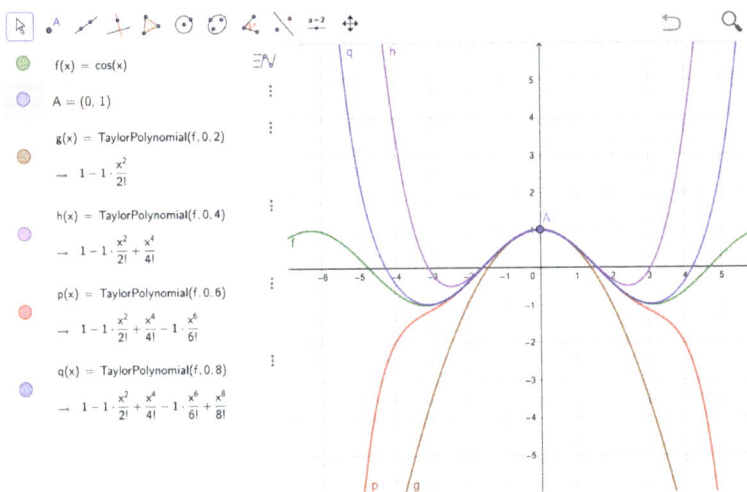

Figure 3. GeoGebra example about Taylor polynomials.

Regarding the calculus of several variables, GeoGebra is a suitable option given that it allows students to rotate three-dimension images in any direction. As an example, Figure 4 shows how to represent the intersection of two surfaces.

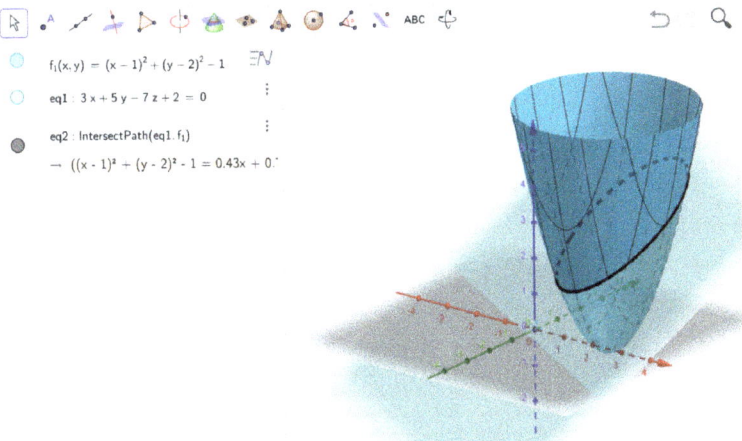

Figure 4. Intersection of two surfaces using GeoGebra.

Switching to WolframAlpha, it is possible to perform calculations such as performing the second derivative of a function and specializing the resulting expression at a point with a single command.

WolframAlpha can also be very convenient in some instances where, together with the requested calculation, the engine also provides a graphic representation of the solution, as in the case of Figure 5.

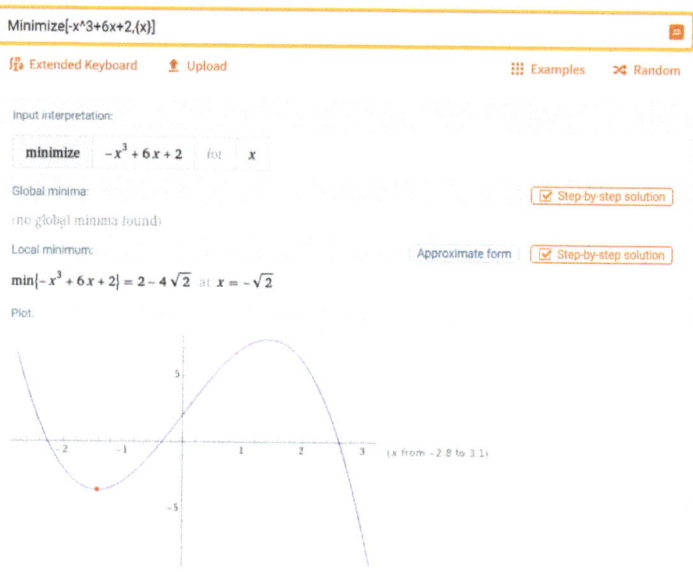

Figure 5. Search for function minimum points in WolframAlpha.

Both GeoGebra and WolframAlpha are supported by a large number of developers who make available their work, so it is possible to access many great online demonstrations and practical examples. This feature is particularly interesting when teaching theorems and their applications, as it is a topic where many students face some difficulties. Some examples are [40,41], where Lagrange's theorem and the Ingetral Mean Value theorem are described using WolframAlpha resources.

4.2. Algebra

This section shows how to use Python for solving different algebra problems using parts of the code developed by Javier García Algarra [42]. In order to correctly execute the following examples, it must be taken into account that NymPy and SymPy modules must be imported through the following commands:

```
import numpy as np
import sympy as sp
from sympy.matrices import import Matrix
```

For convenience, figures included hereafter have been executed as a worksheet in CoCalc. The first example shows how to represent a polynomial and to obtain its roots (see Figure 6).

```
expression = sp.sympify('x**2-4*x-1')
roots = sp.solve(expression)
print("Roots of ",expression)
sp.pprint(roots)

Roots of  x**2 - 4*x - 1
[2 - √5, 2 + √5]
```

Figure 6. Polynomial manipulation.

If we need to solve a system of linear equations, we can use the code displayed in Figure 7.

```
x,y,z = sp.symbols('x,y,z')
c1 = sp.Symbol('c1')
eq1 = sp.Eq(2*x**2+y+z,1)
eq2 = sp.Eq(x+2*y+z,2)
eq3 = sp.Eq(-2*x+y+z,0)

sp.solve([eq1,eq2,eq2],(x,y,z))
```

$$\left[\left(\frac{1}{8} - \frac{\sqrt{1-16z}}{8}, \frac{z}{2} - \frac{\sqrt{1-16z}}{16} + \frac{15}{16}, z \right), \left(\frac{\sqrt{1-16z}}{8} + \frac{1}{8}, -\frac{z}{2} - \frac{\sqrt{1-16z}}{16} + \frac{15}{16}, z \right) \right]$$

Figure 7. Solving a system of equations.

In Python, it is possible to define matrices either directly or through a lambda expression, which can be useful sometimes (see Figure 8).

```
A = Matrix([[-1,0,-3], [3,2,3],[-3,0,-1]])
A
```

$$\begin{bmatrix} -1 & 0 & -3 \\ 3 & 2 & 3 \\ -3 & 0 & -1 \end{bmatrix}$$

```
P = Matrix(2,3,lambda x,y:x*10+y)
P
```

$$\begin{bmatrix} 0 & 1 & 2 \\ 10 & 11 & 12 \end{bmatrix}$$

Figure 8. Matrix definition.

Once we have defined matrix A, Figure 9 shows how to obtain its determinant, inverse matrix, and associated eigenvalues in an easy way.

```
determinant = A.det()
print("Determinant of A: ",determinant)
inverse = A.inv()
print("Inverse of A: ",inverse)
eigenvalues = A.eigenvals()
print("Eigenvalues of A: ",eigenvalues)

Determinant of A:  -16
Inverse of A:  Matrix([[1/8, 0, -3/8], [3/8, 1/2, 3/8], [-3/8, 0, 1/8]])
Eigenvalues of A:  {-4: 1, 2: 2}
```

Figure 9. Matrix operations.

It is also possible to define and operate matrices with symbolic content, as shown in Figure 10.

```
B = Matrix(3,3,lambda x,y:sp.symbols('b'+str((x+1)*10+y+1)))
B
```

$$\begin{bmatrix} b_{11} & b_{12} & b_{13} \\ b_{21} & b_{22} & b_{23} \\ b_{31} & b_{32} & b_{33} \end{bmatrix}$$

```
C = Matrix(3,3,lambda x,y:sp.symbols('c'+str((x+1)*10+y+1)))
C
```

$$\begin{bmatrix} c_{11} & c_{12} & c_{13} \\ c_{21} & c_{22} & c_{23} \\ c_{31} & c_{32} & c_{33} \end{bmatrix}$$

```
B+C
```

$$\begin{bmatrix} b_{11}+c_{11} & b_{12}+c_{12} & b_{13}+c_{13} \\ b_{21}+c_{21} & b_{22}+c_{22} & b_{23}+c_{23} \\ b_{31}+c_{31} & b_{32}+c_{32} & b_{33}+c_{33} \end{bmatrix}$$

Figure 10. Matrices with symbolic contents.

4.3. Coding Theory

In this section, we will demonstrate how to operate with linear codes using SageMath. In the first example, we will define a generator matrix with coefficients defined over the Galois field with three elements, GF(3), as shown in Figure 11.

```
MS = MatrixSpace(GF(3),5,8)
G  = MS([[2,1,0,1,0,0,0,0],
         [0,2,1,0,1,0,0,0],
         [0,0,2,1,0,1,0,0],
         [0,0,0,2,1,0,1,0],
         [0,0,0,0,2,1,0,1]])
print("G=")
G
```

```
G=
[2 1 0 1 0 0 0 0]
[0 2 1 0 1 0 0 0]
[0 0 2 1 0 1 0 0]
[0 0 0 2 1 0 1 0]
[0 0 0 0 2 1 0 1]
```

Figure 11. Generator matrix definition.

Then, we can use G as the generator matrix of a $(8,5)$ code and request information such as the length, dimension, mininum distance, and weight distribution of the code, as shown in Figure 12.

```
C = LinearCode(G)
len = C.length()
print("length = ", len)
dim = C.dimension()
print("dimension = ", dim)
dmin = C.minimum_distance()
print("mininum distance = ", dmin)
wd = C.weight_distribution()
print("codeword weight distribution = ", wd)
```

```
    length =  8
    dimension =  5
    mininum distance =  3
    codeword weight distribution =  [1, 0, 0, 16, 60, 48, 64, 48, 6]
```

Figure 12. Information about the code.

Quite conveniently, we can obtain the generator matrix in systematic form as well as the code's parity check matrix (see Figure 13).

```
C.systematic_generator_matrix()
    [1 0 0 0 0 2 1 0]
    [0 1 0 0 0 0 2 1]
    [0 0 1 0 0 1 2 2]
    [0 0 0 1 0 2 2 2]
    [0 0 0 0 1 2 0 2]
```

```
C.parity_check_matrix()
    [1 0 0 1 0 2 1 1]
    [0 1 0 2 1 1 1 0]
    [0 0 1 0 2 1 1 1]
```

Figure 13. Systematic generator matrix and parity check matrix.

We are also able to check if a received vector is a proper codeword or not, in which case its syndrome will be different from the zero vector, as shown in Figure 14.

```
r1 = vector(GF(3), (2,0,1,1,1,0,0,0))
r1 in C
C.syndrome(r1)
```

```
    True
    (0, 0, 0)
```

```
r2 = vector(GF(3), (1,0,1,1,1,0,2,0))
r2 in C
C.syndrome(r2)
```

```
    False
    (1, 2, 2)
```

Figure 14. Checking if a received vector is a codeword.

In the case of cyclic codes, in addition to matrices, it is also possible to work with polynomials, as can be seen in Figure 15.

```
F.<x> = GF(2)[]
n = 15
g = x ** 4 + x + 1
C = codes.CyclicCode(generator_pol = g, length = n)
C
C.minimum_distance()

    [15, 11] Cyclic Code over GF(2)
    3

h = C.check_polynomial()
h

    x^11 + x^8 + x^7 + x^5 + x^3 + x^2 + x + 1

h*g

    x^15 + 1
```

Figure 15. Defining a cyclic code.

5. Experimental Study

In the experiment performed at the University of Salamanca, two groups of students were selected. The first group, with 57 students studying for a Chemical Engineering degree, represented the experimental group, while the control group was made up of 63 students studying for an Industrial Engineering degree. In the experimental group, students were allowed to use WolframAlpha (or, alternatively, Mathematica, with the same commands). In both cases, students attended a numerical analysis course with comparable contents, so conclusions could be obtained from the comparison. The study took into account the performance of students during the academic years 2018–2019 and 2019–2020.

For the experimental group, three questionnaires, two software exercises in the computer room, and two exams were conducted during the first year associated with this analysis. In contrast, during the second year, two questionnaires, three software exercises, and two exams were monitored. Students from the control group did not participate in software seminars and their only assessment activity was a final written exam at the end of the semester.

The goal of the statistical study presented in this section is to analyse, firstly, the relation between the different assessment activities and the results obtained when using mathematical software instead of traditional problem-solving methods and, secondly, to compare the results with students that did not participate in similar activities.

5.1. Chemical Engineering Degree

The numerical analysis course in the Chemical Engineering degree has 7.5 credits, and the final mark was calculated over 10 points, where 5% corresponds to questionnaires, 10% to software activities, 15% to team work and solving additional problems, and the remaining 70% corresponds to the grades associated with the written exams.

As has been mentioned before, for this study, data from two academic years was collected. In the case of the 2018–2019 course, out of 57 students, seven students that did not attend the different assessment activities have been discarded. Since the Bologna Accord was put into effect, the number of drop-out students has reduced and every year a fewer number of students leave mathematics courses.

Figure 16 shows the box plot representation for the student marks associated with the different assessment activities, where the stars represent the extreme values.

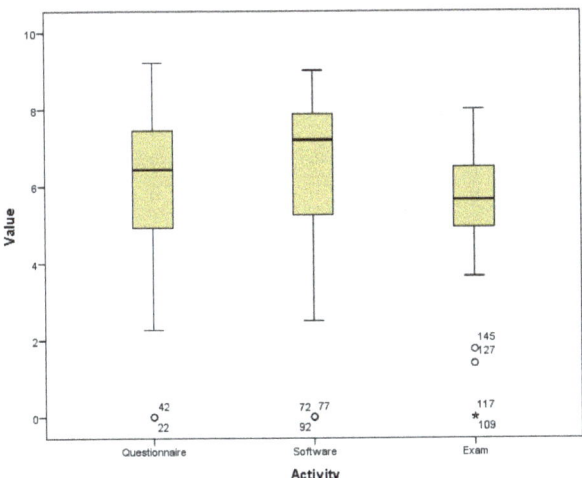

Figure 16. Box plot for the assessment activities results in the academic year 2018–2019.

In the case of software practices, Median > Mean, and Kurtosis = 1.468. In the case of questionnaires and exams, these values are different, as can be seen in Table 2.

Table 2. Descriptive statistics for the academic year 2018–2019.

Concept		Questionnaire	Software	Exam
N	Valid	50	50	50
	Missing	0	0	0
Mean		6.0322	3.2210	5.4018
Median		6.4350	7.2000	5.6400
Mode		0.00 [a]	7.75	5.25
Standard Deviation		2.12767	2.40013	1.73700
Variance		4.527	5.761	3.017
Skewness		−1.038	−1.465	−1.465
Standard Error of Skewness		0.337	0.337	0.337
Kurtosis		0.993	1.468	2.723
Standard Error of Kurtosis		0.662	0.662	0.662
Range		9.22	9.00	8.00
Minimum		0.00	0.00	0.00
Maximum		9.22	9.00	8.00

[a] Multiple modes exist, the smallest value is shown.

Software activity is clearly what suits students the best. Engineering students usually like to work with their hands, in the laboratory or with computers. Moreover, this activity is typically accomplished by collaborating with their fellows, which usually is not the case of exams and to a lesser extent of questionnaires. A consequence of this fact is the absence of a correlation. The biggest one is between software and questionnaires (the Pearson correlation coefficient is equal to 0.720).

We conducted an ANOVA to check the relation between the three activities: questionnaires, software and exams. We have found out that the data meet the homogeneity of variances (the Levene statistic has a significance that is equal to 0.068), they are random samples (the test significance is equal to 0.568), and variables follow a normal distribution (the Kolmogorov–Smirnov test significance is equal to 0.086).

Table 3 shows the results of the analysis of variance that indicates that the hypothesis of equal means is accepted, i.e., the means for questionnaires, software, and exams are equal.

Table 3. ANOVA test results for the academic year 2018–2019.

Concept	Sum of Squares	df	Mean Square	F	Significance
Between Groups	18.402	2	9.201	2.075	0.129
Within Groups	651.933	147	4.435		
Total	670.336	149			

In the 2019–2020 academic year, out of 48 students, eight alumni were discarded as they did not fully participate in all the assessment activities. The statistical analysis is quite similar to the one developed for the previous academic year. In this case, the correlation between activities has been reduced: 0.197 between software and questionnaires. The final average mark of students is 5.93 compared to 5.43 obtained the previous year.

With the goal to avoid the duplication of information we have included Figure 17, where histograms and normal curves for the assessment activities during the 2019–2020 course are displayed.

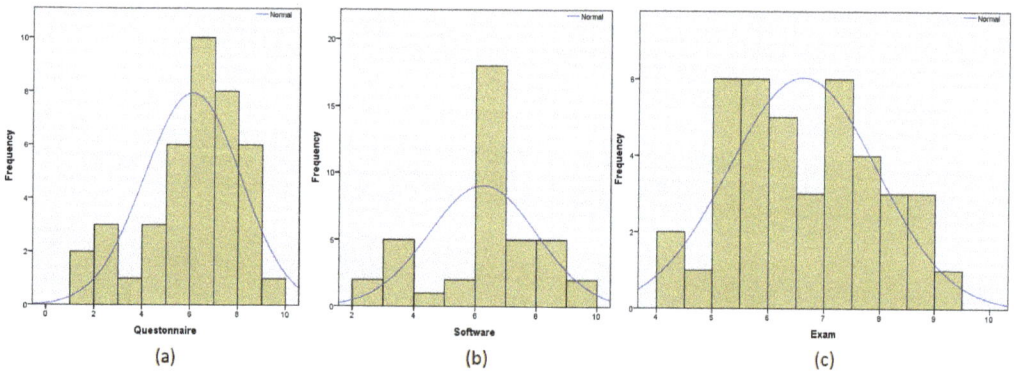

Figure 17. Histogram and normal curve for (**a**) questionnaires, (**b**) software, and (**c**) exams.

5.2. Industrial Engineering

The Industrial Engineering mathematics course has six credits and the final mark, which corresponds to the final exam, is calculated over 10 points. In this instance, the marks from 63 students were collected from the 2019–2020 academic year. For the control group, the final marks obtained in the final exam are shown in Table 4.

Table 4. Final results of Industrial Engineering students.

Mark	Percentage
Not attending	25.40
Between 0 and 4.99	31.75
Between 5 and 6.99	19.05
Between 7 and 8.99	17.46
Between 9 and 10	6.35

In this case, only 42.86% of students passed the exam, and the qualification's mean was 4.74.

5.3. Analysis of the Results

The analysis derived from the data obtained in the Chemical Engineering courses is presented in Table 5.

Table 5. Mean and Standard Deviation for questionnaires (Q), Software (S) and Exams (E).

Course	Mean			Standard Deviation		
	Q	S	E	Q	S	E
2018–2019	6.0322	6.2210	5.4018	2.12767	2.40013	1.73700
2019–2020	5.5441	5.9195	6.5961	2.60838	2.14294	1.32373

The independent samples test was performed in order to obtain the relation between the grades in different assessment activities grouped by year. As a result, we found out that the same variance appears in both courses (the Levene's test for equality of variances coefficient is equal to 0.709) and the t-test for the equality of means returns the 95% confidence interval of the difference equal to $(-1.83377, -0.55491)$ assuming equal variances and to $(-1.82308, 0.56559)$ when equal variances are not assumed, with a significance value of 0.00 in both cases.

Compared to the data obtained from Industrial Engineering students, it can be seen that the mean is lower than the mean for Chemical Engineering students. This could be interpreted as an indication that, when mathematical software is used at class, students improve their understanding of the contents and obtain better results compared to students that are being taught in the traditional way.

6. Conclusions

In this contribution, we have shown how to use some of the best-known free computational packages in order to enhance the learning process for mathematical courses in engineering studies. The usage of engines such as the ones implemented by GeoGebra, WolframAlpha, Python or SageMath allows students to grasp the key concepts seen in class and to practice problems at their leisure, resulting in better learning outcomes and grades.

Using free software has the additional benefit of allowing educators to choose the best option for each topic inside a course, as they are not tied to a specific product that can be optimal for some subjects but inadequate in some other instances. Some of the examples shown throughout this article could even be used in high schools and academies, which are two institutions less likely to commit themselves to investments in things such as mathematical software licences.

An observation made by the authors of this paper during the elaboration of the research is that first-year students are less inclined to use a programming language than a computational engine in order to solve engineering problems, even if they were previously familiar with the programming language in question. Some reasons for this are that students (incorrectly) do not try to interrelate the knowledge obtained in different subjects and that they prefer to use a command instead of coding a small application to obtain fast results. Another conclusion is that students prefer not to install applications if they can obtain the same results by connecting to a remote service providing online compilers or calculators.

Additionally, an analysis of the results obtained when using mathematical software to an engineering mathematics course is included. This analysis allows one to derive some conclusions about the application of mathematical software to different activities such as questionnaires, problems to be solved, and exams. In our study, students that participated in a course that allowed the completion of some activities with the help of mathematical software obtained better marks than students that attended to a more "traditional" course, composed of master classes and problem solving sessions. In general, we believe that students are able to achieve a better understanding of the contents of mathematical subjects if they are allowed to use computational engines, which benefits both students and teachers.

Author Contributions: Conceptualization, methodology, validation, investigation, and writing—review and editing, V.G.M., L.H.E., A.M.M. and A.Q.D.; software and writing—original draft, V.G.M. and A.Q.D.; funding acquisition, L.H.E. and A.Q.D. All authors have read and agreed to the published version of the manuscript.

Funding: This work was supported by the Spanish State Research Agency (AEI) of the Ministry of Science and Innovation (MCIN), project P2QProMeTe (PID2020-112586RB-I00/AEI/10.13039/501100011033), co-funded by the European Regional Development Fund (ERDF, EU).

Institutional Review Board Statement: Ethical review and approval were waived for this study, due to the anonimity of the grades used in the study.

Informed Consent Statement: Student consent was waived due to the anonimity of the grades used in the study.

Data Availability Statement: Data supporting results can be obtained by request from the authors.

Acknowledgments: Víctor Gayoso Martínez would like to thank CSIC Project CASP2/201850E114 for its support. Araceli Queiruga Dios would like to thank Universidad de Salamanca for its support.

Conflicts of Interest: The authors declare no conflict of interest.

Abbreviations

The following abbreviations are used in this manuscript:

2D/3D	2D/3D plotting
AgC	Algebraic combinatorics
AlC	Algebraic computations
AlG	Algebraic geometry
Alg	Algebra
ArG	Arithmetic geometry
Art	Arithmetic
CAl	Computational algebra
CAS	Computer Algebra System
CDm	Distributed computation
CmA	Commutative Algebra
CSIC	Consejo Superior de Investigaciones Científicas
DAl	Discrete algebra
EAHE	European Area of Higher Education
EcA	Econometric analysis
FPA	Floating-Point Arithmetic
InF	Integer factorization
ITEFI	Instituto de Tecnologías Físicas y de la Información
LiP	Linear programming
MDPI	Multidisciplinary Digital Publishing Institute
NTh	Number theory
NuC	Numerical computation
ODE	Ordinary Differential Equation
SAl	Symbolic algebra
SyC	Symbolic computation

References

1. Bologna Process Secretariat. European Higher Education Area and Bologna Process. Available online: http://www.ehea.info/ (accessed on 30 August 2021).
2. Lavicza, Z. Integrating technology into mathematics teaching at the university level. *ZDM-Math. Educ.* **2010**, *42*, 105–119. [CrossRef]
3. M. Tamur, D.J.; Kusumah, Y. The Effectiveness of the Application of Mathematical Software in Indonesia; a Meta-Analysis Study. *Int. J. Instr.* **2020**, *13*, 867–884. [CrossRef]
4. Ahmad, O. Review of symbolic equation solving for engineering problems. In Proceedings of the 8th Asian Conference on Engineering Education (ACEE 2019), Kota Kinabalu, Malaysia, 24–26 June 2019; pp. 39–45.
5. Vick, B. *Applied Engineering Mathematics*; CRC Press: Boca Raton, FL, USA, 2020.

6. University of Salamanca. The University at a Glance. Available online: https://www.usal.es/en/university-glance (accessed on 30 August 2021).
7. U-tad. Centro Universitario de Tecnología y Arte Digital. Available online: https://www.u-tad.com/en/ (accessed on 30 August 2021).
8. European Comission. Europe 2020. A European Strategy for Smart, Sustainable and Inclusive Growth. Available online: https://ec.europa.eu/eu2020/pdf/COMPLET%20EN%20BARROSO%20%20%20007%20-%20Europe%202020%20-%20EN%20version.pdf (accessed on 30 August 2021).
9. Plaza, P.; Martín, S.; Sancristobal, E.; Blázquez, M.; Castro, M.; Díaz, G.; Pérez, C. Science and technology educational quality scaling in Spain. In Proceedings of the 2020 IEEE Frontiers in Education Conference (FIE), Uppsala, Sweden, 21–24 October 2020; pp. 1–8.
10. Hillmayr, D.; Ziernwald, L.; Reinhold, F.; Hofer, S.I.; Reiss, K.M. The potential of digital tools to enhance mathematics and science learning in secondary schools: A context-specific meta-analysis. *Comput. Educ.* **2020**, *153*, 103897. [CrossRef]
11. Kurz, T.L.; Middleton, J.A.; Yanik, H.B. A taxonomy of software for mathematics instruction. *Contemp. Issues Technol. Teach. Educ.* **2005**, *5*, 123–137.
12. Kilicman, A.; Hassan, M.A.; Husain, S.S. Teaching and Learning using Mathematics Software "The New Challenge". *Procedia Soc. Behav. Sci.* **2010**, *8*, 613–619. [CrossRef]
13. Kusbeyzi, I.; Hacinliyan, A.; Aybar, O.O. Open source software in teaching mathematics. *Procedia Soc. Behav. Sci.* **2011**, *15*, 769–771. [CrossRef]
14. Saadon, S.; Rambely, A.S.; Suradi, N.R.M. The Role of Computer Labs in Teaching and Learning Process in the Field of Mathematical Sciences. *Procedia Soc. Behav. Sci.* **2011**, *18*, 348–352. [CrossRef]
15. Botana, F.; Abánades, M.A.; Escribano, J. Using a Free Open Source Software to Teach Mathematics. *Comput. Appl. Eng. Educ.* **2012**, *22*, 728–735. [CrossRef]
16. Berežný, Š. What software to use in the teaching of mathematical subjects? *Acta Didact. Napoc.* **2015**, *8*, 75–85.
17. Ochkov, V.F.; Bogomolova, E.P. Teaching Mathematics with Mathematical Software. *J. Humanist. Math.* **2015**, *5*, 265–286. [CrossRef]
18. Sattar, F.; Tamatea, L.; Nawaz, M. Freeware and Open Source Software Tools for Distance Learning in Mathematics. *Online J. Distance Educ. E-Learn.* **2015**, *3*, 26–32.
19. Joshi, D.R. Useful Applications/Software for Mathematics Teaching in School Education. *Int. J. Inf. Technol.* **2016**, *1*, 29–34.
20. Alonso Izquierdo, A.; González León, M.A.; Martín-Vaquero, J.; Dias Rasteiro, D.M.; Kováčová, M.; Richtáriková, D.; Rodríguez-Gonzálvez, P.; Rodríguez-Martín, M.; Queiruga-Dios, A. Specific mathematical software to solve some problems. In *Calculus for Engineering Students*; Elsevier: Cambridge, MA, USA, 2020; Chapter 15, pp. 327–347. [CrossRef]
21. Flegg, J.; Mallet, D.; Lupton, M. Students' perceptions of the relevance of mathematics in engineering. *J. Math. Educ. Sci. Technol.* **2012**, *43*, 717–732. [CrossRef]
22. Wigderson, A. Mathematics and Computation. Available online: https://www.math.ias.edu/avi/book (accessed on 31 August 2021).
23. Sazhin, S.S. Teaching Mathematics to Engineering Students. *Int. J. Eng. Educ.* **1998**, *14*, 145–152.
24. Chonacky, N.; Winch, D. 3Ms for instruction: Reviews of Maple, Mathematica, and Matlab. *Comput. Sci. Eng.* **2003**, *7*, 7–13. [CrossRef]
25. Abichandani, P.; Primerano, R.; Kam, M. Symbolic scientific software skills for engineering students. In Proceedings of the 2010 IEEE Transforming Engineering Education: Creating Interdisciplinary Skills for Complex Global Environments, Dublin, Ireland, 6–9 April 2010; pp. 1–26.
26. Fangohr, H. A comparison of C, MATLAB, and Python as teaching languages in engineering. In Proceedings of the International Conference on Computational Science, Krakow, Poland, 6–9 June 2004; pp. 1210–1217.
27. Abichandani, P.; Primerano, R.; Kam, M. Recent usage of software tools in the teaching and learning of engineering mathematics. Advances in innovative engineering and technologies. In Proceedings of the International Conference on Innovative Engineering and Technologies (CAASR-ICIET'15), Bangkok, Thailand, 27–28 November 2015; pp. 162–174.
28. GeoGebra. GeoGebra Math Apps. Available online: https://www.geogebra.org (accessed on 30 August 2021).
29. Dimitrov, D.M.; Slavov, S.D. Application of GeoGebra software into teaching mechanical engineering courses. In Proceedings of the 22nd International Conference on Innovative Manufacturing Engineering and Energy (IManEE 2018), Chisinau, Republic of Moldova, 31 May – 2 June 2018; p. 07008.
30. WolframAlpha LLC. WolframAlpha Computational Intelligence. Available online: www.wolframalpha.com (accessed on 30 August 2021).
31. Python Software Foundation. Welcome to Python.org. Available online: https://www.python.org (accessed on 30 August 2021).
32. Python Software Foundation. Sunsetting Python 2. Available online: https://www.python.org/doc/sunset-python-2/ (accessed on 30 August 2021).
33. Fangohr, H. Python for Computational Science and Engineering. Available online: https://fangohr.github.io/teaching/python/book.html (accessed on 30 August 2021).
34. NumPy. The Fundamental Package for Scientific Computing with Python. Available online: https://numpy.org/ (accessed on 30 August 2021).

35. SymPy Development Team. Python Library for Symbolic Mathematics. Available online: https://www.sympy.org/ (accessed on 30 August 2021).
36. SciPy Developers. Python-Based Ecosystem of Open-Source Software for Mathematics, Science, and Engineering. Available online: https://www.scipy.org/ (accessed on 30 August 2021).
37. Sagemath, Inc. Open Source Mathematical Software System. Available online: https://www.sagemath.org (accessed on 30 August 2021).
38. Sagemath, Inc. Collaborative Calculation and Data Science. Available online: https://cocalc.com (accessed on 30 August 2021).
39. Yavuz, I. What does a graphical representation mean for students at the beginning of function teaching? *Int. J. Math. Educ. Sci. Technol.* **2010**, *41*, 467–485. [CrossRef]
40. Kumar, R. Lagrange Mean Value Theorem. Available online: https://www.geogebra.org/m/jyYQM5ZH (accessed on 30 August 2021).
41. Boucher, C. Integral Mean Value Theorem. Available online: https://demonstrations.wolfram.com/IntegralMeanValueTheorem (accessed on 30 August 2021).
42. García Algarra, J. Python Para Matemáticas. Available online: https://github.com/jgalgarra/PythonMatematicas (accessed on 30 August 2021).

MDPI
St. Alban-Anlage 66
4052 Basel
Switzerland
Tel. +41 61 683 77 34
Fax +41 61 302 89 18
www.mdpi.com

Axioms Editorial Office
E-mail: axioms@mdpi.com
www.mdpi.com/journal/axioms

www.ingramcontent.com/pod-product-compliance
Lightning Source LLC
LaVergne TN
LVHW070727100526
838202LV00013B/1187